Gladiolus in Tropical Africa

SYSTEMATICS BIOLOGY & EVOLUTION

Peter Goldblatt

line drawings by
John C. Manning

Timber Press
Portland, Oregon

Library of Congress Cataloging-in-Publication Data

Goldblatt, Peter.
Gladiolus in tropical Africa : systematics, biology & evolution / Peter Goldblatt ; line drawings by John C. Manning.
p. cm.
Includes bibliographical references (p.) and index.
ISBN 0-88192-333-8
1. Gladiolus—Africa, Sub-Saharan. I. Title.
QK391.G64 1996
584'.24—dc20 95-18375
CIP

Timber Press, Inc
The Haseltine Building
133 S.W. Second Avenue, Suite 450
Portland, Oregon 97204, U.S.A.

Printed in Hong Kong

Contents

Preface

Preparation of a thorough monographic treatment of the tropical African species of *Gladiolus* presented more than the normal range of problems associated with this type of project. *Gladiolus,* with more than 250 species, is a large genus by any standards, and although centered in southern Africa, there are more than 82 species in tropical Africa alone. *Gladiolus* is also a genus in which many of the characters on which species are based are floral. The shape of the flower, the relative lengths of the tepals, their orientation, and color are often the primary or the only features on which species are defined. Yet some or all of these characters preserve poorly in herbarium specimens, the basic working units used by plant systematists. As a result, although there are many preserved specimens available in collections throughout the world, field research was vital to achieving an understanding of species, their natural patterns of variation, and ecological associations. Tropical Africa is a huge area, and extensive travel there would not only have been prohibitively expensive, but often dangerous. Moreover, at the time that I was actively engaged in field studies of *Gladiolus,* war and civil unrest prevented effective travel in Mozambique and Angola, the latter an important center of speciation for *Gladiolus.* Zaire and Ethiopia were both politically unstable and travel to the botanically interesting parts of these countries was not practical.

These limitations were partly overcome by the generous assistance of botanical colleagues and informed amateurs who had traveled or lived in these countries and had photographs of many of the wild species of *Gladiolus.* Both their personal knowledge of *Gladiolus* in native habitats and their photographs made it possible to understand more fully new or problematic species and to a large extent made up for the difficulties in traveling to some countries.

Despite the economic importance of *Gladiolus* as an ornamental garden plant and in the cut flower industry, the genus had never been studied in its entirety. The species of *Gladiolus* in southern Africa are relatively well understood, but the tropical African species have not been surveyed since J. G. Baker's treatment of the genus for the *Flora of Tropical Africa,* published in 1898, a time when the continent had barely begun to be explored botanically. Thus, until my research began, the application of names of species already described was confused, and there was no way even to determine the true number of species that occur in tropical Africa. As a rule, tropical African species have relatively wide ranges, and this is true for some species of *Gladiolus* that grow there. Not surprisingly, widely distributed species had been discovered and named in the early phases of botanical exploration of the continent. Difficulties of scientific communication and ignorance of the patterns of variation in individual species made for an unfortunate proliferation of named species and the consequent large number of synonyms for what are now understood to be single widespread species. For example, *G. gregarius,* a very distinctive plant, was given 10 different names based on populations from Nigeria, Zaire, Angola, Tanzania, and Zambia. The even more widespread *G. dalenii,* which also occurs in Madagascar and southern Africa, was given over 40 different names, 27 of them based on collections from tropical Africa. Often such named species actually were virtually identical, but scientists working on plants from as far afield as South Africa, Tanzania, or Ethiopia often did not even consider the possibility that new collections that came into their hands from one country might occur thousands of miles away and had already been described.

Although wide ranges characterize a few species of tropical African *Gladiolus,* smaller ranges are the rule, and the majority of species occur in either western, northeastern, or southern tropical Africa. Even fewer have very narrow distributions, evidently having evolved under peculiar local ecological conditions. Shaba Province of Zaire is one of the areas where local speciation is evident. There, small outcroppings of soils characterized by high concentrations of heavy metals, especially copper, are restricted to a few hills and these support some very localized species of *Gladiolus.* Isolated high mountain areas often also have species of limited ranges. Thus the highlands of southwestern Angola, eastern Zimbabwe, eastern Shaba, and southwestern Tanzania, for example, each also have one or more endemic species of *Gladiolus.*

In this book I have tried to present an analysis of the species of tropical African *Gladiolus,* illustrating as many species as possible in color and in botanically accurate line drawings, made from living plants or from photographs supplemented by preserved herbarium specimens. As in the books I have written dealing with other genera of the *Iris* family, I have attempted to produce a thorough scientific study that is at the same time attractive, readable, easy to understand, and completely accurate botanically. The type specimens of all the species of *Gladiolus* described from tropical Africa were tracked down. The specimens were then critically examined and related to living plants. The huge collections of herbarium specimens of African plants at major herbaria such as Kew, Brussels, and Paris were consulted, as were smaller but important collections in Harare, Nairobi, Pretoria, Zomba, and elsewhere. In this way the distributions of each species across the continent were worked out and mapped. The preserved collections also yielded a surprising number of new species, sometimes misidentified or simply waiting to be described. My research has resulted in the naming of 27 new species. Eight of them have already been described in my 1993 treatment of the Iridaceae for the *Flora Zambesiaca*. The remainder are described here for the first time. This book on tropical African *Gladiolus* is complemented by a study of the genus in Madagascar, published in 1989. Only eight species occur there, seven of them endemic. A second book on *Gladiolus* is planned, treating the southern African species. Currently in preparation, this volume will deal with all the remaining African species of *Gladiolus* and will bring our knowledge of the genus to a comparable level for the entire continent.

I thank the following for allowing me to use their photographs as an aid to understanding some species and for preparing the illustrations: Paul Bamps, F. Billiet, Sheila Collenette, M. G. Gilbert, E. A. S. la Croix, D. C. H. Plowes, Michel Schaijes, Maire Spurrier, Duncan Thomas, and Mats Thulin. Selected photographs are reproduced here with their kind permission. I also thank Isobyl la Croix, Jon Lovett, Michel Schaijes, Maire Spurrier, and Sylvester Chisumpa for generously sharing their knowledge of species in the field, and Jill Lovett and Mary Clarke for hospitality in the course of field work. I acknowledge with gratitude the line drawings made by John C. Manning. Support for preparation of the art work was provided by the Stanley Smith Horticultural Trust, and research was funded by grant DEB 89-06300 from the U.S. National Science Foundation. My sincere thanks are also extended to the following people who have made contributions to the publication expenses of this volume: Maurice Boussard, Verdun, France; John Cook, Whithes, California; Sam and Phil Dominello, S & P Dominello Pty., Ltd., Peats Ridge, New South Wales, Australia; the Dutch Gladiolus Group, Royal General Bulbgrowers Association, Hillegom, Netherlands; H. B. Hay, Tadworth, Surrey, England; Satoshi and Namie Komoriya, Komoriya Nursery, Ltd., Chiba-shi, Chiba-ken, Japan; Fred W. Meyer, Escondido, California; Thom and Marlene Meyer, New World Plants, Bonsall, California; Angus Stewart, Narara, New South Wales, Australia; H. J. Paul Wülfinghoff, Wülfinghoff Freesia BV, Rijswijk, Netherlands.

Introduction

Gladiolus, with more than 250 species, is one of the largest genera of the petaloid monocot plant family, Iridaceae, and is the largest genus of the African and Eurasian subfamily Ixioideae. It is one of the most important plants in horticulture today, both as an ornamental garden subject and as a cut-flower horticultural crop plant. Yet it is surprisingly poorly understood in the wild. Basic knowledge, such as how many species there are, has not until now been known with any certainty. Even the geographical ranges of many of the species were not well established for more than a few species. *Gladiolus* is centered in southern Africa, the countries of South Africa, Swaziland, Lesotho, Namibia, and Botswana, but the genus ranges through all of tropical Africa as well as Madagascar and many of the small offshore islands of the continent and into the Mediterranean basin, southern Europe, and the Middle East, as far east as Afghanistan. *Gladiolus* is fairly well known in southern Africa, where there are at least 150 species, the distribution of which exhibits a pronounced concentration in the southwestern Cape region.

The genus was revised for southern Africa by Lewis, Obermeyer, and Barnard in 1972, at a time when four small genera, *Homoglossum* (= *Petamenes,* 12 species), *Anomalesia* (3 species, including *Kentrosiphon*), and *Oenostachys* (c. 5 species), were regarded as separate from *Gladiolus* although generally acknowledged to be closely related to it. All four of these small genera are now considered to be congeneric with *Gladiolus.* They are understood to have been no more than minor segregates of the larger genus that were recognized because of their unusual floral adaptations for pollination by sunbirds (Goldblatt & de Vos, 1989). The transfer of the 20 species in these genera to *Gladiolus* has resulted in the addition of some 15 species to the 103 recognized by Lewis et al. (1972) in southern Africa. Several more new species, added to *Gladiolus* after the publication of their revision (Hilliard & Burtt, 1983a, b; Goldblatt & Vlok, 1989), have brought the number of species there to more than 125.

The number of species of *Gladiolus* in tropical Africa has until now, however, been so uncertain that the few estimates that had been made had little foundation. The British specialist in Iridaceae, J. G. Baker, in his touchstone revision of *Gladiolus* in the *Flora of Tropical Africa* (1898), admitted 49 species to the genus (including 7 in the addenda and 15 more assigned to two more genera, *Antholyza* and *Acidanthera*), for a total of 63 species for all of tropical Africa. Baker's understanding of *Gladiolus* was tenuous at best, partly because he knew some species only from the literature or poor material, and partly because several more were based on just one or two specimens. Under these circumstances, it should be no surprise that, for example, he recognized the widespread tropical African species, *G. unguiculatus,* by four separate names, three in *Gladiolus* and one in *Antholyza*! And what is now believed to be a single but admittedly polymorphic species, *G. dalenii,* was described by Baker under more than 12 different names, each based on plants from different parts of Africa. More than 100 additional tropical African species of *Gladiolus* have been added to the literature since the publication of Baker's 1898 study. But because most of these are species already known by other names, any estimate of the number of *Gladiolus* species in tropical Africa had to be no more than a rough guess. In this book 82 species of *Gladiolus* are recognized in continental tropical Africa, all but 76 occurring nowhere else. Eight species occur in Madagascar, seven of which are endemic there. (See the Systematic Treatment, Additional or Incompletely Known Species, for two other species in tropical Africa, one of which is too poorly known yet to be described.)

The total number of *Gladiolus* species in sub-Saharan Africa and Madagascar is now believed to be more than 240. To this number can be added the approximately 10 species in Eurasia, including North Africa (which belongs there phytogeographically, rather than to the rest of the African continent). None of the Eurasian–North African species is shared with the rest of Africa so that the best current estimate of the size of *Gladiolus* is more than 250 species. This makes the genus large by any measure, and the second in size in Iridaceae after the North temperate genus *Iris.*

The reasons for the tremendous diversity and success of *Gladiolus* are complex. Obviously, several factors have contributed to this remarkable example of adaptive radiation. These include unusual floral diversity that over time has allowed the species to become adapted to almost all the important floral pollinators in Africa, including various bees, long-tongued flies, sphinx and noctuid moths, and sunbirds. Vegetative adaptation is also significant. Variation in leaf morphology, ranging from the broad soft-textured leaves of some species to the tough narrow leaves with thickened margins and midribs of others, has permitted radiation from moist, high-rainfall habitats to semiarid sites. Other significant adaptations are the ability to flower after the season's leaves have withered and before the next season's vegetative shoot has sprouted, and control of the flowering response to early, middle, or late in the season. Flowering times in *Gladiolus* species, unlike most petaloid monocots, thus extend almost throughout the year. This, like floral diversity, makes available more adaptive niches to the genus and in turn leads to greater species diversification.

History

Gladiolus is a plant of classical antiquity, so named by the Romans. The name, meaning little sword, alludes to the flattened, sword-shaped leaves, actually a characteristic of the entire *Iris* family. In 1753, *Gladiolus* entered the era of modern botany with the introduction of the binomial system of botanical nomenclature by Carl Linnaeus, the Swedish scientist. In *Species Plantarum,* the work in which he laid the foundations of the modern system of naming plants, Linnaeus recognized six species of *Gladiolus,* among which were plants from both Eurasia and the Cape of Good Hope in southern Africa. Outside Europe, the existence of *Gladiolus* was first established at the Cape of Good Hope, at the southwestern tip of southern Africa, where botanical exploration began at the end of the 17th century and was later actively promoted by Linnaeus. Botanical exploration was further encouraged by the 18th century European fascination with exotic and new plants, especially those found at the Cape. Dutch, English, and French patrons continued to support botanical exploration so that by the end of the 18th century, more than 30 African species of *Gladiolus* had been described. Many of these species were discovered and named by Carl Peter Thunberg, Linnaeus's pupil and successor, who completed the first *Flora Capensis* at this time (Thunberg, 1823).

At this early period, *Gladiolus* comprised species that are now assigned to several other southern African genera, including *Watsonia* and *Lapeirousia,* so that the figure of 51 species of *Gladiolus* recorded in the *Flora Capensis* is misleading. Conversely, species of *Gladiolus* with large red flowers and elongated cylindrical perianth tubes were assigned to a second Linnaean genus, *Antholyza. Gladiolus* itself comprised only plants with fairly short tubes and bilabiate flowers. *Antholyza* is now known to be an arbitrary assemblage of species (Brown, 1932; Lewis, 1954a) that are today placed in five different genera (Goldblatt & de Vos, 1989).

The first record of *Gladiolus* in the tropics was the publication in 1788, by the French biologist, Jean Lamarck, of the new species *G. luteus* from Madagascar. The species was described from specimens collected on the east coast of the island in 1770 or 1771 by the French naturalist and explorer, Philibert Commerson. The discovery of the genus on Madagascar, before any of the more than 80 species of tropical African *Gladiolus* became known, is not so surprising, since *G. luteus* is a strand species that is common along the Madagascan east coast. This was close to the ports that were visited by the first explorers and traders to reach Madagascar and it thus required little effort to find the plant. It took nearly 100 years before another of the eight species of Madagascan *Gladiolus* was discovered, and by this time botanical exploration of tropical Africa was well under way.

Early Botanical Exploration in Tropical Africa, 1800–1880

Botanical exploration gradually extended into the African interior from a southern African base at the Cape, but before significant progress had been made the first tropical African species was discovered during the epic expedition to Ethiopia made by James Bruce in 1771. The artist who accompanied Bruce, Luigi Balugani, drew and painted the plant currently known as *Gladiolus abyssinicus* in October 1771. Regrettably, much of Balugani's work remained unknown to the scientific community, and his notebooks, including his drawings of *G. abyssinicus* (which he called "*Antheoliza*"), were only published more than 200 years later (Hulton et al., 1991). This early discovery is thus no more than a footnote to the history of *Gladiolus.*

Until the mid 19th century, botanical exploration in tropical Africa remained focused in Ethiopia, where the German botanist, Georg Heinrich Schimper, arrived in 1838. During the 30 years that he lived in Ethiopia, Schimper made a major contribution to the knowledge of the northeastern African flora. At the beginning of his years in the country, Schimper made the type collection of the plant we now know as *Gladiolus murielae.* Because of its elongate perianth tube, more than 12 cm long, and nearly equal tepals, the plant was considered to be a new genus, *Acidanthera,* by C. F. Hochstetter, who described it as *A. bicolor* in 1844. Under this name *G. murielae* was to become a fairly well known garden plant, often still known as *Acidanthera* to gardeners today. In 1867 the German botanist and expert on Iridaceae, F. W. Klatt, transferred *A. bicolor* to *Sphaerospora,* which he defined as having a long perianth tube and spherical seeds. Unknown to Klatt, however, *G. murielae* has winged seeds quite typical of *Gladiolus.*

Gladiolus abyssinicus was re-collected in 1839 by the biologists of the Lefebvre Expedition to Abyssinia, modern Ethiopia. Both the botanist, Antoine Petit, and the zoologist, Richard Quartin-Dillon, died during the expedition (Petit was killed by a crocodile, a particularly gruesome fate) but not before they had collected the specimens that were to become the types of *Antholyza abyssinicus* (now *G. abyssinicus*), *G. quartinianus,* and *Ixia quartiniana.* Unknown to Achille Richard (1851), who described these species, *I. quartiniana* had already been named, having been described six years earlier as *Acidanthera bicolor,* based on Schimper's collection made a year earlier. Only in 1968 was *G. quartinianus* finally considered conspecific with the southern African *G. dalenii* (then known *G. psittacinus*), described in 1830.

Schimper also made the first gathering of *Gladiolus roseolus,* but his collection, annotated *G. minor* by E. G. von Steudel, was overlooked by the botanical community. A later gathering made by the Italian botanist, Achille Chiovenda, in 1909, was described by him in 1911. At the same time that Chiovenda made his collections of *G. roseolus,* the species was recorded from Cameroon in West Africa, and these plants were named *G. heterolobus* shortly thereafter. In 1862, the German explorer Georg Schweinfurth collected *G. roseolus* near Galabat in western Ethiopia, close to the Sudanese border. This collection, too, was overlooked by contemporary botanists. Schweinfurth also established the presence of *G. dalenii* across southern Sudan in 1869 during his exploration of the region. Traveling in Eritrea, 1890–1894, Schweinfurth also discovered the rare northern Eritrean endemic, *G. mensensis,* and made the first collections of a plant closely related to *G. abyssinicus,* the small-flowered *G. schweinfurthii.* Like its taller relative, this species was at first assigned to the genus *Antholyza* when it was named in Schweinfurth's honor in 1894 by the English botanist and specialist on Iridaceae, J. G. Baker.

The history of *Gladiolus* in tropical Africa becomes rather intricate later in the 19th century, as exploration of the continent accelerated. The West African relative of the Ethiopian *G. murielae, G. aequinoctialis,* which has similar long-tubed and white flowers, was described at a surprisingly early date for this part of the continent. Collected in Sierra Leone, it was grown and flowered in England in 1841 and was described in 1842 by the British expert on bulbous plants, William Herbert, who confidently referred the plant to *Gladiolus.* Details relating to the original collection of *G. aequinoctialis* do not appear to have been recorded. Then, in 1846, during his travels in Mozambique from 1843 to 1847, the German explorer Wilhelm Peter collected a *Gladiolus* later described in the report of the expedition, *Reise nach Mossambique,* as *G. luteolus* by F. W. Klatt (1864). This is the first record of the yellow-flowered form of *G. dalenii* that occurs throughout sub-Saharan tropical Africa, a plant that became widely known in horticulture as *G. primulinus* when recollected after 1890 in central Tanzania and Zimbabwe.

More significant than these scattered discoveries was the systematic exploration of Angola by the Austrian, Friedrich Welwitsch. An energetic biologist, Welwitsch was commissioned by the Portuguese government to explore the botanical resources of that colony (Hiern, 1896). He arrived in Loanda, the capital, in September 1853 and spent the next 8 years traveling widely across Angola, amassing a major biological collection. Welwitsch returned to Portugal in 1861, where he began to work on his specimens. In 1863 he took his entire collection to England, where the only comparative material was available. After Welwitsch's death in 1872, specimens of the plant families not yet adequately studied were entrusted to contemporary botanists. The young expert on bulbous monocot families, J. G. Baker, was given responsibility for the Iridaceae. In 1878, some 25 years after Welwitsch's first collections were made, Baker published the descriptions of 26 new Iridaceae from Angola, among which were 11 species of *Gladiolus*. Of these, *G. gregarius, G. benguellensis, G. huillensis* (referred by Baker to *Antholyza*), *G. laxiflorus, G. welwitschii,* and *G. andongensis* are still recognized, the last two treated here as subspecies of *G. dalenii*.

In West Africa, penetration of the interior along the Niger River by Charles Barter, plant collector on W. F. Baikie's Niger Expedition (1857–1858), resulted in the discovery of the plant now known as *Gladiolus gregarius,* at about the same time that Welwitsch found this species in Angola. Barter's plants from "near Jebba on the Kworra," in southwestern Nigeria, were described by F. W. Klatt in 1867. Unfortunately, the name chosen by Klatt for the species, *G. spicatus,* is a homonym for the southern African *G. spicatus* Linnaeus (1753), now *Thereianthus spicatus,* and is thus nomenclaturally illegitimate. Renamed *G. klattianus* in 1936 by John Hutchinson, the species remained recognized in West Africa for another 30 years. In the second edition of *Flora of West Tropical Africa,* F. N. Hepper (1968) reduced *G. klattianus* to synonymy under *G. gregarius,* until then regarded as a strictly Angolan plant.

Malaria and other diseases as well as the inhospitable climate made West Africa difficult for botanical exploration, and significant discoveries were not made there until after the turn of the 20th century. In East Africa, with its more amenable climate, exploration of the interior from the east coast was more successful and yielded significant discoveries after 1860. David Livingstone's Zambezi Expedition (1858–1863) up the Zambezi River through Mozambique, and up the Shire River into the Shire Highlands of modern Malawi, included the botanist and administrator, John Kirk, and the surgeon, Charles Meller. The first plant collections made by the expedition were shipped to Great Britain in 1860, but owing to some confusion they were delayed in reaching their destination, the Royal Botanic Gardens at Kew. Collections made by Kirk and by Meller included the first records of *Gladiolus atropurpureus, G. decoratus,* and *G. melleri.* These initial collections, all from near Morrumbala, a station at the confluence of the Shire and Zambezi Rivers in Mozambique, were not described until 1876 by J. G. Baker.

In the same year, Baker described *Gladiolus cochleatus.* This species was based on an early collection from Sierra Leone made in about 1800 by the collector Morson, about whom nothing more seems to be known. A homonym for a Cape species of *Gladiolus,* the tropical African *G. cochleatus* was renamed *G. unguiculatus* by Baker the following year. *Gladiolus unguiculatus* is superficially so similar to *G. atropurpureus,* both when alive and preserved, that they have continually been confused and have sometimes been regarded as conspecific. They are, however, separate species and are quite distinct in their vegetative form and growth habit, although less so in their flowers.

The Speke and Grant Expedition (1860–1863), to find the source of the Nile, also included botanical exploration (Grant, 1875). Plants gathered along the route from Zanzibar to Lake Victoria included several species of *Gladiolus,* three of which were thought to be new. Of these, only *G. grantii,* which was named by J. G. Baker (1892) in honor of Colonel James Grant, who was responsible for the botanical activities of the expedition, is still recognized. The two others have been relegated to synonymy: both *G. corneus* and *G. saltatorum* are believed to be minor color variants of *G. dalenii.*

The Scramble for Africa, 1880–1900

The "Scramble for Africa" by the major European powers in the 1880s and 1890s, led to the creation of colonies across all of tropical Africa, spawning numerous organized scientific expeditions, none of which was complete without its botanist. Among the more important of these expeditions, in terms of their botanical results, were the German-sponsored exploration of East Africa. Adolf Engler, botanist on an expedition to the Usambara Mountains of modern Tanzania, discovered *Gladiolus rupicola,* a rare high-altitude streamside species that is restricted to Tanzania and adjacent Kenya. In the highlands north of Lake Malawi, in what became German East Africa, the British explorer Joseph Thomson discovered the plant described by J. G. Baker as *G. thomsonii.* This was the first report from tropical Africa of plants here treated as conspecific with the southern African species, *G. crassifolius.*

Scientific expeditions to Tanzania resulted in the discovery of *Gladiolus oliganthus* and several more species thought to be new at the time. Karl Münzner, the botanist on the Fromm Expedition of 1901 to Lake Tanganyika, discovered *G. muenzneri*. The species was named in his memory by François Vaupel, who described many of the new species collected on German expeditions to both East and West Africa. A. F. Stolz, stationed at Kyimbila in the high mountains north of Lake Malawi, contributed a number of important collections in the 1890s. Among these was the first record of *Acidanthera goetzii,* now renamed *G. curtifolius*. This poorly known species was not re-collected until 1964, when it was found in northern Malawi.

In northern East Africa, the fairly widespread, dry-country species, *Gladiolus candidus,* was discovered in 1876 by J. M. Hildebrandt near Mombasa, in coastal southeastern Kenya. The checkered history of this species, which was given five different specific epithets in the course of just 6 years, three of them by J. G. Baker, is discussed in more detail below. The high-mountain East African endemic, *G. watsonioides,* was discovered on Mt. Kilimanjaro by Joseph Thomson and Henry H. Johnson in 1884. A small-flowered variant of this species, collected on Kilimanjaro in 1887 and 1889, formed the basis for the species first named *Antholyza gracilis* by Ferdinand Pax in 1892. Currently, this rather puzzling variant of *G. watsonioides* is not recognized taxonomically.

At the British settlements established in the Shire Highlands at Blantyre and Zomba, Malawi, interest in documenting the natural history of the area continued. *Gladiolus buchananii* (conspecific with *G. zambeziacus*) was named by Baker in 1892 in honor of James Buchanan, the agriculturalist at the settlement at Zomba. Buchanan also sent back living plants to Kew, among which was *G. brachyandrus,* named by Baker in 1879, and now known to be conspecific with *G. melleri*. Alexander Whyte, British naturalist and administrator, commemorated in *G. whytei* (a white-flowered variant of *G. atropurpureus*), was one of the first people to climb Mount Mulanje, the highest mountain in southern central Africa, where the local endemic, *G. bellus,* was discovered in 1891. Whyte was the first botanist, and one of the first Europeans, to reach the Nyika Plateau in the far north of Malawi in 1895. Among the plants that Whyte collected there was the type of *G. nyikensis*. Specimens of the same species from Chitipa (then Fort Hill), further north in Malawi, that Whyte collected in July 1896 were described by Baker as *G. venulosus*.

Both these northern Malawian species are conspecific with *Gladiolus erectiflorus,* which was discovered in 1894 by the engineer, Alexander Carson, in northern Zambia close to the southern shore of Lake Tanganyika. The first person to collect plants in this part of tropical Africa, Carson made several notable discoveries, including *G. oligophlebius, G. gracillimus,* and what J. G. Baker described as *G. tritonioides,* a later name for *G. laxiflorus,* first recorded by Welwitsch in Angola some 40 years earlier.

In southern Africa, early 19th century exploration had slowly extended outward from the Cape into the interior. In 1815, William Burchell collected *Gladiolus edulis* (now *G. permeabilis* subsp. *edulis*) near Kuruman, in the northern Cape, close to the Botswana border. Then, in 1827, the Danish botanist, C. F. Ecklon, who collected extensively in southern Africa, described what we now know as *G. dalenii,* although he thought the species belonged in the genus *Watsonia.* The plants he saw were cultivated in Cape Town but known to be native to Natal, hence the name he chose, *W. natalensis.* Corms of this species were sent to Holland, where they flowered in 1829, and the plant was formally described by Cornelius van Geel as *G. dalenii* the same year. The same species was also grown in Britain, where J. D. Hooker named it *G. psittacinus* some months after the publication of *G. dalenii.* Meanwhile, corms had been distributed to gardens in Europe by the Dutch botanist, C. G. C. Reinwardt, under the manuscript name *G. natalensis.* When Hooker (1831) learned this, he retracted his *G. psittacinus,* assuming that the name *G. natalensis* was valid and had priority. The confusion over the early nomenclature of this widespread and common tropical African plant was not resolved until 1978.

Southern tropical Africa had been in the British sphere since the mid 19th century, and it was not exposed to the colonial competition that took place further north. Possibly as a result, it was slow to be explored botanically. Only Botswana was deliberately explored at this early period, by the Czech, Emil Holub, a medical doctor as well as naturalist and explorer (Gunn & Codd, 1981). Starting from the northern Cape in 1873, Holub traveled to Shoshong near Mahalapye in northeastern Botswana. On a second expedition in 1875 and 1876, he traveled as far as Sesheke in southern Zambia. Holub discovered the striking species named *Antholyza zambeziacus* by J. G. Baker, now *Gladiolus magnificus.* He also discovered the mysterious *G. micranthus,* named by Baker in 1892. The specimens on which the name was based have been lost, and the species, subsequently renamed *G. bakeri* by F. W. Klatt, cannot be identified (see Incompletely Known Species). The flora of adjacent Zambia and Zimbabwe received little attention until the 20th century. There, most exploration was done by amateurs, in contrast to the almost systematic botanical exploration by the Germans and British in central, eastern, and western Africa.

Nineteenth century botanical exploration in Africa was brought to a climax with the publication of *Flora Capensis* and *Flora of Tropical Africa,* in both of which the Iridaceae, including *Gladiolus,* were treated by J. G. Baker (Baker, 1896, 1898). *Flora Capensis* dealt with the plants of southern Africa (all of South Africa as well as Lesotho and Swaziland), and *Flora of Tropical Africa* treated the plants of the rest of sub-Saharan Africa. In these works *Gladiolus* had a somewhat different circumscription than it has today. Species with extremely zygomorphic flowers, with the upper tepal substantially larger than the upper laterals, and the lower three tepals reduced in size and more or less vestigial, were assigned to *Antholyza.* The genus then included species that are today referred to five different genera. The type of flower that characterized *Antholyza* is believed to be the result of a series of adaptations for sunbird pollination. These adaptations have evolved several times in African Iridaceae and have led to a good deal of nomenclatural confusion (Goldblatt & de Vos, 1989).

The circumscription of *Antholyza* was altered and expanded in 1892 by Ferdinand Pax when he described *A. labiata* from Togo in West Africa. This plant is conspecific with *Gladiolus unguiculatus* and differs from it in no significant way. The inclusion of small-flowered and short-tubed species of *Gladiolus* in *Antholyza* was ignored by most botanists, but not the Belgian, Emile de Wildeman, who in 1902 described *A. descampsii* from Shaba, a long-bracted variant of *G. gregarius,* and *G. gilletii* from the lower Congo, a plant conspecific with *G. unguiculatus* and Pax's *A. labiata.* This expanded concept of *Antholyza* did not find support, although Auguste Chevalier followed this tradition in his studies of West African Iridaceae (Chevalier, 1920).

The other genus that included species of *Gladiolus* was *Acidanthera,* which was recognized on the basis of its white flowers, extremely long perianth tube, and nearly equal tepals. Most well known of the tropical African species of *Acidanthera,* and the type of the genus, *A. bicolor* (i.e., *G. murielae*) was first collected in Ethiopia by G. H. Schimper. *Acidanthera* later came to include a disparate array of species, some now referred to *Hesperantha* and *Geissorhiza.* The genus is now understood to be founded solely on floral adaptations for sphinx-moth pollination, and all the species of *Acidanthera* have been transferred to other genera. *Acidanthera bicolor* and its close relatives, transferred to *Gladiolus* by Marais (1973), are regarded here as constituting a section within *Gladiolus* subgenus *Ophiolyza.*

Although Baker's *Flora of Tropical Africa* treatment of *Gladiolus* summarized knowledge of the genus gathered through the century, *Gladiolus* was still too poorly known for his work to have lasting value. Several species were known from just one or two specimens, and vast parts of the continent had not or had hardly been explored botanically. In fact, nearly half the species we recognize today had not been discovered by 1900, let alone named.

The Turn of the Twentieth Century, 1901–1945

The vast Congo River basin and surrounding highlands were late in receiving botanical attention. The lower Congo is poor in geophytes, but the upper Congo, especially the mineral-rich region of Katanga, now the Zairian province of Shaba, is the most species-rich area for *Gladiolus* in tropical Africa. Sporadic collecting there in the late 19th century resulted in the discovery of a handful of species of *Gladiolus.* In 1891, a certain Captain Descamps collected specimens that Emile de Wildeman described in 1901 as *Antholyza descampsii. Gladiolus verdickii,* discovered in 1900 and named in honor of the collector, Edgard Verdick, was described by de Wildeman in 1901, and the long-misunderstood *G. debeerstii,* collected by the missionary, Gustave Debeerst, in 1894 or 1895, in "Tanganyika" (*sic*), the name given to eastern Katanga, was described by de Wildeman in 1913. Knowledge of the wealth of *Gladiolus* species in Shaba thus began to emerge after 1900, but it was not until after World War II that Shaba was realized to be a major center for the evolution and radiation of *Gladiolus* in tropical Africa.

In Angola, the 1903 Cunene–Zambezi Expedition, led by Otto Warburg into botanically unknown southern Angola, yielded significant new collections of species that were until then only known from Friedrich Welwitsch's collections made almost 50 years earlier. Although four of these were thought to be new by Hermann Harms, only *Gladiolus magnificus* (described in *Antholyza*; Harms, 1903) remains recognized. This was only because the earlier names for the plant, *A. zambeziacus* and *A. spectabilis,* cannot be transferred to *Gladiolus,* in which these epithets have already been used. Harms's *G. longanus* is conspecific with *G. benguellensis,* and *G. kubangensis* and *G. baumii,* the last named in honor of the botanist on the expedition, Hermann Baum, are later synonyms of *G. pallidus* (Baker, 1898).

Particularly noteworthy in this period in the history of *Gladiolus* were the discoveries of German naturalist, Theo (or Theodore) Kassner, who traveled from Cape Town to Alexandria in Egypt (Kassner, 1911). Kassner collected numerous specimens along the way, most significantly in northern Zambia and Zaire. Reaching the Zambian Copperbelt in the summer of 1907 and southern Zaire early in 1908, he made the first collections of *G. velutinus, G. puberulus,* and *G. linearifolius* (Vaupel, 1913, 1920). This last species was re-collected only in 1970 and is still poorly known. Kassner's collections also included several notable range extensions, among which was the discovery of the western Angolan *G. huillensis* in northern Zambia. His collection of this plant was actually thought to represent a new species, described as *G. pubescens* by Vaupel in 1920.

Exploration in the West African colony of Kamerun (part of modern Cameroon) yielded *Gladiolus mirus,* the type of which was collected in 1909 although it was first collected in 1895 (Vaupel, 1913). Also discovered at this time was the plant that Vaupel named *G. heterolobus,* here regarded as conspecific with the Ethiopian *G. roseolus.* The extensive range of *Gladiolus* in West Africa only gradually became known later in the 20th century as British and French botanists gradually explored this vast area. *Gladiolus iroensis* was discovered by Auguste Chevalier in 1903 in southern Chad. The French botanist, H. Jacques-Felix, made the first collections of *G. praecostatus* in 1936, and *G. chevalieranus* in 1937. Both species are endemic to West Africa, *G. praecostatus* restricted to the Nimba Mountains and the adjacent highlands of interior Liberia, Ivory Coast, and southern Guinea, and *G. chevalieranus* to the Fouta Djallon in western Guinea. The last of the three local West African endemic species of section *Acidanthera, G. leonensis,* restricted to interior Sierra Leone, was only discovered in 1949, and it was not described until 1973. The first species of *Gladiolus* to be described from continental tropical Africa (Herbert, 1842), also from Sierra Leone, was collected there before 1840, long before the botanical exploration of the continent began in earnest.

In 1911, Emile de Wildeman described two more new species of *Gladiolus* from Zâire, *G. hockii* and *G. luembensis,* both minor variants of the widespread *G. dalenii.* De Wildeman described another five new species of *Gladiolus* in 1913, based largely on collections made by Theo Kassner, mentioned above, and the Belgians, Adrien Hock and Henri Homblé. Of these, *G. velutinus* and *G. debeerstii* are still recognized.

In British South Africa, later Northern and Southern Rhodesia, and now Zambia and Zimbabwe, largely ignored botanically until after 1900, the only notable collection of *Gladiolus* made there before the turn of the century was the plant recorded by the naturalist, Frank Oates, near Bulawayo in 1874. Described by R. A. Rolfe in 1889 as *G. oatesii,* the species has remained poorly understood. It is, I believe, conspecific with the widespread tropical African *G. unguiculatus,* although it is an unusual form of the species. The fairly common Zimbabwean *G. sericeovillosus* subsp. *calvatus* was first collected there in 1906 by the South African collector, H. G. Flanagan. It was not, however, associated with *G. sericeovillosus* and was either treated as a separate species, *G. dehnii,* or was referred to another South African plant, *G.*

elliotii (Lewis et al., 1972). C. E. F. Allen, Forester for the British South Africa Railways, made the first collection of *G. serapiiflorus* in 1906 near Kalomo in southern Zambia, but it remained undescribed until 1993.

The most significant early collections made early in the 20th century were those of Richard Eyles, beginning in 1906. Eyles was to make collections throughout much of Zimbabwe in the following 30 years. The plant treated in this book as *Gladiolus crassifolius* was evidently first collected in Zimbabwe in the Chimanimani District of eastern Zimbabwe in 1907 by C. F. M. Swynnerton. His collections were the basis for *G. gazensis,* described by A. B. Rendle in 1911. *Gladiolus magnificus,* a species discovered in northern Botswana in 1875 and first recorded in 1901 in Angola, was only found in Zimbabwe in 1941, where it has since been found to be common. None of the eastern Zimbabwean and adjacent Mozambican endemic *Gladiolus* species was collected until after 1945.

The first collections of *Gladiolus dalenii* from Zimbabwe that reached H. M. L. Bolus in Cape Town shortly after World War I attracted the attention of this prolific botanist. She described two new species from Zimbabwean specimens, and one more from plants from Kenya. These were only minor variants of this common tropical African species (Bolus, 1928, 1929). Bolus (1933) named a fourth plant *G. xanthus,* which is conspecific with *G. velutinus,* named 20 years earlier by de Wildeman.

In 1921 the remarkable red-bracted *Gladiolus dichrous* was discovered on Mt. Elgon in eastern Uganda. It is so unusual a species that its correct affinities to *Gladiolus* were not initially evident and it was assigned to a new genus, *Oenostachys* (Bullock, 1930), so named for its enlarged and red-purple bracts that dominate the flowering spike and all but conceal the small whitish flowers. The circumscription of *Oenostachys* was expanded by the British botanist, N. E. Brown (1932), to include *G. abyssinicus* as *O. abyssinicus.* Brown was aware of the heterogeneous composition of *Antholyza,* and he proposed that the genus be restricted to include just one South African species. Brown referred the erstwhile tropical African species of *Antholyza* to *Homoglossum* and *Petamenes.* The types of both these genera are Cape species now believed to be only distantly related to the species in tropical Africa that were included in the same genera (de Vos, 1976; Goldblatt & de Vos, 1989).

The years 1930–1945 saw no major botanical activity in tropical Africa. B. D. Burtt made the only known collection of *Gladiolus canaliculatus* in central Tanzania in 1933, however, a plant described here for the first time. After World War II, new collections began to accumulate so that known distribution ranges were gradually expanded. New collections also have the result of bridging some of the variation between some species, thus showing some of the redundancy in species numbers.

The Modern Era, 1945–1968

After World War II, Shaba Province of Zaire began to receive close botanical attention. The Belgian biologist, Paul Duvigneaud, traveled extensively in the province in 1959 and 1960, especially in the virtually unexplored southwest toward the Angolan border. Duvigneaud also systematically collected at mine sites where ore bodies close to the surface produced heavy-metal-enriched soils. Among a substantial number of edaphically restricted species of several plant families documented at these sites were *Gladiolus actinomorphanthus, G. robiliartianus, G. tshombeanus, G. ledoctei* (Duvigneaud & Denaeyer-de Smet, 1963), and *G. microspicatus* (Còrdova, 1990). Duvigneaud also made the first and only known collections of *G. salmoneicolor* and *G. pungens,* both of which he considered new species but which were only described much later by S. Còrdova (1990). Duvigneaud also amply documented another species that he thought was undescribed, here named *G. pusillus* (to which he gave the manuscript name *G. opheliae*). *Gladiolus pusillus* was actually first recorded by the Belgian botanist, Paul Quarré, in 1933 near Elisabethville (now Lubumbashi), but is was subsequently found in a wide area of northern Zambia as well as Shaba.

Zimbabwe also began to receive careful botanical study after 1945. The fairly common *Gladiolus sericeovillosus* was not at this time associated with this southern African species, and collections made by Gretel Dehn were described as the new *G. dehnianus* by H. Merxmüller in 1951. Lewis et al. (1972) believed that *G. dehnianus* was conspecific with the Transvaal species *G. elliotii,* and it is only as a result of my studies that *G. dehnianus* has been associated with the glabrous variant, *G. sericeovillosus* subsp. *calvatus* (Goldblatt, 1993). The first of the eastern Zimbabwean and western Mozambican highland *Gladiolus* endemics, *G. flavoviridis,* was discovered in 1949 by N. C. Chase. Then, *G. juncifolius* was found in 1950 by the South African, Sheila Thompson, and *G. zimbabweensis* was first collected in 1955 by the collecting team of A. W. Exell, F. A. Mendonça, and Hiram Wild, who were making collections for the new flora project, *Flora Zambesiaca.*

Angola still provided novelties, for although it was comparatively well collected by Welwitsch, additional botanical exploration there has brought to light a handful of new species described here, including *Gladiolus scabridus,* discovered by John Gossweiler in 1906 but inadequately documented until the 1960s. The southwestern Angolan *G. fenestratus* was first collected by the missionary Eugene DeKindt in 1898, but the species was only properly documented in 1969 by the Swedish botanist, Lars Kers. Kers also made the first collection of *G. stenosiphon,* a species restricted to the Chela Mountains, which form the western edge of the Huila Plateau of southwestern Angola. The unusual *G. amplifolius* is still known from just one gathering, made near Huambo by Manuel da Silva in 1970.

The extraordinary work of Mary Richards in northern Zambia and southwestern Tanzania in the late 1950s and early 1960s, and of Jean Pawek in Malawi in the 1970s, meticulously documenting the flora of this part of central Africa, deserves special mention. Richards discovered the central Tanzanian endemic, *Gladiolus richardsiae,* here named in her honor. Both Richards and Pawek documented the distributions of several tropical African species, some at the time known only from the type specimens collected the previous century. Their ample and well-preserved collections have contributed substantially to the understanding of a number of species, notably *G. intonsus, G. muenzneri, G. oliganthus, G. oligophlebius,* and *G. pusillus.*

The Synthetic Period, 1968–Present

The first of a series of major regional floristic studies of tropical Africa, begun in West Africa, resulted in the publication of *Flora of West Tropical Africa* in the period 1927–1936 by John Hutchinson and J. M. Dalziel. A second, much revised edition followed in the 1960s, in which the Iridaceae were treated by F. N. Hepper (1968). In both of these studies *Acidanthera* was regarded as a genus separate from *Gladiolus,* but Hepper's treatment of *Gladiolus* was notable for treating the tropical African *G. quartinianus* as conspecific with a southern African *G. psittacinus,* the earliest name for which has since been found to be *G. dalenii.* In 1972, the Belgian, D. Geerinck, published a revision of the genus for Zaire, Rwanda, and Burundi. Geerinck entertained a very broad species concept, and his inclusion of *G. atropurpureus* in *G. unguiculatus, G. tshombeanus* in *G. gracillimus,* and *G. melleri* (as a subspecies) in *G. dalenii* seem to me unwarranted and hence are not followed.

Also in 1972, the South African botanist, Amelia Obermeyer, with the help of T. T. Barnard, British *Gladiolus* grower and breeder and expert on the genus, completed G. J. Lewis's revision of *Gladiolus* in southern Africa, left unfinished at her untimely death in 1967. This monograph of *Gladiolus* for all of southern Africa, including southern Mozambique and Swaziland (but not Botswana), recognized 103 species in the subcontinent. It brought order to a large part of this diverse genus, with its complex and confused taxonomy and hundreds of names, many never typified or properly documented. Although published relatively recently, the monograph is outdated as a result of nomenclatural and taxonomic changes, the discovery of several new species in southern Africa, and the inclusion in *Gladiolus* of the small segregate genera *Homoglossum* and *Anomalesia.* There are now thought to be 150 species of *Gladiolus* in southern Africa, including Botswana. Just five species in southern Africa are shared with tropical Africa (excluding southern Mozambique) and Madagascar, bringing the total species in the genus to more than 250 with the addition of the seven species endemic to Madagascar and some 10 Eurasian members of the genus.

While modern floristic and revisionary studies were being produced, one long-neglected part of tropical Africa, southern Ethiopia was explored more thoroughly. This area had first been botanized by P. O. Bally and J. B. Gillett in the 1950s. Bally and Gillett established the occurrence of *Gladiolus boranensis* in southern Ethiopia, a plant first recorded from southern Sudan, and Bally made the first collection of *G. calcicola,* endemic to the region around Harar in 1954. The first gatherings of *G. balensis* and *G. negeliensis* were, however, not made until after 1970, when M. G. Gilbert, Mats Thulin, Kai Vollesen, and Mesfin Tadesse collected extensively in Sidamo and Bale Provinces. These two species are described here.

Gladiolus is now relatively well known throughout its range, and I have every confidence that the conclusions reached in this work are by and large accurate. Much collecting does, however, remain to be done. Southern Ethiopia and Somalia remain poorly known. Central Tanzania also merits more botanical attention. Three species that occur there, *G. canaliculatus, G. richardsiae,* and *G. stenolobus,* are each known from single gatherings. Perhaps there are still unrecorded species there. Angola, especially the western escarpment, has received botanical attention since the 1850s, yet species such as *G. amplifolius, G. fenestratus,* and *G. harmsianus* are all too little known. Southern Zaire merits additional exploration. Species there such as *G. camilae, G. curtilimbus, G. debeerstii,* and *G. linearifolius* need additional collecting, something that may add to our understanding of their relationships. Likewise, *G. puberulus* and *G. verdickii* have unusual variant populations in Shaba, the significance of which remains to be determined. New information may well show them to be worthy of taxonomic recognition. An unusual population of *G. permeabilis* on the Great Dyke in Zimbabwe may also be a new species, and the significance of the plant from the Mafinga Mountains in Malawi, believed on the basis of a poor specimen to be *G. pretoriensis,* otherwise only known from South Africa, needs to be investigated. The list of additional work on tropical African *Gladiolus* could be enlarged considerably. This book will make it possible to focus on particular regions and particular problems for others to address.

Morphology

CORM

Like most Ixioideae, the corms of *Gladiolus* are more or less globose to depressed-globose and produce roots in an unorganized fashion from the lower half. As well as assimilatory roots (Figure 1A–C), a corm will often produce one or more thick, fleshy contractile roots (Figure 1B). These serve to pull the corms deeper into the ground. The contractile roots thus ensure that corms remain buried in the soil at a level suitable for the plant. Corms vary considerably in size, a feature often directly related to plant size. The largest corms are found in some species of the *G. dalenii* complex (section *Ophiolyza*), often 30–40 mm in diameter. The smallest corms are found in subgenus *Gladiolus*. They are particularly small in species of section *Hebea*, for example, in *G. gracillimus* and *G. pusillus*. Corm surface and internal tissue may be white or red to orange, the latter most common in the *G. dalenii* complex, and there is a weak correlation between corm size and color, larger corms most often having orange flesh. Perhaps this is related to tannin content and defense against predation. Outgroup comparison suggests that small corms with white flesh are ancestral in *Gladiolus*. The production of small cormlets on short, threadlike stolons is often characteristic of species or species groups. *Gladiolus dalenii* is notable for the production of many cormlets; fewer cormlets are produced in all the species of section *Decoratus*, in which the cormlets are typically concealed in the fibrous corm tunics. In section *Blandus*, *G. verdickii* is notable for the production of numerous tiny cormlets on stiff ascending fascicled stolons (Figure 2E). In the closely related *G. erectiflorus*, one to three cormlets are produced on slender stolons. Cormlet production has limited taxonomic value, however, because species are often variable for the character. Both *G. verdickii* and *G. erectiflorus* can, however, be recognized consistently by the way their cormlets are formed.

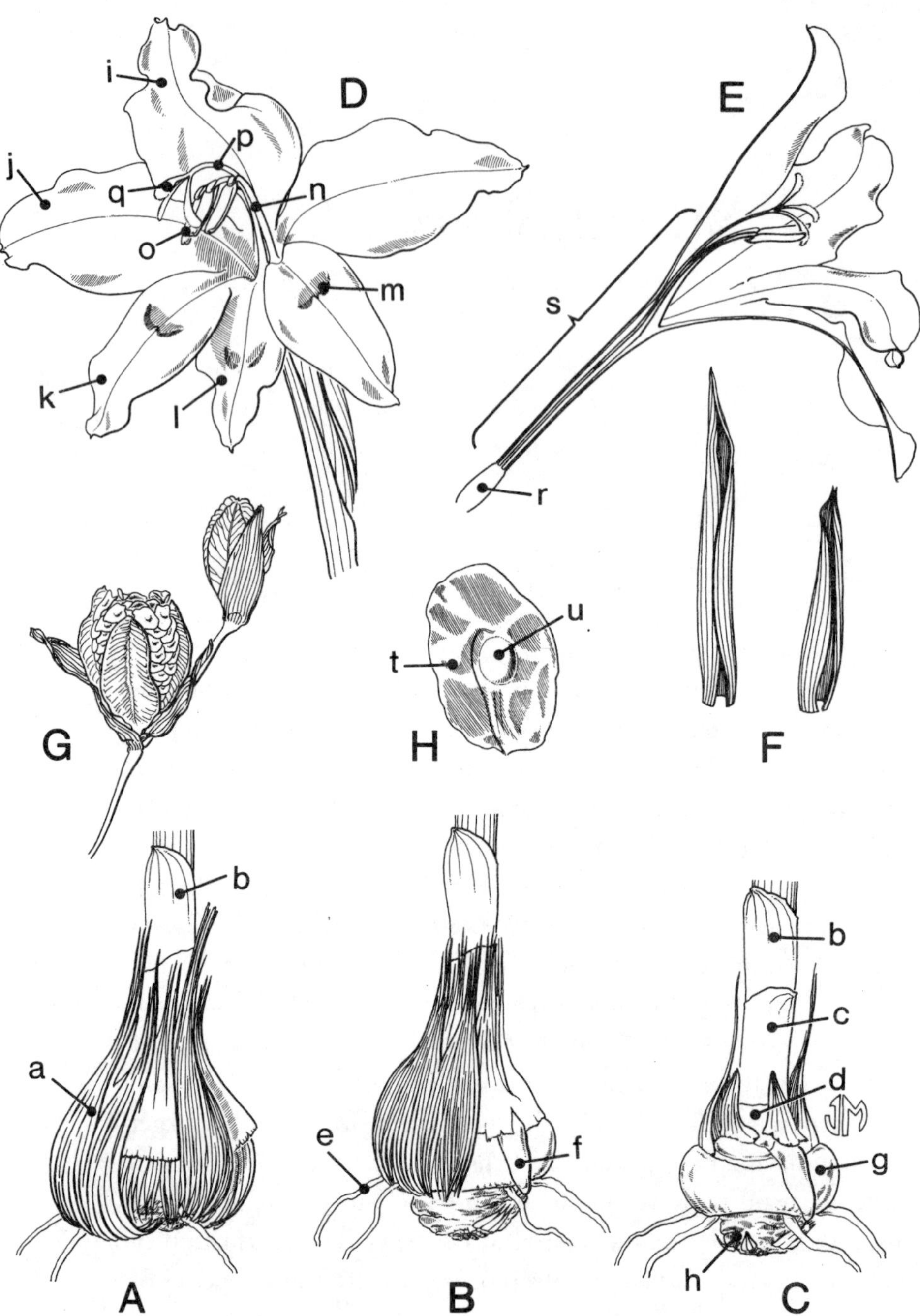

FIGURE 1. *Gladiolus carneus,* type species of section *Blandus*, illustrating the various reproductive and vegetative organs of the *Gladiolus* plant. A–C, Base of plant, showing corm tunics (a), upper (inner) cataphyll (b), middle cataphyll (c), and lower (outer) cataphyll (d), roots (e), contractile root (f), current (g) and previous season's (h) corm. D, Whole flower, dorsal tepal (i), upper lateral tepal (j), lower lateral tepal (k), lower median tepal (l), nectar guide (m), filaments (n), anthers (o), style arching over the stamens (p), style branch (q). E, Vertical section of flower, with details of ovary (r), perianth tube (s), consisting of a slender lower part and wider upper part or throat, with stamens attached at the top of the lower part of the tube. F, Outer (larger) and inner (smaller) floral bracts. G, Portion of spike with an intact capsule and a dehiscing capsule with the seeds visible. H, Single seed with broad wing (t) and small round seed body in the center (u).

Corm coverings, more commonly called corm tunics, derived from the bases of the cataphylls and lower leaves, are often characteristic of species groups of *Gladiolus* (Figure 2). Least specialized are probably the coriaceous to firmly papyraceous, brown tunics such as those found in most members of subgenus *Ophiolyza* (Figures 1A, 2C–G). These layers closely resemble dry foliage and they decay to irregularly broken fragments that sometimes become more or less fibrous, because the vascular tissue is more resistant to decay than the mesophyll. Similar plesiomorphic corm tunics occur in some tropical African members of subgenus *Gladiolus* (Figure 7), for example, *G. crassifolius*. More typical of the subgenus are specialized, soft-papery tunics, and the more common tunics composed of reticulate fibers of medium to fine texture. Such tunics are characteristic of most of the endemic species of *Gladiolus* in Madagascar, and in tropical Africa of *G. atropurpureus* (Figure

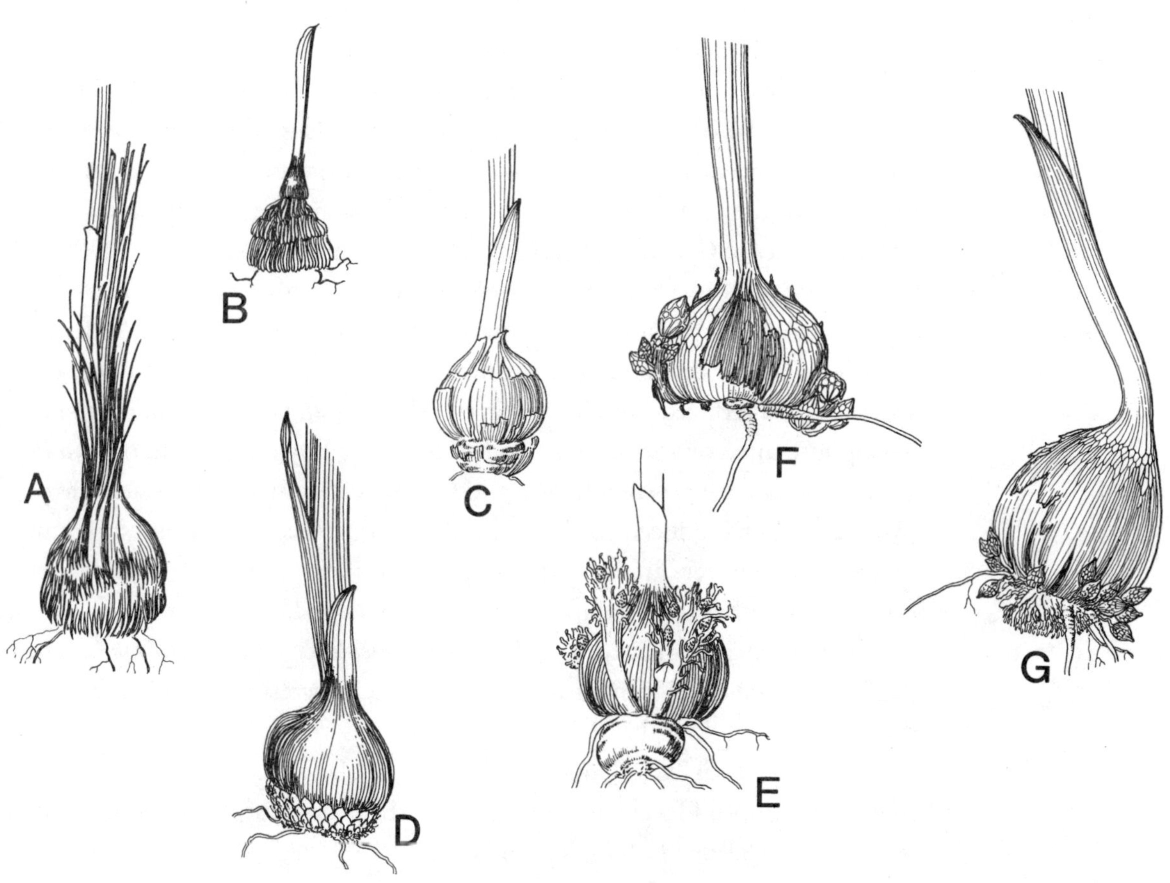

FIGURE 2. Corm types in tropical African species of *Gladiolus*. A, *G. tshombeanus*. B, *G. atropurpureus*. C, *G. rupicola*. D, *G. dalenii* subsp. *dalenii*. E, *G. verdickii*. F, *G. aequinoctialis*. G, *G. longispathaceus*.

2B), its close allies of section *Gladiolus,* section *Heterocolon,* and section *Decoratus* of subgenus *Ophiolyza.* In *G. atropurpureus* the tunic fibers are characteristically thickened into clawlike ribs below, an apomorphy that makes it possible to recognize this and its close ally, *G. serapiiflorus.* In section *Heterocolon* the tunics typically accumulate in a thick mass of coarse fibers that extend upward around the base of the stem, apomorphic for the alliance (Figure 2B). More specialized corm tunics are found among some Cape species of *Gladiolus* in which the hard, more or less woody texture is probably associated with the coarse-grained nutrient-poor soils and hot, dry summers.

LEAVES

Apart from developed foliage leaves, plants have three (occasionally only two remain in herbarium specimens) cataphylls that sheath the underground part of the shoot and always lack blades (Figure 1A–C). These foliar organs are the first ones produced by the shoot, and they grade from pale and membranous to firm in texture. The inner (and uppermost) cataphyll generally extends shortly above the ground and then becomes green or sometimes purplish. The distinction between the upper (inner) cataphyll and the lowermost leaf can be obscure, particularly when the foliage leaves have vestigial blades. The lower cataphylls are usually truncate, and the upper acute. Above the ground, the cataphylls are sometimes pubescent and downy, notably in *Gladiolus dalenii, G. roseolus,* and *G. velutinus,* and their close allies.

Foliage leaves vary in number from a maximum of eight to just one. Based on outgroup comparison, the production of several leaves must be regarded as plesiomorphic, and reduction to few apomorphic. Among the species with a reduced number of leaves, *Gladiolus unguiculatus* stands out in having just three or sometimes two leaves on the flowering stem, *G. canaliculatus* consistently has two leaves, and *G. curtilimbus* may have either one or two leaves, both reduced to short sheaths. Reduction in leaf number seems to have occurred repeatedly in different lineages within the genus, and there is no reason, other than leaf number, for example, to believe that *G. canaliculatus* is immediately related to either *G. unguiculatus* or *G. curtilimbus.* The degree of blade development is an important characteristic in *Gladiolus* and the presence of long-bladed foliage leaves inserted on the stem below ground level (basal leaves), and smaller shorter-bladed leaves inserted on the stem above ground level (cauline leaves), is the plesiomorphic condition not only of *Gladiolus* but of Ixioideae in general.

Another apomorphic trend in leaf organization is the reduction of the blades of the cauline leaves. Either the uppermost or all the cauline leaves may be elaminate (without blades). All that remains of these leaves are the sheathing bases, and they are termed sheathing leaves. Suppression of the leaf blades on the flowering stem is a repeated specialization in *Gladiolus* and appears to be an important ecological adaptation. In subgenus *Ophiolyza* the two early-flowering subspecies of *G. dalenii,* subspecies *andongensis* and *welwitschii,* leaf blades on both cauline and basal leaves are much reduced or absent, as they are in *G. melleri* and *G. roseolus.* In these taxa the foliage leaves are produced on separate shoots later in the season from the same corm as the flowering stem (Figures 49, 50). In *G. huillensis* (section *Ophiolyza*) and *G. iroensis* (section *Acidanthera*), the leaf blades are usually short but plants do not produce separate leafy shoots.

Similar adaptations occur in subgenus *Gladiolus,* but only in *G. unguiculatus* do plants regularly produce long leaf blades on separate shoots after flowering at the beginning of the wet season. In other species with reduced leaf blades (e.g., *G. atropurpureus, G. curtifolius, G. curtilimbus, G. muenzneri, G. serapiiflorus*), all photosynthetic activity takes place in the leaf sheaths, stems, and vestigial blades. These species do not produce separate leafy shoots. The degree of reduction of the leaf blades is generally variable in a species, and individual populations are characterized by having vestigial to moderately developed blades. This is particularly so in *G. atropurpureus,* populations of which usually have very short leaf blades, but in some late-flowering populations plants often have leaves with blades as long or somewhat longer than the sheaths. This has led to the naming of plants with differences in the degree of blade development as separate species (see discussion under the species).

Additional leaf features of taxonomic importance include shape in outline and transverse section, texture, and pubescence. Equitant, narrowly lanceolate, and firm-textured (more or less coriaceous) leaves are the rule in Ixioideae. Linear leaves, as in the related *Gladiolus linearifolius* and *G. sulcatus,* and in *G. permeabilis* among others, are apomorphic, as are the soft-textured leaves of *G. decoratus, G. mirus,* and *G. oligophlebius.* Typical leaf specializations in Ixioideae are the thickening of the midrib and the margins by the enlargement of the fibrous caps of the primary veins and subepidermal marginal tissue. These areas then usually appear semitransparent or hyaline when alive due to the absence of green tissue. On drying, the difference in coloration becomes more pronounced. Rarely, as in *G. sericeovillosus* subsp. *calvatus,* only the margins become thickened and the midrib is hardly distinguishable from one or more secondary veins. More elaborate development of the margins and midribs into ridges or wings extended at right angles to the leaf blade are well known in Cape species of *Gladiolus* and also occur in *G. pallidus* and *G. sulcatus* in tropical Africa.

Alternatively, the midrib and margins may be so heavily thickened that the nonthickened blade surface becomes restricted to narrow grooves or sulci (Figure 6D, F). An equal degree of marginal and midrib thickening results in a centric (terete) blade in which the four longitudinal sulci may only be visible by microscopic examination (Figure 6C, G, H). This degree of blade specialization is characteristic of *Gladiolus pretoriensis*, its close allies in southern Zaire (section *Heterocolon*), and a few other tropical African species, including the Angolan *G. pallidus* and the eastern Zimbabwean *G. juncifolius*. Clearly a xeromorphic adaptation, leaf thickening has evolved repeatedly in *Gladiolus* as it has in the related *Geissorhiza* (Goldblatt, 1985), and occasionally in other genera (*Hesperantha*), whereas this feature is ancestral in *Romulea* and *Devia* (Goldblatt & Manning, 1990).

Leaf pubescence occurs occasionally and is a very useful taxonomic character. *Gladiolus puberulus* and *G. intonsus* (subgenus *Gladiolus*) consistently have heavily pubescent leaves. A light, scattered pubescence on the leaf sheaths and sometimes the lower part of the blades is characteristic of *G. laxiflorus, G. zambesiacus,* and *G. zimbabweensis,* and a few others. In *G. velutinus* the leaf surface has a velvet texture, a result of the elongation of the papillae of the epidermal cells.

STEM

For convenience, the flowering stalk is called the stem here although this is not strictly correct, the corm also being a part of the stem system. All species of *Gladiolus* have an aerial flowering stem bearing leafy organs along its length, usually three basal cataphylls, a cluster of longer leaves inserted near the ground, and one or more smaller leaves inserted on the stem above the ground. Stems may be simple or branched, but in many species they may be consistently unbranched (simple), a common specialization in the genus as it is in other genera of subfamily Ixioideae. Species of section *Ophiolyza,* including the *G. dalenii* complex, seldom produce branches unless individuals are particularly robust. In subgenus *Gladiolus* branching is suppressed in species with reduced leaf blades and in several species with laminate leaves as well. Distinctive short branches are produced in the Shaban *G. ledoctei,* always in the axil of the penultimate cauline leaf.

Stems are terete throughout the genus and pubescent only in some southern African populations of *Gladiolus sericeovillosus.* Stems are typically flexed outward and inclined above the sheath of the uppermost leaf or just below the base of the spike in subgenus *Gladiolus,* or are erect throughout as in nearly all species of subgenus *Ophiolyza*. Stem thickness is directly related to plant size, and diameter at the base of the spike, given in all descriptions, is a useful measure.

INFLORESCENCE

The flowers of all species of *Gladiolus* are arranged in a spike, as they are in most genera of subfamily Ixioideae (Figure 3). The flowers are thus sessile (without pedicels) and each flower is subtended by a pair of opposed bracts, a larger outer and smaller inner bract (Figure 1F). In *Gladiolus* the spikes are characteristically secund (probably plesiomorphic in the genus) and either inclined (subgenus *Gladiolus*) or erect (subgenus *Ophiolyza*). Spirally arranged flowers occur in some species with actinomorphic flowers (probably a derived condition in *Gladiolus*). Distichous and erect spikes characterize *G. sericeovillosus* and the southern African *G. elliotii* and *G. oppositiflorus*. Number of flowers per spike is particularly variable even within populations and depends largely on health and age of the plants. A particular range is, however, characteristic of a species. This has resulted in the use of epithets such as "*multiflorus*" (a synonym of *G. gregarius*) and "*pauciflorus*" for species. Variability of flower number within species limits the usefulness of the character

The bracts are typically green and firm in texture in *Gladiolus* and are fairly constant in size for species and even species groups, thus they are a useful taxonomic character. The outer bracts are typically slightly longer than the inner, and the latter is usually apically forked for a short distance. Unusually attenuate bracts characterize *G. erectiflorus* and its close relatives, *G. stenosiphon* and *G. verdickii* (section *Tenuibracteus*). In most species of section *Ophiolyza* the bracts lightly clasp the stem in the lower half, and in *G. sericeovillosus* and the southern African *G. elliotii* the inner bract margins are united below, forming a tube around the flower bud and the perianth tube of the open flower. In species of xeric habitats the bracts often become membranous or dry above, and in a few members of section *Heterocolon* the bracts often become entirely dry and tawny-colored as flowering progresses.

Unusually large floral bracts relative to the flowers and internodes of the spike occur in *Gladiolus gregarius* and its close relatives, *G. harmsianus* and *G. macrospathus,* and to a lesser extent, in *G. microspicatus* (Figures 21–23). In these species the bracts are firm and straight, clasping the stem and at least two and sometimes three internodes long. These bracts and the associated narrow capsules are apomorphic for this small clade.

Large, purplish to bright red floral bracts occur in a few northeast African species, notably *Gladiolus abyssinicus* and *G. dichrous.* Here the bracts seem to be part of the attraction for pollinators, and their flowers, thought

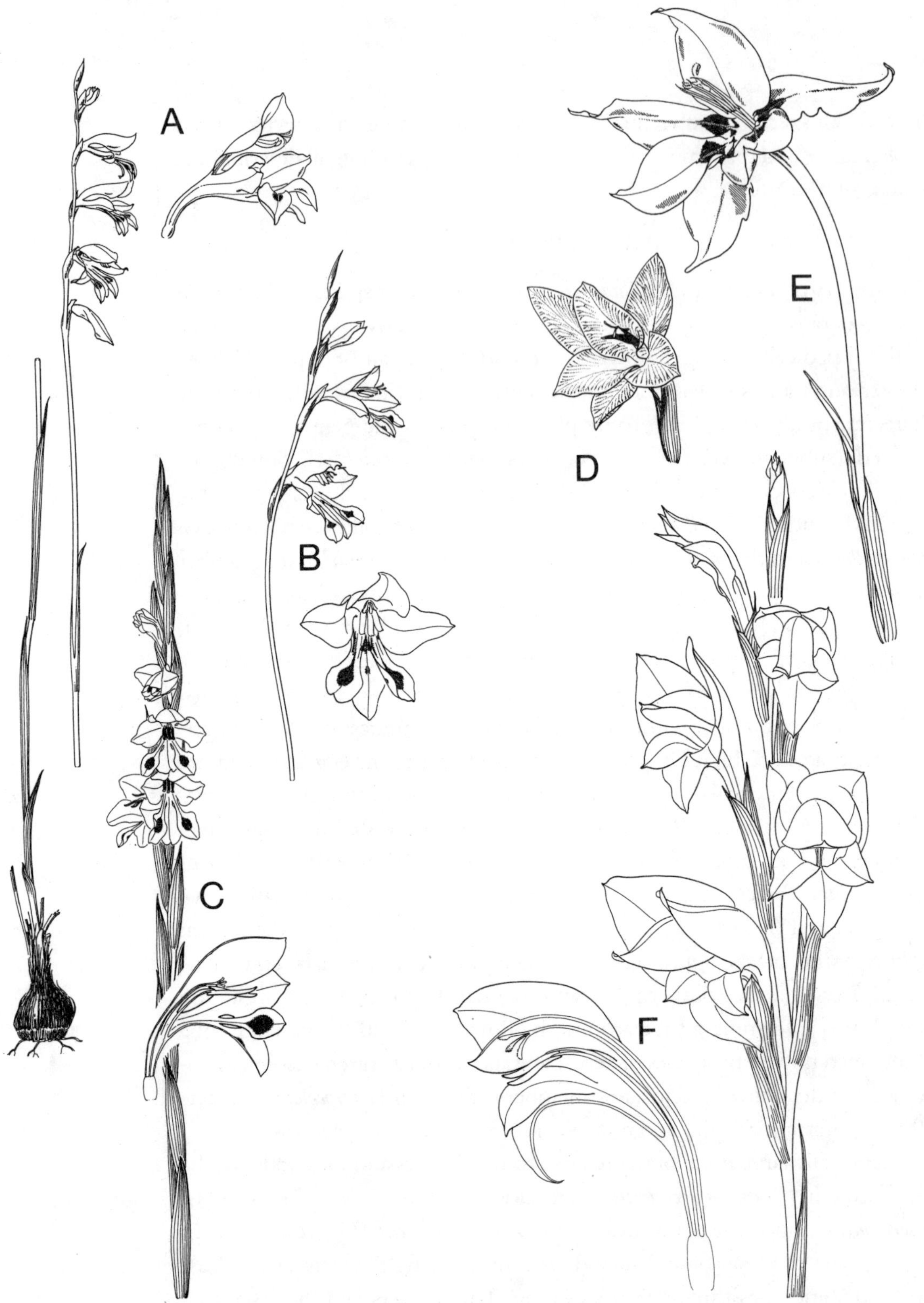

FIGURE 3. Flower types in subgenus *Gladiolus* (A–C) and subgenus *Ophiolyza* (D–F). A, *G. unguiculatus.* B, *G. atropurpureus.* C, *G. gregarius* (note the large bracts, unusual for the subgenus). D, *G. erectiflorus.* E, *G. aequinoctialis.* F, *G. dalenii* subsp. *dalenii.*

to be pollinated by sunbirds, are themselves fairly small. So large and bizarre are the bracts of *G. dichrous* that the species was for a time considered a separate genus, *Oenostachys,* named for the wine-colored bracts (Bullock, 1930).

PERIANTH

More variable than any other organ in the genus, the perianth of *Gladiolus* is the key to understanding the evolutionary success of the genus. The perianth ranges from actinomorphic in a few scattered species across the entire genus, including *G. actinomorphanthus* (Figure 25) from Shaba and *G. bojeri* from Madagascar (Goldblatt, 1989). In both these species the actinomorphic condition is secondary. In *G. actinomorphanthus* the spike is inclined and the short tube curved, neither feature compatible with ancestral actinomorphy. In *G. bojeri* the actinomorphic flower is unusually small and unique in the genus in being truly pendent on a drooping peduncle. The species also has derived, finely fibrous corm tunics and remarkably long cylindric capsules. All these features suggest that *G. bojeri* is a specialized species and therefore that its actinomorphic flower is secondary.

Outgroup comparison, the recommended method for determining ancestral traits, is not useful in the case of perianth symmetry in *Gladiolus* because the immediate generic relationships are uncertain. All that can be said is that it seems likely that floral zygomorphy is the plesiomorphic (ancestral) condition, as is relatively small flower size and a fairly short perianth tube. Comparison with other genera of Ixioideae in which a range of floral types occur suggests that zygomorphy probably evolved as an integrated unit, a bilabiate perianth with a larger dorsal (upper or posterior) tepal inclined to arched over unilateral stamens, combined with closely aligned lower tepals marked in contrasting colors to form a nectar guide (Figure 1D).

Perianth coloration is remarkably varied in *Gladiolus,* indicating adaptation to a variety of pollinators. One repeated specialization is the development of a white or cream flower, generally with purple or reddish streaks on the lower tepals. This specialization is frequently accompanied by an elongation of the perianth tube. In tropical Africa this type of flower characterizes the species of section *Acidanthera,* and the tube in this group is remarkably long, 10–12 cm. In southern Africa similar perianth coloring and a relatively long tube characterize *G. tristis,* an important species in early *Gladiolus* breeding, as well as *G. longicollis,* and these species are also known to be pollinated by moths. Among other tropical African species, white, long-tubed flowers occur in *G. harmsianus* (allied to *G. gregarius*), *G. stenosiphon* (allied to *G. erectiflorus*), and several more species of both subgenera. The repeated development of this type of flower, associated with pollination by sphinx moths, demonstrates the plasticity of the *Gladiolus* genome and its ability to respond to changing adaptive pressures.

Despite the variability in the *Gladiolus* flower, there appear to be two fundamentally different basic types. A first group of species, subgenus *Gladiolus,* has small flowers, generally (18–)25–40 mm long (Figure 3A–C). Associated with this small flower, the lower tepals are typically united below for a short distance and narrowed in the lower half into claws, and the nectar guides are often on the upper third of the tepals. Except for species with obviously modified long perianth tubes, the tube is (6–)10–15 mm long and about half as long as the dorsal tepal. The lower tepals are often channeled below, and usually more or less directed forward below and sharply flexed downward distally. In profile, the lower tepals exceed the dorsal tepal although they are seldom longer than the dorsal in linear measurement. The term *prognathous* is sometimes applied to the flowers of such species.

A second group of species, here treated as subgenus *Ophiolyza*, is characterized by substantially larger flowers, 35–80 mm long, not including long-tubed moth-pollinated species in which the flower may be up to 150 mm long (Figure 3D–F). Associated with the larger flower, the lower three tepals are hardly if at all united to one another, are not or only weakly narrowed below but never clawlike, and are generally either gently or strongly curved from the base. The nectar guides in the subgenus usually take the form of longitudinal streaks on the lower part of the tepals. Viewed in profile the lower tepals often seem as long or shorter than the upper. *Gladiolus dalenii,* probably the most important species in *Gladiolus* breeding, and one of the most well known species, is typical of subgenus *Ophiolyza.*

Adaptive floral radiation in both subgenera has led to a blurring of the limits of the two groups. For example, based on flower size alone the relationships of *Gladiolus sericeovillosus* and *G. ecklonii* are not immediately clear. Although the flowers are fairly small and accord better with subgenus *Gladiolus,* the spotted or streaked distribution of dark pigment, shape and orientation of the lower tepals, longitudinal nectar guides, and comparatively large bracts convincingly place them in subgenus *Ophiolyza.* Similarly, the fairly large-flowered tropical African *G. laxiflorus* nevertheless seems better placed in subgenus *Gladiolus* in view of its clawed and distally marked lower tepals, short bracts, and even shorter perianth tube. In so large and diverse a genus as *Gladiolus* such apparent inconsistencies should not be surprising. The affinities and correct classification of problematic species requires careful analysis and may in some cases be arbitrary, depending on the intuition of the systematist.

Another important series of floral adaptations, most common in tropical Africa in subgenus *Ophiolyza,* is that for pollination by sunbirds (Figure 4). Perianth color is bright red, and the perianth tube is usually cylindric but sharply divided into upper and lower parts. The tepals may be subequal but more often the dorsal tepal is much enlarged, or the lower three are much small than the upper. Prime examples of this flower type are found in the Ethiopian species *Gladiolus abyssinicus, G. longispathaceus,* and *G. schweinfurthii* (Figure 4D–F), which probably form a single lineage with *G. watsonioides.* In southern tropical Africa, *G. magnificus* has a similar type of flower (Figure 4B), but this species is most likely independently derived from *G. dalenii,* itself thought to be pollinated by sunbirds although it lacks some of the adaptations usually associated with this type of pollination (Figure 4A). Species of *Gladiolus* with red flowers, elongate perianth tubes, and somewhat to extremely reduced lower tepals also occur in southern Africa (Figure 4G–I) and these have in the past been regarded as belonging to separate genera, *Homoglossum* (= *Petamenes*) and *Anomalesia.* Some of the tropical Afri-

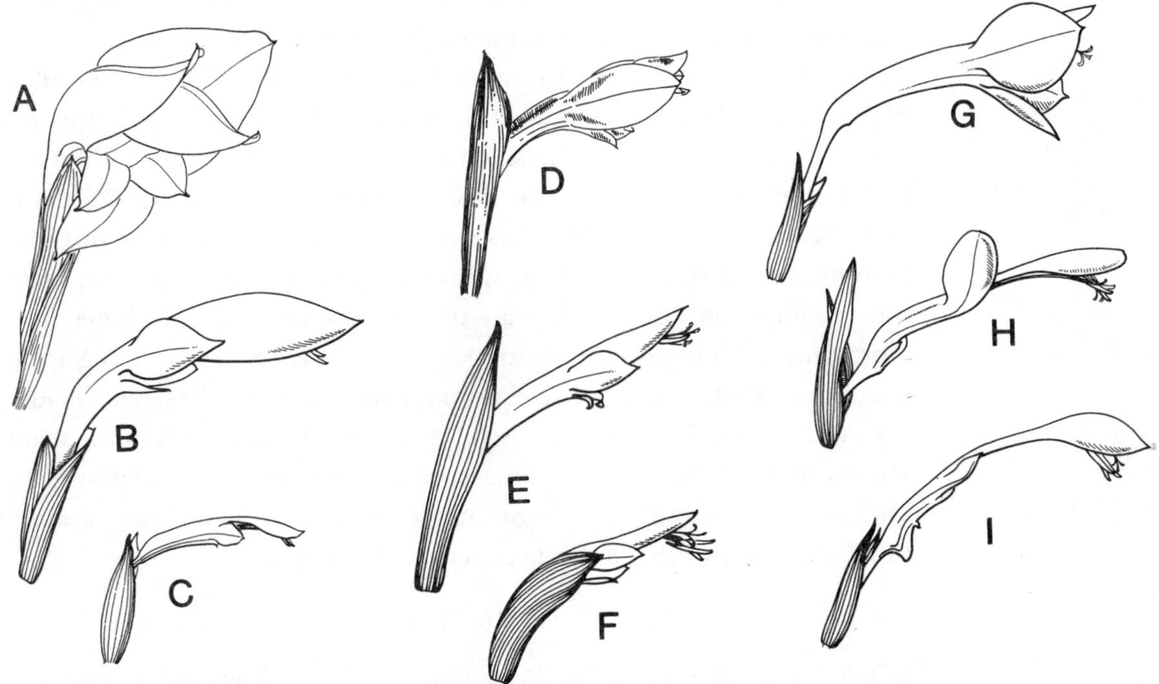

FIGURE 4. Examples of flowers of species of *Gladiolus* adapted for sunbird pollination. A, *G. dalenii* subsp. *dalenii.* B, *G. magnificus.* C, *G. huillensis.* D, *G. longispathaceus.* E, *G. abyssinicus.* F, *G. schweinfurthii.* G, *G. bonaspei.* H, *G. cunonius.* I, *G. saccatus.* Species A–F are tropical African; G–I are southern African and belong to two separate clades. The repeated convergence is striking.

can species with similar flowers have at times been included in *Homoglossum*. There seems no doubt that these bird-pollinated flowers have evolved independently at least twice in both tropical and southern Africa, and species with such flowers are best included in *Gladiolus* instead of being treated as one polyphyletic genus or two to four smaller genera, each derived from different ancestors within *Gladiolus*.

ANDROECIUM

Like all Ixioideae and most Iridaceae there are three stamens in *Gladiolus* (Figure 1D). The filaments are inserted at the base of the upper part of the perianth tube, usually at the point where the tube widens (Figure 1E), but the distinction is not evident in section *Acidanthera*. The stamens are unilateral and arcuate to horizontal in most species but symmetrically arranged around the style in the few species with actinomorphic flowers. Filaments are normally exserted for some distance from the tube, although more or less included in the flower, but in several species the filaments are included. This is often a useful diagnostic character, allowing the immediate recognition of *G. benguellensis* and *G. melleri* from their relatives in section *Ophiolyza*. In subgenus *Gladiolus* included filaments characterize *G. canaliculatus, G. curtifolius,* and *G. muenzneri,* possibly a synapomorphy for these three species.

The anthers lie parallel to one another in species with zygomorphic flowers and are usually contiguous (Figure 1D). Variation in size is a useful taxonomic character in distinguishing allied species. Fairly large anthers are typical of subgenus *Ophiolyza,* and smaller anthers of subgenus *Gladiolus*. Anthers are subbasifixed in all the tropical African species, and most of those from southern Africa, where a few species related to *G. saccatus* have sagittate anthers. Although obtuse apices are the rule, the connective is drawn into short acute appendages in some species of section *Tenuibracteus*. Long acute apiculate appendages characterize sections *Decoratus* and *Acidanthera* (Figure 3E). Anther color is sometimes useful taxonomically. Pale yellow to cream anthers are most common, but blue to violet anthers occur in some species. The character is useful in distinguishing *G. erectiflorus* from *G. zambesiacus,* although there are other differences between these species.

GYNOECIUM

Simple but apically expanded to spathulate style branches are apomorphic characteristics in *Gladiolus*. Most other genera of Ixioideae have either filiform style branches, or the branches are divided apically for some distance. Variation in style type is rare within genera of the subfamily. The only genus with style branches comparable to those of *Gladiolus* is *Babiana,* a largely southern African genus with one species on Socotra and one ranging across southern tropical Africa. Whether the style branches of *Gladiolus* and *Babiana* are structurally identical and homologous awaits investigation. In bud, the margins of the style branches of *Gladiolus* are conduplicate, so that the stigmatic surfaces are concealed. After anther dehiscence the margins unfold, exposing the minute stigmatic hairs. There are few useful taxonomic characters in either the ovary or style (Figure 1E). The ovary is ovoid to oblong and varies in size in relation to overall flower size. The style is slender and arches over the stamens, except in the few species with actinomorphic flowers when the style is erect. The point of division of the style into its three stigmatic branches relative to the anthers is often characteristic of a species, although the feature is somewhat variable. Comparatively short styles with very short branches characterize the *G. gregarius* complex (except *G. harmsianus*), in which the style divides opposite the lower half of the anthers and the style branches are less than 2 mm long. More typically, the style divides near or just beyond the anther apices, and the branches are at least 3 mm long (Figure 1E).

FRUITS & SEEDS

The *Gladiolus* fruit is a capsule, and the genus can normally be recognized by its inflated, almost ellipsoid or ovoid-ellipsoid shape (Figure 1G), not found elsewhere in Iridaceae. The capsules are also typically fairly coriaceous, although they become brittle when completely dry. In a few species, notably *G. abyssinicus* and its allies, *G. longispathaceus* and *G. schweinfurthii* (section *Ophiolyza*), the capsules are more or less ovoid or sometimes nearly globose. Unusually small capsules characterize *G. erectiflorus* and *G. verdickii* (section *Tenuibracteus*) and perhaps other members of the section, the capsules of which are not known. Species of *Gladiolus* with globose seeds also have ovoid to barrel-shaped capsules, a shape found in many other genera of Ixioideae. This suggests that capsule shape is at least partly a result of the shape of the seeds. *Gladiolus gregarius* and its immediate allies, *G. harmsianus, G. macrospathus,* and *G. microspicatus* (section *Gladiolus*), have distinctively narrow capsules, also unusual in their firm, almost woody texture. The species all have very long bracts that enclose the capsules and perhaps influence the narrow shape.

Fairly large, light brown, broadly winged seeds are characteristic of *Gladiolus.* The seed body itself is globose, but part of the seed coat extends in halo around the seed as a broad wing (Figure 1H). The seed surface is lightly reticulate and shows the outline of the epidermal cells. Seeds are oval to round and 8–12 × 5–7 mm in most species of subgenus *Ophiolyza*, although they may be substantially smaller in some species, especially in section *Tenuibracteus*. In subgenus *Gladiolus* the seeds, when winged, are 4–7 × 2–5 mm but are occasionally up to 8 × 6 mm, for example, in *G. laxiflorus*. The wing is vestigial or lacking in section *Heterocolon* and some populations of *G. gregarius* and *G. microspicatus* (section *Gladiolus*). In some populations of *G. candidus* (section *Acidanthera*) the wings are reduced laterally or entirely.

Other Characteristics

LEAF ANATOMY

Although the leaves of *Gladiolus* vary considerably in their number and in the shape and degree of development of the blade, they all conform to the presumed plesiomorphic type for Ixioideae (Rudall & Goldblatt, 1991) in basic organization. They are equitant (isobilateral and unifacial), the marginal epidermal cells are of the same type as those of the blade surface, a marginal vein is well developed, and there is a strand of subepidermal sclerenchyma associated with the marginal vein (Figures 5, 6).

Outgroup comparison supports the assumption that the plesiomorphic leaf (blade) shape for *Gladiolus* is the plane, narrowly lanceolate form most common in the genus, especially among species that have several (more than three) to many leaves. Important specializations of the leaf blade are the following:

1. Development of a thickened margin, usually paralleled by thickening of the pseudomidrib (Figures 5, B, D, 6). This is reflected anatomically by the enlargement of the sclerenchyma bands between the vascular bundles and the epidermis. Lacking pigment, the sclerenchyma bands appear as transparent or hyaline ridges and on drying assume a yellowish color very different from the intercostal areas.
2. Reduction of blade width to a linear shape. This is not necessarily associated with hyperdevelopment of the vascular sclerenchyma caps, as, for example, in *Gladiolus permeabilis.* More often, however, narrow leaf blades are accompanied by strong and nearly equal development of midrib and margins (Figure 6C–H). These are typically raised and arched over the intercostal areas (e.g., *G. juncifolius, G. sulcatus, G. zimbabweensis*). In the

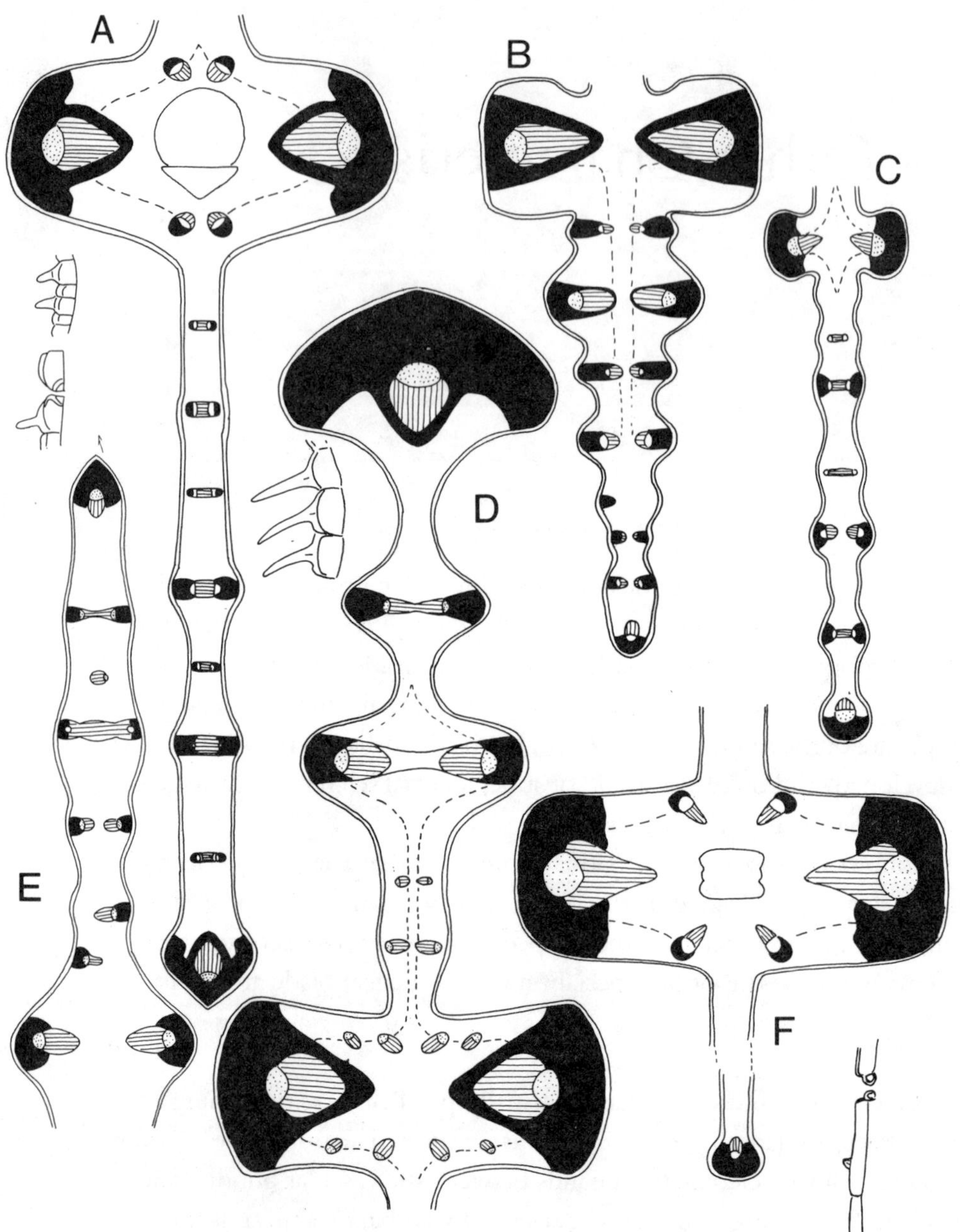

FIGURE 5. Transverse sections of leaves of selected tropical African species of *Gladiolus*. Half of the leaf is drawn, from margin to midvein, for all except F, in which only the margin and midvein area is shown (xylem tissue hatched, phloem dotted, sclerenchyma black). A, *G. zambesiacus*. B, *G. unguiculatus*. C, *G. serenjensis*. D, *G. sericeovillosus* subsp. *calvatus* (note the long, hairlike papillae in the detail of the epidermal cells). E, *G. erectiflorus*. F, *G. murielae*.

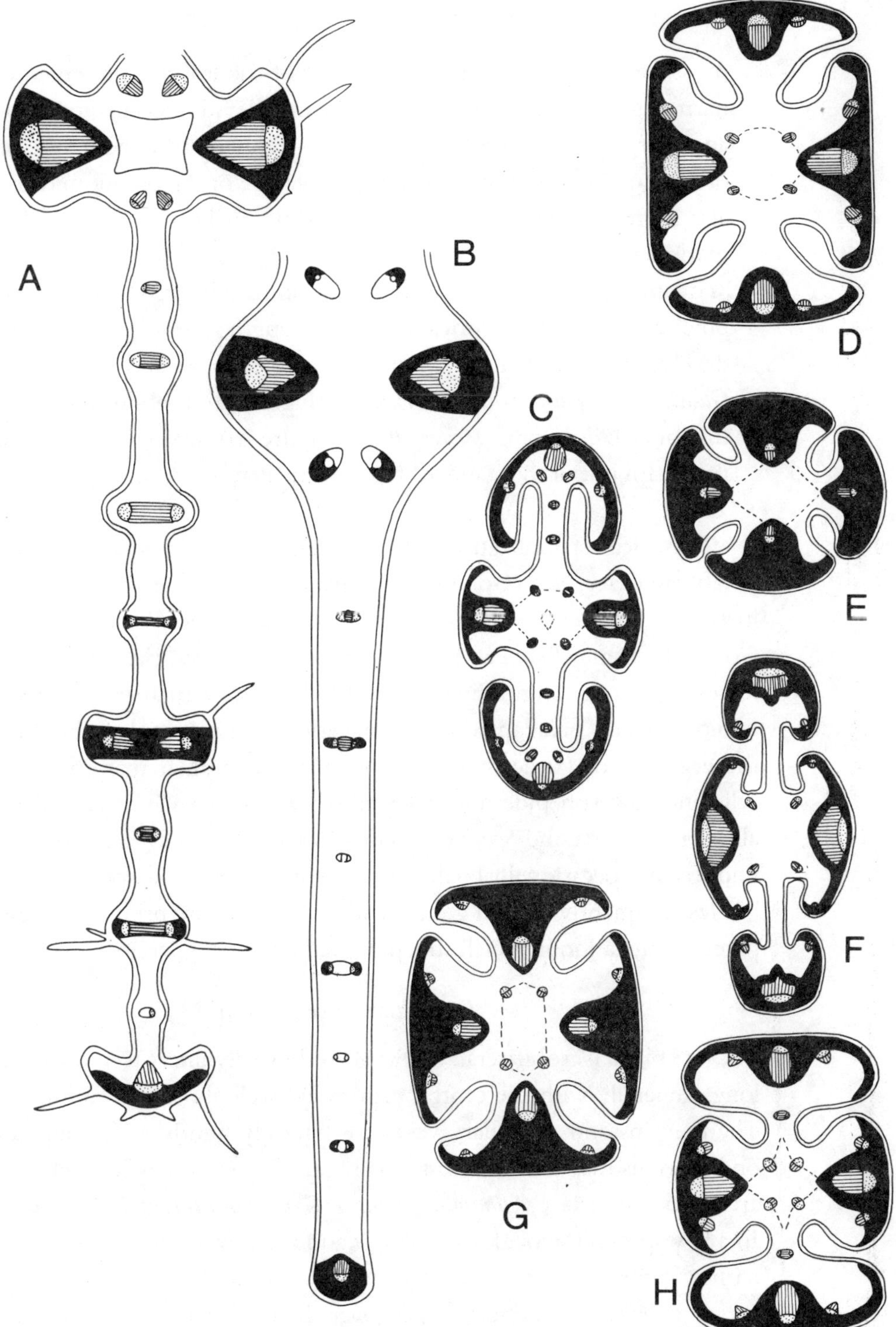

FIGURE 6. Transverse sections of leaves of selected tropical African species of *Gladiolus.* Only half-leaves are drawn for A and B, showing the portion from the margin to just beyond the midvein region (xylem tissue hatched, phloem dotted, sclerenchyma black). A, *G. intonsus.* B, *G. horombensis* (a Madagascan species with unspecialized plane leaves). C, *G. zimbabweensis.* D, *G. robiliartianus.* E, *G. tshombeanus.* F, *G. juncifolius.* G, *G. ledoctei.* H, *G. actinomorphanthus.*

most extreme state the intercostal part of the leaf is represented only by narrow grooves (sinuses) between the raised and thickened pseudomidrib and margins. Such leaves are essentially centric and either terete or more or less square in transection. The vasculature of the pseudomidrib is usually more strongly developed and is thus easily distinguished from the marginal veins.

3. Reduction of the leaf thickness and accompanying reduction of xeromorphic features such as cuticle thickness, length of epidermal papillae, and development of fiber caps. For comparable length and width the leaf of *Gladiolus decoratus* and *G. serenjensis* (Figure 5C) is about half as thick as that of *G. dalenii* or *G. sericeovillosus* (Figure 5D), which have leaves of similar width but of the standard type for the genus.

These leaf specializations are paralleled in other genera of Ixioideae, notably in *Geissorhiza* (Goldblatt, 1985) and *Romulea* (de Vos, 1972) as well as in several southern African species of *Gladiolus* (de Vos, 1976; Rudall & Goldblatt, 1991). *Romulea* has almost exclusively centric, oblong to terete leaves with four longitudinal grooves. The basic structure of the leaves of Ixioideae seems preadapted to specializations of the type described above. In those genera of Ixieae that have specialized columnar marginal epidermal cells and lack subepidermal marginal sclerenchyma (*Tritonia* and its close allies, and *Freesia* and *Sparaxis*), centric leaves or heavily raised margins and midribs may occasionally be developed (e.g., *Devia,* some species of *Tritonia*) but less frequently than in genera with leaves of more primitive (plesiomorphic) organization in which subepidermal marginal sclerenchyma is present.

Leaf Transverse Section

The leaf blade is isobilateral, unifacial, and usually linear to more or less oblong, generally with the central vein raised well above the laminar surface. The margins and secondary veins are typically lightly raised, and the secondary veins may also be substantially raised, sometimes so thickened as to arch over the surface (*Gladiolus intonsus, G. praecostatus,* and *G. sericeovillosus*). In a few species the blades are centric and then oval to terete or square with narrow sinuses between the thickened margins and midribs (Figure 6). The cuticle is usually thick and domed over the epidermal cells above veins but thinner over intercostal areas and within sinuses, especially so in species with thin blades (*G. decoratus, G. murielae,* and most species of section *Acidanthera*). The epidermal cells are normally two to three times wider than high, to four to five times wider than high in species with thin-textured leaves (*G.*

decoratus, G. murielae). In the intercostal areas of the blade, the epidermal cells bear three to five papillae, each about as high as depth of cell. The papillae are somewhat shorter in section *Acidanthera* and about twice as high as the cell depth in *G. elliotii* and *G. sericeovillosus.* Long simple trichomes usually associated with the veins characterize *G. intonsus* and *G. puberulus.* Similar trichomes are confined to lower part of blade and sheath in *G. laxiflorus, G. zambesiacus, G. zimbabweensis,* and a few other species. Epidermal cells at the base of the sinuses are very large and about as wide as high. Stomata are restricted to intercostal areas, or in the case of centric leaves, to the sinuses. The stomata are usually sunken but barely so in section *Acidanthera* and *G. decoratus.* The guard cells are much smaller than neighboring epidermal cells. The vascular bundles are arranged in two opposite rows, and they vary considerably in size. The primary bundles are more or less median in position, either alone forming the pseudomidrib or often associated with pairs of much smaller bundles. As in all Iridaceae with isobilateral leaves, the xylem pole of the vascular bundles is oriented toward the interior, and the phloem toward the exterior. The phloem of the primary and marginal bundles has well-developed fibrous caps that reach the epidermis. In centric leaves the fiber caps of associated veins are confluent with the primary fiber cap, and the submarginal sclerenchyma extends to the edge of the sinuses. Secondary and tertiary bundles are usually much smaller than the primary, and the xylem of opposed bundle pairs is often confluent. The vascular bundle sheaths sometimes have thickened walls, especially those of the primary and marginal veins. This feature is most developed in centric leaves. Occasionally, the bundle sheaths of opposed veins are thickened and continuous and form girders (*G. intonsus, G. zambesiacus*). The chlorenchyma generally extends from the intercostal epidermis to leaf center except at the pseudomidrib. In centric leaves the chlorenchyma lines the sinus and extends laterally to the edges of the vascular bundle sheaths. Tannin cells are often interspersed with chlorenchyma, or sometimes tannin cells are restricted to the inner layer of chlorenchyma in centric leaves of *G. robiliartianus* and *G. tshombeanus.* Elongate styloid crystals with pointed ends, and square or rectangular in transverse section, are scattered in the chlorenchyma. These crystals are especially common in *G. zimbabweensis* but sparse in other species.

POLLEN

So far investigated only in the broadest way, pollen grains of *Gladiolus* conform to the plesiomorphic pollen type of subfamily Ixioideae. They are uniform in their boat shape with a broad aperture and double-banded operculum (Goldblatt et al., 1991). I have examined the exine sculpturing only under the light microscope and it appears to differ among the species in no discernably different way, being tectate and microperforate with small supratectal scabrae. The same exine sculpturing is present in the opercular bands. In addition to the five species examined by Radelescu (1970) and Schulze (1971), I have examined pollen grains of 13 tropical African species, including examples from all major tropical African species groups. Size is the only apparent variable (Table 1), but this shows no correlation with species relationships, and a weak correlation with length of perianth tube. Pollen grains range from 83 μm in length in *G. grantii* and *G. melleri* to more than 135 μm in *G. sulcatus*. There is no apparent relationship between pollen grain size and species relationship. Minor details of the exine sculpturing show no significant variation, and the opercular bands are of comparable length and width in the species examined. There appears to be no reason to examine the pollen grains of *Gladiolus* in more detail for the purpose of achieving a better understanding of the taxonomy and phylogeny of the genus.

TABLE 1. Pollen grain length in selected tropical African species of *Gladiolus*.

SPECIES	GRAIN LENGTH (μm)
Subgenus *Gladiolus*	
G. amplifolius	95.2
G. curtifolius	114.3
G. sulcatus	135.7
Subgenus *Ophiolyza*	
G. abyssinicus	109.5
G. aequinoctialis	133.3
G. dalenii	95.2
G. erectiflorus	123.8
G. grantii	83.3
G. gunnisii	95.2
G. melleri	83.3
G. mirus	95.2
G. murielae	119.0
G. sericeovillosus	119.0

CHROMOSOME CYTOLOGY

Thought to be relatively uniform cytologically (Goldblatt, 1971, 1989), *Gladiolus* is now known to exhibit an unexpected degree of variability among the tropical African members of the genus (Goldblatt et al., 1993). The genus as a whole is fairly conservative in its chromosome cytology. All 50 southern African species, 6 Madagascan species, and 14 of 20 tropical African species studied cytologically have basic chromosome numbers of $x = 15$. The majority of these are diploid, $n = 30$. A number of moderately to highly specialized tropical African members of subgenus *Gladiolus* have what are assumed to be derived base numbers. The only representative of the specialized section *Heterocolon,* the southern Zairian endemic *G. actinomorphanthus,* has $n = 14$. *Gladiolus atropurpureus,* specialized both in its reduced leaf blades and thickened corm tunics, has $n = 12$. The closely related and vegetatively similar *G. serapiiflorus* has $n = 11$. *Gladiolus unguiculatus,* which like *G. atropurpureus* also has leaf blades largely reduced or absent but produces foliage leaves on separate shoots, has $n = 13$ and, possibly, 12. The two species of the *G. gregarius* complex, apomorphic in having long floral bracts, *G. gregarius* and *G. microspicatus,* both have $n = 11$. Because all these species are specialized and thus appear derived in subgenus *Gladiolus,* their chromosome numbers are assumed also to be derived.

The current belief that the ancestral base number in the genus is $x = 15$ remains an acceptable hypothesis. Chromosome number in tropical African members of subgenus *Gladiolus* appears from available evidence to have undergone dysploid reduction in four separate lineages. Until more counts are obtained for subgenus *Gladiolus,* the systematic and phylogenetic significance of dysploidy will remain difficult to understand except in the broadest terms. Difficulties of collecting in central Africa, especially in Zaire, make progress in expanding our knowledge difficult. It also seems important that southern African species related to the dysploid tropical African taxa be examined cytologically. Among those that require study are *G. woodii,* apparently closely allied to *G. atropurpureus* and *G. pretoriensis,* a likely ancestor of the dysploid Zairian *G. actinomorphanthus.*

In tropical African members of subgenus *Ophiolyza,* two species, *Gladiolus dalenii* and *G. bellus,* are polyploid. Populations of *G. dalenii* subsp. *dalenii* have either $2n = 60$ or 90, but the subspecies is diploid in Madagascar (Goldblatt, 1989). The only counts for *G. dalenii* subsp. *andongensis* are also diploid, $2n = 30$. *Gladiolus bellus,* restricted to Mt. Mulanje in southern Malawi, is tetraploid, $2n = 60$, based on a sampling of one population. Counts for two populations of *G. decoratus* have both yielded $2n = 39$, probably triploid plants, and from this evidence the species appears to be dysploid with $x = 13$. That all four plants from the two populations examined were apparently triploid is puzzling.

Clearly, the chromosome cytology of *Gladiolus* in tropical Africa needs more study. No examples of dysploidy are known in the genus outside tropical Africa, and polyploidy is equally rare in southern Africa. In Eurasia all the species counted are polyploid with numbers $2n$ = 60, 90, or 120 (Raamsdonk & de Vries, 1989; Goldblatt et al., 1993). In the past, the chromosome record for *Gladiolus* has made is seem that polyploidy was significant in the genus, particularly so in tropical Africa. All of the older reports of polyploidy in tropical African species were, however, for species now considered conspecific with the widespread *G. dalenii*. The occurrence of four separate dysploid reduction series in subgenus *Gladiolus* and one more likely example in subgenus *Ophiolyza* emphasizes the complexity and evolutionary depth of the genus in tropical Africa, already evident from the morphological diversity and speciation across the area.

The ancestral basic chromosome number in *Gladiolus*, x = 15, indicates a polyploid event in its ancestry, and the genus is thus evidently paleotetraploid. This number is shared only with the mono- or ditypic genus *Radinosiphon* (Goldblatt, 1993) and is almost certainly derived in Iridaceae. It presumably represents a significant synapomorphy for *Gladiolus*. Apart from the shared base number, there is no compelling reason to believe that *Gladiolus* and *Radinosiphon* are immediately related. Most other genera of Ixioideae have lower basic numbers, x = 11 and 10 being the most common, but also x = 26, 18, 16, and 14. Two genera believed to be closely related to *Gladiolus*, *Geissorhiza* and *Hesperantha*, both have x = 13 and they, too, must be regarded as paleotetraploid. The southern African genus *Tritoniopsis*, x = 16 (Goldblatt, 1971, 1981), appears to be only distantly related to *Gladiolus*. *Babiana*, a genus possibly fairly closely related to *Gladiolus* (see Generic Relationships, next chapter), has a base number x = 7 (Goldblatt, 1971).

The only other departure from the standard chromosomal pattern in tropical African *Gladiolus* is the occurrence of B chromosomes in a few species. The significance of these supernumeraries is uncertain in most plants, and *Gladiolus* is no exception. In *Gladiolus*, in which all the chromosomes are small, the B chromosomes cannot be distinguished at mitotic metaphase and their existence is usually inferred by the departure from the basic number of the species in question. A population of *G. atropurpureus* from Mt. Malosa in southern Malawi has $2n$ = 36. Counts from elsewhere suggest the base number for the species is x = 12. These plants may either be triploid, $2n$ = $3x$ = 36, or they may have an unusually large number of B chromosomes, $2n$ = 24 + 12B. Other reports of B chromosomes in *Gladiolus* include the record of $2n$ = 30 + 0–2B in the Madagascan *G. decaryi* (Goldblatt, 1989), and $2n$ = 26 + 2B in *G. unguiculatus*. Clearly, much remains to be learned about the chromosome cytology of *Gladiolus,* and the discovery of dysploidy in three apparently unrelated lineages of the genus in tropical Africa should provide the stimulus to devote more energy to this aspect of future research in *Gladiolus.*

REPRODUCTIVE BIOLOGY & POLLINATION

Little is known about the pollination of *Gladiolus* in tropical Africa. *Gladiolus* species are normally outcrossing and probably self-incompatible, thus requiring animal-mediated transfer of pollen to accomplish pollination. Adaptation to different pollination strategies is one of the driving forces for evolution and species diversification. In view of this, it is surprising how poorly known *Gladiolus* is in this respect. Pollination strategies are somewhat better known in the southern African species of the genus (Vogel, 1954; Johnson, 1992; Johnson & Bond, 1994) than in the tropical African members of the genus.

Species with white flowers with tubes exceeding 50 mm (usually also sweetly scented) are widely thought to be pollinated by sphinx moths. This mode of pollination has been reported in *Gladiolus longicollis* from Natal, South Africa, and is generally assumed for a few other southern African species, including *G. tristis* and even *G. liliaceus,* which has light brown to purplish flowers. Both these species have long perianth tubes and produce a strong sweet fragrance in the late afternoon and evening. In tropical Africa, sphinx-moth pollination is assumed, solely from floral morphology, for all the species of section *Acidanthera* as well as for isolated species of section *Ophiolyza* (*G. gunnisii, G. richardsiae*), section *Blandus* (*G. bellus, G. usambarensis*), and section *Tenuibracteus* (*G. stenosiphon*). In subgenus *Gladiolus,* sphinx-moth pollination is assumed for two species of section *Gladiolus, G. curtifolius* and *G. harmsianus.* These two species are not closely related and the pollination syndrome is assumed to have arisen independently in each from ancestors with more conventional pollination, probably by bees.

Short-tubed species of subgenus *Gladiolus* are probably pollinated by anthophorid bees. In southern Africa this has been established for a few species, including *G. scullyi* and *G. orchidiflorus* (Goldblatt & Manning, unpublished; V. B. Whitehead, pers. comm.) and probably occurs in many more. The southern tropical African *G. zambesiacus,* which has pale pink flowers with yellow nectar guides, appears to be pollinated by anthophorid bees, according to my observations in Malawi. Short-tubed and small-flowered members of subgenus *Ophiolyza* also appear to be pollinated by anthophorid bees, based on my observations in eastern Zimbabwe of *G. sericeovillosus,* which has a dull greenish perianth with pale yellow nectar guides.

Species of *Gladiolus* with moderate to large-sized pink or whitish flowers with red nectar guides and moderately long tubes, typical of section *Blandus* in South Africa, are thought to be pollinated by long-tongued flies of either the family Nemestrinidae or Tabanidae (Vogel, 1954; Johnson, 1992). My observations of *G. crassifolius* in eastern Zimbabwe suggest that pollination by the genus *Prosoeca* (Nemestrinidae) may also occur in this species.

Another pollination strategy known to be important in *Gladiolus* is bird pollination. Sunbirds (*Nectarinia*) pollinate red-flowered and long-tubed *G. bonaspei, G. saccatus,* and *G. watsonius* (and several more species) in southern Africa (Vogel, 1954). Sunbirds are also believed to pollinate tropical African species with flowers that have a similar shape and coloration. Bird pollination appears to have evolved at least three times in southern African species of *Gladiolus* (Goldblatt & de Vos, 1989). It probably also evolved independently at least three more times in tropical Africa, in the *G. abyssinicus* complex (including also *G. longispathaceus, G. schweinfurthii,* and perhaps *G. watsonioides*), in the taxonomically isolated *G. huillensis,* and in *G. dalenii* and the closely allied *G. magnificus.* Direct observations are lacking for all these examples. It is interesting to note that the vernacular name for *G. dalenii* in eastern Kenya translates to "coconut of the sunbird," presumably an example of a pollination strategy giving rise to a common name.

The entire field of pollination biology for tropical African *Gladiolus* is virtually unexplored. Investigation of plant-animal relationships will no doubt be a valuable source of new information about the biology and adaptive radiation of *Gladiolus.*

Relationships, Phylogeny & Classification

GENERIC RELATIONSHIPS

It is firmly established that *Gladiolus* is a member of subfamily Ixioideae, one of four subfamilies currently recognized in the Iridaceae (Goldblatt, 1990a, 1991). The genus has all the synapomorphies that define Ixioideae, including a basally rooting corm, a flower with a well-developed perianth tube, a long-lived perianth (not deliquescing on the same day that it opens), sessile flowers subtended by a pair of opposed floral bracts, pollen exine with a perforate tectum and scabrate sculpturing, and pollen grain apertures with a two-banded operculum (Goldblatt, 1990a; 1991; Goldblatt et al., 1991). *Gladiolus* also has leaves with a well-defined midvein and anatomical specializations that include mesophyll cells elongated across the leaf axis and epidermal cells with sinuous margins and two or more papillae (Rudall & Goldblatt, 1991). These latter features also define two of the three tribes of Ixioideae, Watsonieae and Ixieae (Goldblatt, 1990a, 1991). Within Ixioideae, *Gladiolus* falls in the Ixieae, in which corm ontogeny is apical-axillary (Goldblatt, 1990a), and the seeds lack chalazal crests, a feature characteristic of other tribes of Ixioideae. *Gladiolus* also has undivided style branches, probably the plesiomorphic condition in Ixioideae, whereas the genera of Watsonieae have deeply divided style branches, the apomorphic condition, and also a corm ontogeny that is entirely axillary (Goldblatt, 1990a).

Within Ixieae, the relationships of *Gladiolus* are more difficult to assess. The tribe, including some 20 genera, is the largest of the three tribes of Ixioideae, and the patterns of generic relationship within it are obscure. At least, however, it seems likely that *Gladiolus* is most closely allied to a group of genera that have lightly angular (or more or less rounded) seeds with secondary sculpturing (the cell outlines are well defined) and intact seed vasculature. In addition, this alliance has unspecialized leaf anatomy, in which the leaf margins have epidermal cells of the same shape and size as those of the blade area, and the tissue under the epidermal margins is heavily thickened into a strand of sclerenchyma that is closely associated with a marginal vein (Rudall & Goldblatt, 1991). Other genera of Ixieae have hard, rounded seeds with a smooth surface in which the cell outlines are obscure or absent (lacking secondary sculpturing) and usually an excluded seed vasculature (Goldblatt & Manning, 1995). This group also has a derived leaf anatomy in which the marginal epidermal cells are columnar and have secondarily thickened walls, and a subepidermal marginal strand of sclerenchyma is absent. It seems likely that both the basic seed features of *Gladiolus* and its leaf anatomy are plesiomorphic for Ixioideae and so say nothing about the mutual relationships of the genera with these characters. It does, however, seem unlikely that *Gladiolus* is related to any genera of the *Tritonia–Crocosmia* alliance, or to *Sparaxis, Ixia,* or *Dierama,* all of which have derived seed features and leaf margin anatomy. The closely allied pair of genera, *Freesia* and *Anomatheca,* can probably also be excluded from the list of genera immediately related to *Gladiolus* because although they have seeds with intact vasculature, the seeds are hard, round, and smooth, and the marginal anatomy is of the derived type.

This leaves just a handful of genera as possible close allies of *Gladiolus.* These include *Radinosiphon,* a genus of one or two species of southern tropical Africa, and *Geissorhiza, Hesperantha, Melasphaerula,* and *Romulea,* all predominantly or exclusively southern African genera that probably constitute a clade defined by woody corm tunics and usually asymmetric corms (not so in *Melasphaerula*) with roots produced from a defined ridge. The tropical and southern African *Babiana,* an apparently taxonomically isolated genus, may also fall in this group of possible relatives of *Gladiolus.* It has specialized plicate leaves, surface pubescence, and smooth seeds with an unusual, apomorphic type of sculpturing. The leaf anatomy accords with that in *Gladiolus,* as does the intact seed vasculature, but these features are both plesiomorphic and say nothing about their mutual relationship. Similar apically expanded style branches alone are synapomorphic for these two genera. An additional two genera, the southern African *Syringodea* and the Eurasian *Crocus,* are probably not immediately related to *Gladiolus* and are not of direct concern here. Species of these two genera are highly specialized, acaulescent plants, most probably allied to *Romulea.* At least *Syringodea* has woody corm tunics and asymmetric corms of the *Romulea* type, and both *Syringodea* and *Crocus* have derived inaperturate pollen grains.

A cladistic analysis of the relationships the genera considered the most likely allies of *Gladiolus* on the basis of the argument developed above suggests that *Radinosiphon* may be the genus most closely related to *Gladiolus.* In the analysis (Figure 7, Tables 2 and 3), in which *Ixia* is used as the outgroup for character polarization, the clade *Geissorhiza–Hesperantha–Romulea* (to which could be added *Melasphaerula, Syringodea,* and *Crocus*) falls out first, closest to *Ixia,* and most distant from *Gladiolus. Babiana* plus *Radinosiphon* and the sections of *Gladiolus* form a clade in which *Babiana* is the sister lineage to the remaining taxa. The support for this tree comes only from only

TABLE 2. Characters used in the cladistic analysis. The plesiomorphic (or ancestral) state is indicated by (0), and the apomorphic (or derived) state by (1), or in the case of multistate characters, by (2) and (3). Multistate characters are unordered.

CHARACTER NUMBER	PLESIOMORPHIC STATE → APOMORPHIC STATE
1	Flowers actinomorphic (0) → flowers zygomorphic (1)
2	Corm tunics coriaceous to fibrous (0) → corm tunics woody (1)
3	Seed surface matte and reticulate or colliculate (0) → surface hard, smooth, and shiny (1)
4	Seed vasculature retained under the seed coat (0) → vasculature excluded (1)
5	Perianth tube shorter than the bracts (0) → perianth tube about as long as the bracts or exceeding them (1)
6	Dorsal tepal longer than the perianth tube (0) → dorsal tepal about as long or shorter than the tube (1)
7	Capsules ovoid and somewhat truncate (0) → capsules more or less ellipsoid (1)
8	Lower tepals not united basally united nor clawed (0) → lower tepals united basally and clawed (1)
9	Nectar guides near the tepal bases (0) → nectar guides on the distal parts of the tepals (1)
10	Flowers and other reproductive organs small (0) → floral organs large (1)
11	Spikes erect (0) → spikes inclined (1)
12	Flowers spirally arranged (0) → flowers secund (1)
13	Floral bracts green and leafy-textured (0) → bracts membranous (1), bracts firm-textured (2), bracts attenuate and dry apically (3)
14	Seed without broad wings (0) → seeds with a broad membranous wing (1)
15	Anther apices obtuse and without appendages (0) → anther apices acute and with apiculate appendages (1)
16	Flowers shades of pink blue or purple (0) → flowers bright red with white markings (1), flowers white with purple markings (2)
17	In profile, the lower tepals appearing as long or shorter than the upper tepals (0) → lower tepals exceeding the upper (1)
18	Corms symmetric (0) → corms asymmetric, roots produced from a ridge (1)
19	Seeds relatively small (0) → seeds large (1)
20	Style branches simple and filiform (0) → style branches each deeply divided (1), branches expanded apically (2)

TABLE 3. Data matrix for the taxa included in the cladistic analysis. The plesiomorphic state is indicated by 0, the apomorphic state by 1, 2, or 3, and an unknown or inapplicable state by ? Multistate characters are unordered.

	CHARACTER NUMBER			
	00000	00001	11111	11112
TAXON	12345	67890	12345	67890
Ixia	00110	00000	00100	00000
Hesperantha	01001	00000	00000	00100
Geissorhiza	01000	00000	00000	00100
Romulea	01100	00000	00200	00101
Babiana	00100	00000	00000	00002
Radinosiphon	10001	10000	11000	00000
Gladiolus	10000	01110	11010	01002
Hebea	10000	01110	11010	01002
Heterocolon	10000	00110	11000	01002
Ophiolyza	10001	01001	01010	00012
Blandus	10000	01001	01010	00012
Tenuibracteus	10000	01001	01310	000?2
Acidanthera	10001	11001	01011	20012
Decoratus	10001	01011	01011	1?012
Ranisia	10001	11001	01010	00002
Cardinales	10001	01001	01010	10012

one character, the shared apically expanded style branches of *Gladiolus* and *Babiana.* This character is absent from *Radinosiphon,* and the plesiomorphic state, filiform style branches, found in the genus is treated as a reversal in the analysis. That there may be a reasonably close relationship between *Gladiolus* and *Babiana* is also suggested by the work of Shneyer (1990). Based on serological comparisons of seed proteins, Shneyer found that of the genera considered here to be possible close allies of *Gladiolus*, *Babiana* seemed more closely related to the genus than were *Geissorhiza, Hesperantha,* or *Romulea.* Genera of the *Crocosmia–Chasmanthe* alliance and *Dierama*, however, appeared to be as close to *Gladiolus* as *Babiana. Radinosiphon* was not included in her analysis.

In the consensus tree (Figure 7A) of 16 equally parsimonious trees obtained using the ie (explicit enumeration) option of the Hennig86 program for parsimony analysis (Farris, 1988), *Radinosiphon* actually forms a trichotomy with the two subgenera of *Gladiolus.* This topology is not, however, one obtained in any of the 16 trees. Instead, *Radinosiphon* falls within in the clade with the sections of either subgenus *Gladiolus* or subgenus *Ophiolyza*. This instability leaves the status and affinities of *Radinosiphon* in doubt, although an

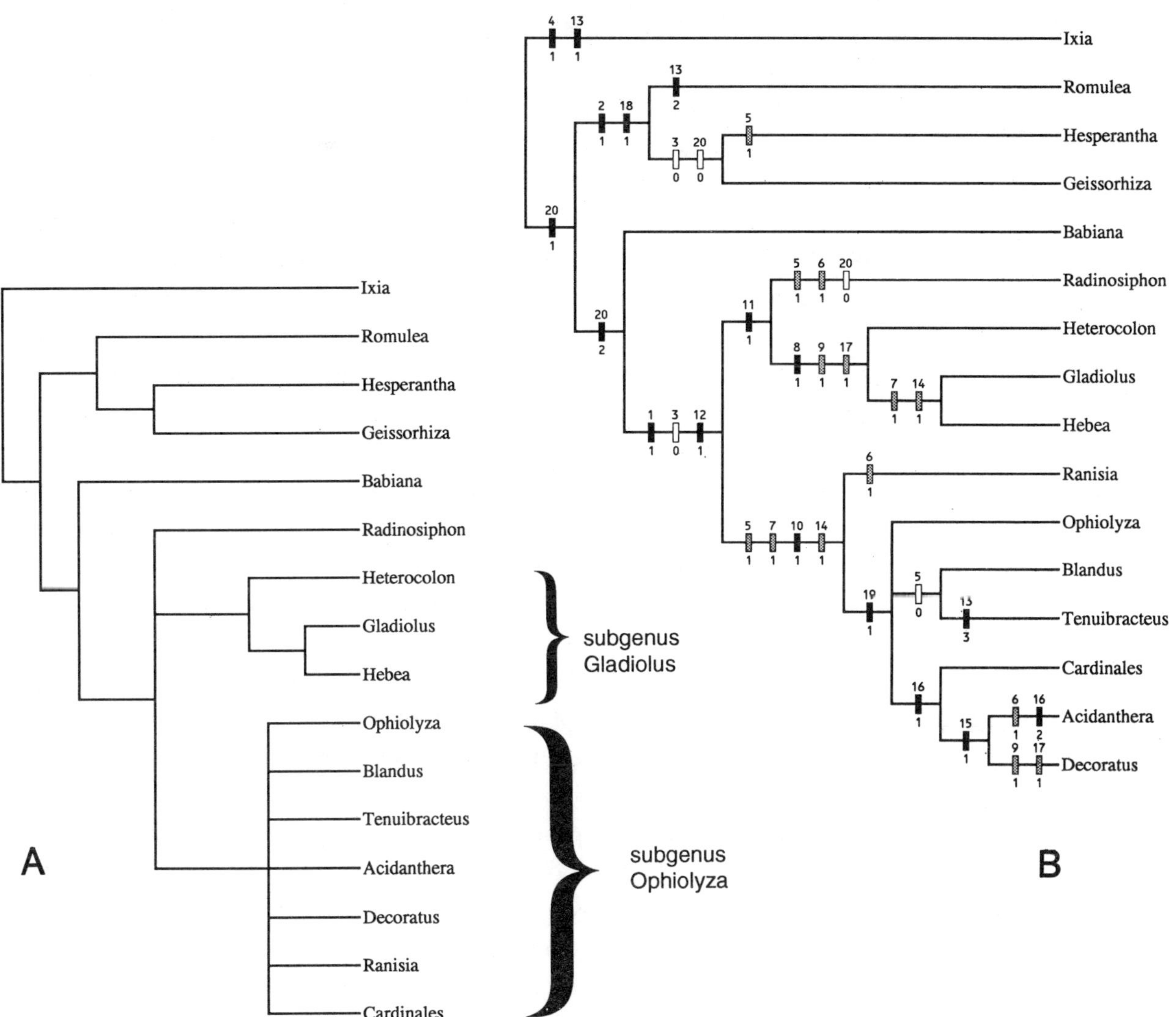

FIGURE 7. Cladograms showing the relationships of *Gladiolus* to closely allied genera of subfamily Ixioideae, and including the sections of *Gladiolus* recognized here. A, The consensus tree (length 35, consistency index 68, retention index 81), showing only those branch topologies present in all the trees—note the position of *Radinosiphon* and the completely unresolved situation among the sections of subgenus *Ophiolyza*. B, One of the sixteen equally parsimonious trees to show the distribution of characters and character changes. Trees were initially generated using the Hennig86 package of programs for parsimony analysis (Farris, 1988), with ie option. The trees were analyzed using CLADOS version 1.2 (Nixon, 1992) and the illustrations were produced using this program. Characters used in the analysis are listed in Table 2, and the character matrix is shown in Table 3. Numbers above the bars on the tree correspond to the character numbers in Table 2; numbers below the bars indicate character state, also from Table 2. Bar color: black color, change to the apomorphic state (i.e., specialized condition); white, reversal to the plesiomorphic state (e.g., loss of a specialized condition); stippled, parallel character change (e.g., character state change shared by more than one taxon in the analysis).

argument can be made on the basis of the results of the cladistic analysis that *Radinosiphon* should be included in *Gladiolus.* An examination of the character changes to produce a topology that includes *Radinosiphon* in clades with the tribes of subgenus *Gladiolus* (Figure 7B), or with those of subgenus *Ophiolyza,* makes this aspect of the cladistic analysis appear unsatisfactory.

Gladiolus and *Radinosiphon* share three apomorphic characters: the zygomorphic flower, secund flower arrangement, and an inclined spike. Two or all three of these characters may well be linked, and in any event this alone is a weak argument at best for concluding that they are closely related. One additional piece of evidence links the two genera, their basic chromosome number. Both are paleotetraploid genera with a unique basic number for Ixioideae, $x = 15$.

More significantly, *Radinosiphon* differs from *Gladiolus* in lacking winged seeds and in having ovoid capsules and filiform style branches. Inclusion of *Radinosiphon* in *Gladiolus* (Figure 7B) implies that the inflated ellipsoid capsules (character 7) and the winged seeds (character 14) of *Gladiolus* arose independently in the two subgenera, something I find unlikely. It also implies that the filiform style branches (character 20) of *Radinosiphon* are a reversal. I find it impossible to accept the assumptions about these three character changes and therefore reject the topologies in the shortest trees obtained here. I recommend that *Radinosiphon* continue to remain a genus separate from *Gladiolus,* pending a more thorough analysis of *Radinosiphon.*

The best that can be said about the relationships of *Gladiolus* is that the genus probably has no close relatives, and this may be taken as an indication that the genus diverged from ancestral ixioid stock at a remote time, probably in the early Tertiary. Other genera of Ixieae are probably more recent in origin, and of these *Radinosiphon* seems most nearly related. A more satisfactory answer to these phylogenetic questions will probably only be achieved when new methods of gene sequencing are used to examine questions about generic relationships in the Ixioideae.

An aspect of the cladistic analysis that I find attractive is the strong support for the two subgenera that I recognize. The three tropical African sections of subgenus *Gladiolus,* sections *Gladiolus, Hebea,* and *Heterocolon,* are linked by several apomorphic states (four in Figure 7B), although section *Heterocolon* appears in the basal position (lacking both winged seeds and ellipsoid capsules). Likewise, the sections of subgenus *Ophiolyza* form a well-supported clade. Relationships of the sections themselves are confusing. Only sections *Cardinales, Acidanthera,* and *Decoratus* are linked, first on the basis of assumed ancestral red flower color, and then the latter two by their peculiar acute anther appendages. The southern African section *Ranisia* falls in subgenus *Ophiolyza,* but its position there is weakly supported. In four of the seven shortest trees obtained in the cladistic analysis it is the basal section of the subgenus, and sister to the remainder. Its placement in subgenus *Ophiolyza* needs to be critically examined. I do not intend here to examine in further detail the intersectional relationships. This must wait for the future, when the southern African species are better known and can be taken into account.

SUBGENERIC CLASSIFICATION

Patterns of variation in *Gladiolus* suggest no obvious discontinuities that delineate major species groupings. Hence, it should be no surprise that there is no subgeneric classification that is currently accepted. The only working divisions of the genus are those of Lewis et al. (1972). Their avowedly artificial system was used to delineate four groups, *Plurifoliatae, Paucifoliatae, Unifoliatae,* and *Exfoliatae,* based on the number of foliage leaves present on the flowering stem. The system is useful for identification purposes, but because leaf number has been reduced independently in several lineages in both subgenus *Gladiolus* and subgenus *Ophiolyza,* it has no value in indicating natural relationships. I propose a novel classification here, in which I recognize two subgenera, *Gladiolus* and *Ophiolyza*, each based largely on flower size and associated features such as bract length, and capsule and seed size (Figure 3). This results in the more or less equal division of the species into the two subgenera, and most of these can be fairly easily placed in one or the other of them. Where size alone fails to place species, a comparison of associated characters usually results in satisfactory assignment of species to one or the other of the subgenera. This classification is supported by the cladistic analysis (Figure 7), in which the main generic groupings of *Gladiolus* as well as the genera most likely allied to the genus are analyzed.

Adaptive radiation into a variety of habitats and utilizing several pollination strategies has led to considerable convergence among members of both subgenera, with the result that differences between them are occasionally blurred and species cannot always be confidently assigned to subgenus. This does not necessarily mean the classification is flawed, but rather that it is not always useful. Within each subgenus there are several species clusters that appear to monophyletic assemblages that can be grouped further into sections. Most are easy to recognize, and some of them have long been recognized, sometimes as separate genera, as is the case with *Acidanthera, Homoglossum,* and *Oenostachys.*

The subgenera of *Gladiolus* may be defined as follows. Subgenus *Ophiolyza* typically has large flowers (32–)40–190 mm long (length of the tube plus dorsal tepal). The dorsal tepal is (18–)35–50 mm long and usually either slightly shorter to about as long as the perianth tube, or considerably shorter in section *Acidanthera*. The tube is rarely less than 25 mm long, and the bracts are (20–)25–60(–80) mm long. Associated with the large flower and large bracts are proportionately large stamens, with anthers 8–12 mm long, and an erect and sturdy stem generally 3–5 mm in diameter at the base of the spike. Likewise, the capsules are usually relatively large, (10–)15–35 mm long, and the seeds are mostly 8–11 × 5–7 mm. Outgroup comparison suggests that the large flower is a derived condition. All genera of Ixieae in the sense of Goldblatt (1990a, 1991) have comparatively small flowers. The large-flowered condition is not associated with polyploidy. Nearly all members of subgenus *Ophiolyza* are diploid, $n = 15$. Most populations of the widespread *G. dalenii* are tetraploid or hexaploid, $2n = 60$ or 90, on the African mainland, but even there some populations are diploid, $2n = 30$, and the species is apparently consistently diploid on Madagascar.

Small-flowered subgenus *Gladiolus* is not merely a residual assemblage of unspecialized species. All members of the subgenus have a number of discrete specializations. The small flower is typically (12–)20–35 mm long (occasionally up to 70 mm), and the dorsal tepal is 10–25 mm long, normally half again as long as the tube. The bracts are mostly (5–)10–20 mm long and usually exceed the perianth tube. In its size and proportions this flower is assumed to be plesiomorphic in the genus. Other features of the flower appear, however, to be derived. These include the generally narrow lower tepals, additionally united below for some distance, and when viewed in profile the lower three tepals seem to be about as long or longer than the three upper ones. The lower tepals, or at least the lower laterals, are also abruptly narrowed below more or less into discrete claws that are usually also channeled.

Typically, the perianth tube is about half to two-thirds as long as the dorsal tepal, always the largest tepal, and the bracts are longer than the tube, which thus emerges below the bract apices or just exceeds the bracts. A few species, evidently adapted to pollination by long-tongued insects, have longer tubes, but in subgenus *Gladiolus* tepal and anther size remain fairly constant. Spikes are almost always inclined in subgenus *Gladiolus,* owing to a bend in the stem at the base of the spike, and stems are typically 1.5–2.5 mm in diameter.

An additional distinction between subgenus *Ophiolyza* and subgenus *Gladiolus* is the nectar guides. These markings are usually in the distal third of the lower three tepals in subgenus *Gladiolus,* thus on the tepal limbs, and additionally the markings often contrast strongly in color, sometimes consisting of transverse bands of pale and dark pigment. In subgenus *Ophiolyza* the nectar guides are located on the lower half of the tepals and are usually uniformly pale in color. In dried specimens the nectar guides of subgenus *Ophiolyza* usually disappear, but they remain evident for years in subgenus *Gladiolus*. The characteristics of the two subgenera may be summarized as follows:

Subgenus *Gladiolus*

1. Flowers small: perianth mostly 15–30 mm long, upper tepal 10–25 mm long and about half again as long as the tube.
2. Tube mostly 10–15 mm, usually half to two-thirds as long as the bracts.
3. Bracts (5–)10–20(–50) mm long.
4. Spike inclined and (1–)1.5–2.5 mm in diameter.
5. Lower tepals united basally for some distance and narrow and channeled below (lower tepals more or less clawed).
6. Nectar guides on the upper half of the tepals, usually strongly contrasting and often transverse.

Subgenus *Ophiolyza*

1. Flowers large: perianth mostly (32–)40–190 mm long, upper tepal (18–)35–45 mm long and slightly longer but more often slightly to much shorter than the tube.
2. Tube mostly 25–50 mm, slightly shorter to about as long or much longer than the bracts.
3. Bracts (20–)25–60(–80) mm long.
4. Spike erect and (2–)3–5 mm in diameter.
5. Lower tepals barely or not at all united basally and not notably narrowed or channeled below (lower tepals not clawed).
6. Nectar guides on the lower half of the tepals, usually weakly contrasting and not transverse.

Early attempts to produce a useful subgeneric classification of *Gladiolus* are of little more than historical curiosity. H. Persoon (1805), for example, divided *Gladiolus* into five groups of undefined rank, some not given names, and three of the five are now regarded as other genera. Persoon must, however, be credited with the name *Hebea*, later used by J. G. Baker (1878b, 1896) and F. W. Klatt (1882) as a subgenus including several southern African species clustered around *G. alatus* that have an enlarged dorsal tepal and both lower and upper tepals abruptly narrowed below, thus more or less clawed. *Hebea* was regarded as a section of *Gladiolus* by G. Bentham and J. D. Hooker (1883), who did not recognize subgenera in *Gladiolus*. The species of *Hebea* fall within my definition of subgenus *Gladiolus* but constitute a natural grouping that keeps its sectional rank. Only two species of section *Hebea* reach tropical Africa, the widespread southern African *G. permeabilis* and the Angolan endemic *G. fenestratus.* The sectional position of the latter is by no means certain and will not be settled until fruits and seeds of the species are discovered. Baker recognized one other subgenus, *Schweiggera,* a synonym of the South African genus, *Tritoniopsis,* and four groups of no stated rank, *Parviflorae, Blandi, Cardinales,* and *Dracocephalae.* The last three fall within subgenus *Ophiolyza,* whereas group *Parviflorae* is equivalent to section *Gladiolus.*

Klatt's (1882) classification included four more subgenera in addition to *Hebea* and subgenus *Gladiolus,* which he called *Eugladiolus,* a formulation no longer nomenclaturally admissible. Subgenus *Ophiolyza,* the name adopted here for the large-flowered subgenus of *Gladiolus,* included the longer-tubed species with large flowers and smaller lower tepals; subgenus *Hyptissa* included larger-flowered species with nearly equal tepals; subgenus *Ranisia* included large flowered species a tube longer than the bracts; and subgenus *Limonia* included both small- and large-flowered species and was poorly defined. The names that Klatt used for his subgenera are taken from the literature, but because he did not cite their sources, even indirectly, they must be treated as new names. Subgenus *Hebea* is obviously Persoon's (1805) group *Hebea,* and subgenera *Ophiolyza, Hyptissa,* and *Ranisia* are names for genera segregated from *Gladiolus* by Salisbury (1866), although *Ranisia* included quite different species in Salisbury's treatment. *Limonia* can be traced to *Lomenia* Pourret, also misspelled *Lemonia* in the literature. Nomenclaturally, *Lomenia* is congeneric with the southern African *Watsonia,* species of which were included in *Gladiolus* by Persoon. Because Klatt designated no types for his subgenera, I have designated lectotype species here for each of them so that the application of the names will in future be fixed. Both Baker and Klatt treated *Acidanthera* as separate genera and included within it several southern African species now transferred to *Hesperantha* and *Geissorhiza.*

English botanist R. A. Salisbury (1866) provided a novel treatment of *Gladiolus.* He divided *Gladiolus* into five genera, the only one of which to receive recognition for some time was *Homoglossum.* Although Salisbury's taxonomy is largely of historical interest only, it provides an insight into species clusters that Salisbury thought were natural units.

Sections of Subgenus *Gladiolus*

Species clusters among the tropical African members of subgenus *Gladiolus* than can conveniently be arranged in sections are difficult to delineate. A cursory survey of the Eurasian species of *Gladiolus* suggests that they form a single lineage, conveniently grouped in section *Gladiolus,* but this needs critical examination, something beyond the scope of this book. There seems to be no fundamental difference between Eurasian species and the majority of the tropical African members of the subgenus. Most thus fall into the same section. *Gladiolus permeabilis,* however, seem to belong in a predominantly southern African alliance, section *Hebea*, defined by having linear leaves, typically with only the midvein area raised, all the tepals fairly strongly narrowed below (thus more or less forming claws), and the sheaths of the upper leaves free to the base. The incompletely known Angolan *G. fenestratus* also falls in the section. The type of section *Hebea* is the Cape species, *G. alatus,* which has very strongly modified tepals, but other Cape species more closely resemble *G. permeabilis.*

I recognize one other tropical African section of subgenus *Gladiolus,* section *Heterocolon.* The section is defined by the terete leaves with thickened margins and midribs, seeds with variously reduced wings, ovoid capsules, and unusual red-brown corm tunics that form a neck around the base of the plants. The section was founded by O. Kuntze (1898) for the southern African *G. pretoriensis,* but a group of central African species of the subgenus match exactly with the definition of the section. These species include *G. ledoctei, G. actinomorphanthus, G. tshombeanus,* and a few more, mostly of southern Shaba Province of Zaire, and restricted to ultrabasic and heavy-metal-enriched soils. Based on the strong vegetative similarity to two of the Shaban species, *G. gracillimus* and *G. pusillus* are also included in the section. These two species, however, have winged seeds and softly textured corm tunics, probably an adaptation to their marshy habitat. Lewis et al. (1972) considered *G. pretoriensis* to be most closely related to *G. permeabilis*, but this seems to me unlikely. *Gladiolus permeabilis* has narrow leaves but without the raised margins that one would expect in an ancestor of a species that has both heavily thickened margins and midribs, and it is, moreover, specialized in the narrow dorsal tepal held well apart from the other tepals and in the peculiar firm, coarsely fibrous corm tunics.

Sections of Subgenus *Ophiolyza*

In contrast to subgenus *Gladiolus,* the tropical African species of subgenus *Ophiolyza* can fairly readily be grouped into distinct clusters. Least specialized of these is section *Blandus,* the flowers of which are usually shades of pink or cream, moderate in size, and have perianth tubes slightly shorter that the dorsal tepals. The tepals are usually more or less equal and often have poorly defined nectar guides. Section *Blandus* is best known from the Cape region of South Africa, where the type species, *G. carneus,* is fairly common. The inclusion of Ethiopian species in section *Blandus* here is provisional. *Gladiolus balensis, G. mensensis,* and *G. negeliensis* could as well be placed elsewhere. If correctly assigned to section *Blandus* they are taxonomically isolated. Closely related is section *Tenuibracteus,* an exclusively tropical alliance, in which the upper leaves always have long blades and the floral bracts are attenuate. As in section *Blandus,* the flowers are usually shades of pink or cream, the tepals are more or less equal, and the perianth tubes are relatively long in relation to the dorsal tepal. Section *Ophiolyza* is more distinctive. The dorsal tepal is usually about as long as the tube and the bracts are strongly developed, sometimes very prominent. The dorsal tepal is enlarged and hoodlike, often horizontal or tilted downward toward the ground, concealing the stamens. Often, the lower tepals are shorter than the upper three and are curved outward and downward for their entire length so that in profile the dorsal tepals much exceed the lower three. This pattern is particularly pronounced in *G. abyssinicus, G. dichrous,* and *G. magnificus,* in which the lower tepals are nearly vestigial.

In the two remaining tropical African sections of subgenus *Ophiolyza,* sections *Decoratus* and *Acidanthera,* the anthers have unusual acute apiculate appendages. Apart from this the two sections are quite different. Section *Decoratus* has red or pink flowers, often with prominent yellow to white nectar guides, and the leaves are soft-textured. The flowers have long tubes, slightly exceeding the bracts. Flowers of section *Acidanthera* are always white, often with dark purple nectar guides, and the perianth tubes are very long, generally twice as long as the bracts and at least 60 mm long, often as much as 120–150 mm. Typically, the inner bracts are also reduced in size, but this is not the case for all the species. Although the suite of perianth characters is clearly an adaptation to pollination by sphinx moths, species of section *Acidanthera,* nevertheless, seem closely allied and have a similar overall appearance. Flowers of *Gladiolus bellus, G. gunnisii, G. richardsiae,* and *G. usambarensis* also seem adapted for sphinx-moth pollination but have somewhat more rigid leaves, and the stamens lack apiculate anther appendages. *Gladiolus bellus* and *G. usambarensis* seem best placed in section *Blandus,* but the remaining two are most probably related to *G. dalenii* and are included in section *Ophiolyza* here. A last section of subgenus *Ophiolyza* is the exclusively southern African section *Cardinales,* which like section *Decoratus* has red flowers with white markings. The anthers lack appendages and the leaves are firm in texture. The floral similarity is probably convergent. The flowers of section *Cardinales* are known to be adapted for butterfly pollination (by *Meneris tulbaghia* according to Marloth, 1917; Johnson, 1992; and Johnson & Bond, 1993). Pollinators in section *Decoratus* are unknown but may be in the same insect group, although not *Meneris,* which does not occur outside South Africa.

PLATE 1. *Gladiolus crassifolius,* pale pink flowers of the summer-flowering form from the Vumba Mountains, eastern Zimbabwe (D. C. H. Plowes).

PLATE 2. *Gladiolus crassifolius,* with the deep pink flowers and branched habit of the spring-flowering form, from eastern Zimbabwe (D. C. H. Plowes).

PLATE 3. *Gladiolus zimbabweensis,* from Mt. Inyangani, eastern Zimbabwe (D. C. H. Plowes).

PLATE 4. *Gladiolus juncifolius,* from the Chimanimani Mountains, eastern Zimbabwe (D. C. H. Plowes).

PLATE 5. *Gladiolus laxiflorus,* found in a dry marsh in southern Malawi (E. A. S. la Croix).

Plate 6. *Gladiolus unguiculatus,* in characteristic habitat, flowering in dry grass in a dambo in Malawi (E. A. S. la Croix).

Plate 7. *Gladiolus unguiculatus,* showing details of the pale pinkish flowers with dark-colored nectar guides, from central Malawi (P. Goldblatt).

PLATE 8. *Gladiolus atropurpureus,* the pale-flowered form with purple nectar guides, growing in southern Malawi (E. A. S. la Croix).

PLATE 9. *Gladiolus serapiiflorus,* the deep yellow flowers and distinctive flower shape are evident in this plant from Shaba Province, Zaire (M. Schaijes).

PLATE 10. *Gladiolus muenzneri,* a typical pale-cream-flowered plant from the Nyika Plateau, northern Malawi (E. A. S. la Croix).

PLATE 11. *Gladiolus gregarius*, showing the grassland habitat in a population from Shaba Province, Zaire (F. Billiet).

PLATE 12. *Gladiolus gregarius*, typical markings in the robust Zairian form of the species (F. Billiet).

PLATE 13. *Gladiolus macrospathus*, showing the light mauve flowers typical of the species in plants from Burundi (F. Billiet).

PLATE 14. *Gladiolus harmsianus,* a spike of this rare species from southwestern Angola, showing the straight axis and long, tightly sheathing bracts (P. Bamps).

PLATE 15. *Gladiolus harmsianus,* showing the pale flower and elongate perianth tube (P. Bamps).

PLATE 16. *Gladiolus gracillimus,* found in its typical habitat, a boggy meadow along a stream in southwestern Tanzania (P. Goldblatt).

PLATE 17. *Gladiolus zambesiacus,* in its characteristic habitat, open rocky sites, on Zomba Mountain, Malawi (P. Goldblatt).

PLATE 18. *Gladiolus zambesiacus,* close-up of the attractive pink flower with pale yellow nectar guides (P. Goldblatt).

PLATE 19. *Gladiolus balensis,* with white flowers with dark yellow nectar guides, growing in southern Ethiopia (M. Gilbert).

PLATE 20. *Gladiolus negeliensis,* showing the long perianth tube, white flower with undulate tepals (M. Thulin).

PLATE 21. *Gladiolus rupicola,* with its striking large red flowers, found in rocky grassland above evergreen forest in western Tanzania (D. Thomas).

PLATE 22. *Gladiolus usambarensis,* the typical pure white flowers of plants from northern Tanzania (anonymous).

PLATE 23. *Gladiolus bellus,* showing the white flower with striking red nectar guides in this Malawian endemic, known only from Mt. Mulanje (J. D. Chapman).

PLATE 24. *Gladiolus erectiflorus,* spikes of cream to pink flowers from Dedza in central Malawi (P. Goldblatt).

PLATE 25. *Gladiolus erectiflorus,* showing details of the flowers with its characteristic deep pink- to red-veined flowers and dark purple anthers (P. Goldblatt).

PLATE 26. *Gladiolus verdickii,* the pale creamy yellow flower in this plant from southern Shaba Province, Zaire, has less conspicuous veining on the tepals than its close relative, *G. erectiflorus* (M. Schaijes).

PLATE 27. *Gladiolus dalenii* subsp. *dalenii,* showing the characteristic hooded flower in an orange-flowered plant from Zomba Mountain, southern Malawi (P. Goldblatt).

Plate 28. *Gladiolus dalenii* subsp. *dalenii,* showing the unusual slender, purple-flowered form of this polymorphic species, from Malawi (E. A. S. la Croix).

Plate 29. *Gladiolus velutinus,* with its pale greenish yellow flowers, growing in a marsh in southwestern Tanzania (J. Lovett).

Plate 30. *Gladiolus magnificus,* spikes of scarlet flowers, growing in light woodland in western Zimbabwe (D. C. H. Plowes).

PLATE 31. *Gladiolus benguellensis,* with scarlet flowers and yellow nectar guides, found near Kolwezi in southern Zaire (M. Schaijes).

PLATE 32. *Gladiolus melleri,* just coming into flower in a dambo in southern Malawi, with leafless spikes rising above dry grass (E. A. S. la Croix).

PLATE 33. *Gladiolus roseolus,* with pure white flowers in a plant from western Ethiopia (M. Thulin).

PLATE 34. *Gladiolus pauciflorus,* showing an unusual form of the species with well-developed markings on the lower tepals in a population from southern Ethiopia (M. Gilbert).

PLATE 35. *Gladiolus abyssinicus,* showing the unusual floral form assumed to be adapted for bird pollination, found growing along a stream in the mountains of western Saudi Arabia (I. S. Collenette).

PLATE 36. *Gladiolus decoratus,* with its striking dark reddish flowers and whitish nectar guides, in a colony in southern Malawi (E. A. S. la Croix).

PLATE 37. *Gladiolus oligophlebius,* showing the characteristic pale pink flower and elongate whitish nectar guides in plants from the shores of Lake Malawi (E. A. S. la Croix).

PLATE 38. *Gladiolus sudanicus,* flowering after good rains in the semiarid Blue Nile Gorge in central Ethiopia (M. Gilbert).

PLATE 39. *Gladiolus sudanicus,* showing the typical red perianth and elongate pale nectar guides typical of section *Decoratus* (M. Gilbert).

PLATE 40. *Gladiolus murielae,* with graceful spikes of long-tubed white flowers, growing on Zomba Mountain in southern Malawi at the southern extremity of its range (E. A. S. la Croix).

PLATE 41. *Gladiolus candidus,* from southern Ethiopia showing the pure white flowers without nectar guides characteristic of the species (M. Thulin).

Development of the Garden *Gladiolus*

Almost all gladioluses seen for sale as cut flowers or for garden culture have a hybrid origin. They are the product of interspecific crosses and subsequent selection and recrossing of the hybrids. The deliberate hybridization of plants is a practice that began early in the 19th century, initially out of sheer curiosity and later with deliberate aims in mind. The first recorded efforts to produce hybrids in *Gladiolus* were those of the English scientist, William Herbert, later Dean of Manchester. He described his experiments (Herbert, 1847), begun in the 1820s, at Spofforth, Yorkshire, where he grew several southern African species. An important result of his work was the discovery that hybrids between the late spring-and early summer-flowering species from the Cape Region of South Africa, including *G. angustus, G. cardinalis, G. carneus,* and *G. tristis,* proved hardy and could be left in the ground all year long. Hybrids between earlier-flowering Cape species, including *G. carinatus* and *G. caryophyllaceus,* which grow throughout the winter, needed to be raised under glass, just as for the parent species. A number of Herbert's hybrids were given names, and they were distributed to many gardeners and nurseries in Britain and elsewhere.

Herbert's example was followed by several nurserymen who grew bulbous plants. At the time the cultivation of southern African *Gladiolus* was quite fashionable, and many different species were available in the nursery trade. In 1823 Colville's of Chelsea flowered hybrid plants that resulted from a cross between *G. cardinalis* and *G. tristis,* and placed the resulting hybrids on the market. This strain was illustrated and described in 1826 by the botanist, Robert Sweet, who had a particular interest in plants cultivated in Britain. Sweet called Colville's plant *G. colvillei.* According to current botanical convention, the name should be written in the form *G.* ×*colvillei,* thus indicating that it has a hybrid origin and is not a naturally occurring plant. Colville's *Gladiolus* is still cultivated today, and it can be found in florist stores in Europe in early summer. In France, it is often seen under the name, "les Colvilles."

Gladiolus breeding took a different direction after the discovery in the 1830s of *G. oppositiflorus,* a species described by Herbert in 1837. This is a summer-flowering and winter-dormant species from eastern South Africa, as opposed to the winter- and spring-growing Cape species (but sometimes summer-flowering) then widely grown. *Gladiolus oppositiflorus* was crossed with the hybrid between two Cape species, *G. cardinalis* × *G. carneus.* The resulting offspring is believed to have been the plant later called *G. ramosus* (its actual parentage is uncertain because of inadequate documentation and some confusion over the application of names of species). Another series of hybrids is the so-called "*nanus*" strain, the exact parentage of which is in doubt. Most likely the strain was the result of back-crossing *G. ramosus* with *G. carneus,* or according to some authorities, with *G. cardinalis.* This attractive line, resembling the early Colville stock, included several named cultivars and is still grown today. Although best known in western Europe, they are available in North America.

More exciting was the introduction of another summer-flowering species, the robust and easily grown *Gladiolus dalenii.* Although one of the tropical African species of *Gladiolus,* the first specimens to be introduced to Europe were from Natal in South Africa. Some of the first hybrids with *G. dalenii* as one of the parents were made in Belgium about 1837 by H. S. Bedinghaus, head gardener of the Duke d'Aremberg at Enghein. The strain was later obtained by Messrs. van Houtte of Ghent and were marketed as "*gandavensis*" hybrids, or *G.* ×*gandavensis* ("from Ghent," in Latin). Many new hybrid strains were produced in the following years by crossing the "*gandavensis*" hybrids with other cultivars. According to T. T. Barnard (in Lewis et al., 1972), the most important of the plants used in the crosses was probably *G. ramosus,* itself of apparently of hybrid origin. As new species were discovered in southern Africa during a period of active botanical exploration in the mid to later 19th century, breeders tried to cross these with their old stock, no doubt hoping to develop new strains for the flower market. In 1870, Victor Lemoine, a nurseryman at Nancy, France, acquired *G. papilio,* another summer-flowering southern African species. This plant was notable on two counts. It is winter-hardy in temperate regions, and has very dark purple blotches on the lower tepals, something lacking in the *Gladiolus* strains then available. Lemoine succeeded in crossing *G. papilio* with *G.* ×*gandavensis* and the resulting seedlings bore the attractive purple blotches of one of its parents. These "*lemoinei*" hybrids, also called *G.* ×*lemoinei,* proved also to be hardy, although this was not considered one of its attractions. The "*lemoinei*" stock was crossed with another southern African species, the stunning red- and white-flowered *G. saundersii,* to yield the plants that became known as *G.* ×*nanceianus.* These large-flowered *Gladiolus* hybrids have flowers in a range of different colors, so well known today in the florist trade. Lemoine also produced the range of brown, green, and dull-spotted varieties by using a green-flowered race of *G. dalenii* (also called *G. dracocephalus*) in some of his crossing experiments. Later this strain was used in Germany to produce purple, lilac, and lavender colors.

Other important developments in *Gladiolus* breeding also took place in Germany. Max Leichtlin of Baden Baden also crossed "*gandavensis*" stock with *G. saundersii* (some authorities say it was *G. cruentus,* but it is unlikely that this rare, red-flowered Natal species was ever available for extensive plant breeding). The result was a large-flowered strain with pale patches on the lower tepals. Leichtlin's hybrids were not exploited commercially and were sold, becoming the property of John Lewis Childs of New York in 1891. Continued hybridization using this strain in North America yielded plants that were the ancestors of the common garden and florist-trade *Gladiolus.* These plants are the result of more than 60 years of crossing and selection and have the genes of four different species.

Conspicuously lacking in the hybrids being developed in the late 19th century was any yellow color. The introduction in the 1890s, probably from near the Victoria Falls in Zimbabwe, of a plant then known as *Gladiolus primulinus* provided the source of the missing color. *Gladiolus primulinus* is now considered conspecific with *G. dalenii* subsp. *dalenii,* but it is a slender variant of the species and has clear yellow flowers. The use of this plant seems to have been the last time wild stock was involved in the breeding of the large-flowered *Gladiolus* cultivars. Back-crossing among the existing hybrid strains now yields the variability and novelty we see today in markets. Details of modern *Gladiolus* breeding are beyond the scope of this book and are the subject of specialized study. Several works are available on the subject, and I refer readers especially to *Growing Gladioli* (Anderton & Park, 1989) and *Gladioli for Everyone* (Garrity, 1975).

Attempts have been made over the years to produce additional hybrid strains using other wild species. Often the results have been striking, but difficulties with propagation, cultivation, or marketing have made little impression on the flower market. Many species of *Gladiolus* have a delightful scent, and repeated efforts have been made to produce a strain of scented hybrids. Wonderfully scented hybrids have been bred, and some have been acclaimed at exhibitions. None, however, have survived the test of time, and none are regularly available in the horticultural trade. T. T. Barnard's *Gladiolus* cultivar Cristabel, a hybrid between two Cape species, *G. virescens* and *G. tristis,* both scented, is winter-growing. Although it has great charm, it suffers from small stature, short-lasting flowers, and the difficulty of raising large stocks. *Gladiolus* breeding continues, both in search of new variants and for disease resistance. Few of the more than 250 wild species have, however, been tapped for their horticultural potential, and only one, *G. primulinus,* as far as is known has been from tropical Africa. Of the great diversity of the genus there, only the East African *G. murielae* (also called *Acidanthera bicolor, A. murielae,* or *G. callianthus*) is currently available in the horticultural trade. Production of *Gladiolus* hybrids from wild species is relatively easy once plants are obtained. Almost all the species can be intercrossed, and providing that chromosome numbers match, offspring are fertile and readily produce quantities of seeds. Perhaps the future will yield new hybrid strains resulting from crosses of a series of species not yet tried by plant breeders. I refer readers to the book by Anderton & Park (1989) and the essay by Barnard in Lewis et al. (1972) for more detailed information about the history and breeding of *Gladiolus* hybrids.

GROWING *GLADIOLUS*

Growing *Gladiolus* is, like that of most plants with bulbs or corms, a relatively easy and rewarding undertaking. All *Gladiolus* species produce corms that are dormant for a part of the year, either winter for the summer-growing species or summer for the winter-growing species. Dormant corms can be obtained from nurseries and through trade catalogs and are usually sold with instructions about planting, including the appropriate season. In most of North America and northern Europe, only the summer-growing species and the large hybrid forms are readily grown in gardens in the open. Planted in the spring, after any danger of a freeze, corms will sprout quickly and produce a fan of handsome sword-shaped leaves, and in time spikes of flowers. In California, the Southwest, the South, and all across southern Europe, *Gladiolus* corms can be left in the ground all year, and a wide range of winter-growing species can be grown in the open ground. Even in milder parts of the Midwest and East Coast, corms may survive the winter, as they have for me in St. Louis. One cannot depend on this, and it is advisable to either lift corms for winter storage or purchase new stock each year.

Ideally, *Gladiolus* should not only be grown at the appropriate season but under conditions that most closely approximate natural conditions. This is particularly true of wild species, many of which grow in limited habitats in nature. Most gladioluses are fortunately fairly adaptable and will grow in almost any soil, provided it is well drained. Sunny conditions are also preferable—shade encourages the production of a lanky, weak stem and a lax spike. Most wild species of *Gladiolus,* but perhaps not the very tall species, also make interesting pot plants. They thrive indoors when given well-lighted growing conditions, and the charming and colorful flowers of the smaller-flowered species can be seen at their best in pots brought into the house as flowering commences. The small-flowered species seldom achieve success in open garden displays, because they tend to be lost among displays of annuals or larger-flowered perennials. Pockets of wild species in the rock garden, however, are likely to make interesting accents.

A great variety of the large-flowered hybrids are available in the trade, and a few "*nanus*" strains are also seen in many bulb dealer's catalogues, often advertised as being winter-hardy. To my mind these lovely plants are seen much too rarely. Only one wild species is routinely offered for sale, the tropical African *Gladiolus murielae,* also known as *G. callianthus,* and it certainly is an interesting garden subject. If possible it should be planted close to a patio, where the delightful scent can be enjoyed in the evenings. A species that is not, to my knowledge, available in cultivation, *G. laxiflorus,* would make a fine addition to any garden. It will produce more than one spike in the early summer, and the large flowers are a fine deep pink or light purple. In the wild it occurs in seeps and marshes, and on riverbanks, and I expect that it would make a good plant for a streamside garden. Coming from southern tropical Africa, it is unlikely to survive harsh winters, even though it would then be dormant. The climates of southern Europe, Australia, and the southern United States are likely to suit its requirements.

Regretfully, sources of any of the wild species of *Gladiolus* are limited. Seeds of a few southern African species are available from specialist nurseries, but few if any of these offer any of the tropical African species, either as corms or seeds. Many of them are worth trying to grow, and perhaps a few will be found amenable to garden conditions. Particularly handsome species include the bright red-flowered *G. rupicola,* the tall white-flowered *G. bellus,* and the lovely red or deep pink *G. decoratus* and *G. oligophlebius.* There is a great variety of flower color in *G. dalenii,* and although the species resembles the large-flowered hybrids that were derived from it, it is more graceful and to my mind a much more attractive plant. The list of desirable species is much longer. If any wild species of *Gladiolus* can be found, I recommend them to the bulb enthusiast. Try them out.

Phytogeography

Eighty-four species of *Gladiolus* (including the two discussed at the end of the Systematic Treatment, under Additional or Incompletely Known Species) in two subgenera, *Ophiolyza* and *Gladiolus,* are recognized here for tropical Africa. For purposes of this monograph, tropical Africa encompasses all of sub-Saharan Africa, south to the borders of the countries normally considered part of southern Africa: Namibia, Botswana, and South Africa. I also arbitrarily excluded temperate southern Mozambique, south of the Save River, from tropical Africa. *Gladiolus* is thus by far the largest genus of the Iridaceae in tropical Africa. *Moraea,* the next most speciose genus of the family there, is represented by just 27 species. Large as *Gladiolus* is in tropical Africa, the genus is substantially eclipsed in southern Africa, where there are estimated to be at least 150 species. There are just eight species of the genus in Madagascar (Goldblatt, 1989) and probably no more than 10 in all of North Africa, southern Europe, and Turkey and the Middle East, excluding southern Arabia.

Species distributions range from the widespread and nearly ubiquitous (e.g., Maps 14, 22, 49) to narrow endemics known from small areas, even from single localities (e.g., Maps 9, 15, 39). These is no apparently fundamental difference in the distributions of species in the two subgenera in tropical Africa. In contrast, subgenus *Gladiolus* predominates in southern Africa and Madagascar, and only species of this subgenus occur in Eurasia–North Africa. Among the widespread species, *G. dalenii* (subgenus *Ophiolyza*) and *G. gregarius* and *G. unguiculatus* (both subgenus *Gladiolus*) have comparable ranges. All three extend from Senegal in West Africa southward to southern central tropical Africa. *Gladiolus dalenii* (including all three subspecies) is the most widespread species of the whole genus. Its range covers all of tropical

Africa, western Arabia, Madagascar, and the eastern half of southern Africa. The ranges of *G. gregarius* and *G. unguiculatus* are comparable, but neither extends beyond the African mainland, nor do they occur in Kenya, Ethiopia, or most of Uganda. There is no apparent reason for their absence from most of northeastern Africa, and the pattern awaits satisfactory explanation.

Seventy-seven of the 82 tropical African species of *Gladiolus* are endemic there. *Gladiolus dalenii,* as mentioned above, extends to western Arabia, where two more tropical African species occur, *G. abyssinicus* and *G. candidus,* both members of subgenus *Ophiolyza. Gladiolus abyssinicus* has its main distribution in northern Ethiopia, whereas the eastern African and southern Ethiopian *G. candidus* has outlying populations in Dhofar in southern Oman. The latter distribution pattern is without parallel in the Iridaceae. Like *G. dalenii,* the widespread *G. unguiculatus* (subgenus *Gladiolus*) also extends southward out of the tropical zone into southern Africa. It is recorded widely in the Transvaal and also occurs in eastern Botswana. *Gladiolus sericeovillosus* subsp. *calvatus* (subgenus *Ophiolyza*), which extends widely across Zimbabwe, also occurs in the Transvaal, South Africa; subspecies *sericeovillosus* is exclusively southern African. *Gladiolus permeabilis* subsp. *edulis* (subgenus *Gladiolus*) has a similar range. This widespread southern African plant extends from the southern Cape to northern Zimbabwe in the east and to northern Namibia in the west. The two remaining subspecies of *G. permeabilis* have smaller ranges in the southern Cape and Natal. Despite favoring semiarid habitats, subspecies *edulis* is evidently rare in Botswana but it does occur in the east of the country. Only one more species, *G. magnificus* (subgenus *Ophiolyza*), a close relative of *G. dalenii,* occurs outside tropical Africa; it has isolated populations in northern Namibia and northern Botswana, but its main range is eastern Zimbabwe and southern Zambia and is known from only a single gathering in southeastern Angola.

Within tropical Africa there are striking differences in the patterns of diversity for *Gladiolus* across the continent, in numbers of species and numbers of endemics (Map 1). Species favor open sites in areas of moderate to high rainfall. Thus it is no surprise that *Gladiolus* is best represented in highlands where grassland and savanna vegetation predominates, and conversely, is poorly represented in lowlands where the natural vegetation is a closed-canopy forest. Even admitting this limitation, the concentration of species is remarkably variable. The area of greatest diversity for *Gladiolus* in tropical Africa is Shaba Province in Zaire, where 28 species have been recorded, 12 of which are endemic. This dwarfs the next two most diverse centers, the southwestern Angolan highlands, with 14 species and six endemics, and south-

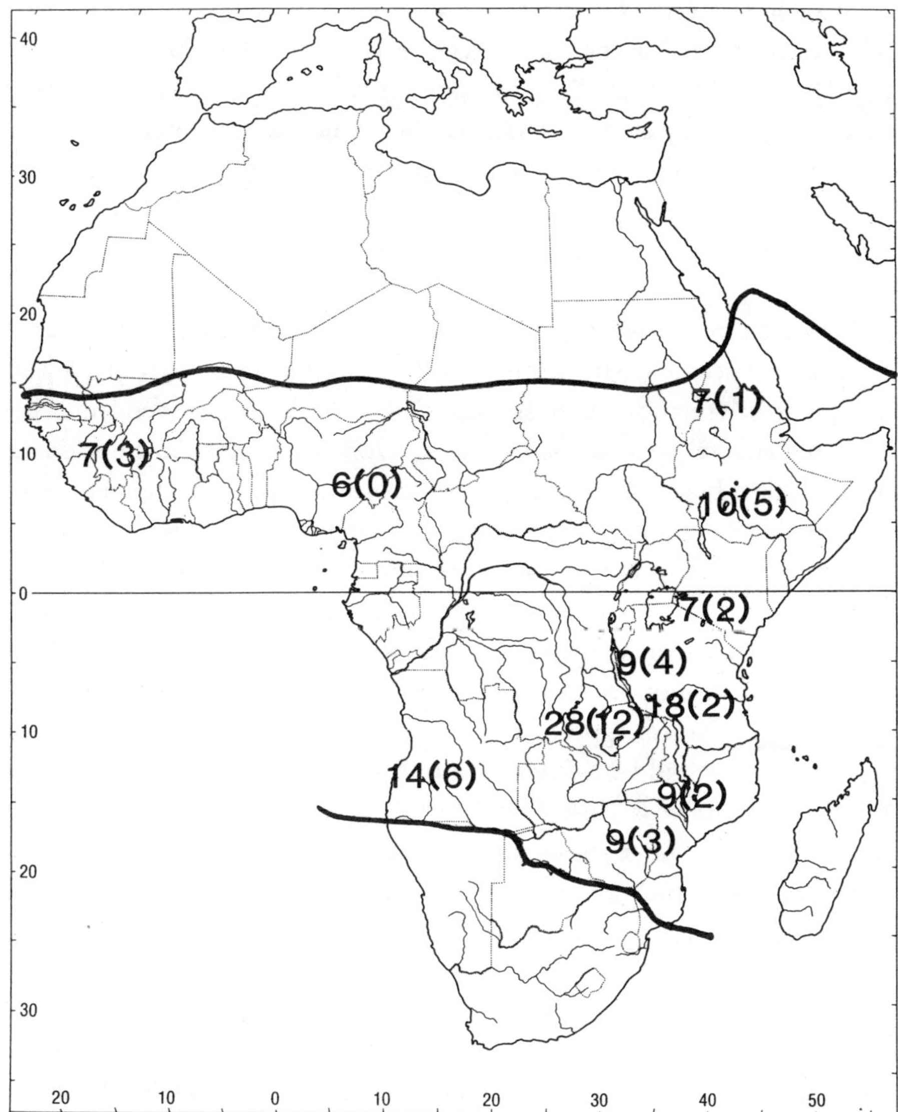

MAP 1. Centers of diversity and endemism in tropical African *Gladiolus.* The first number indicates the total number of species for each center, and in parentheses, the number of species endemic to the center.

western Tanzania–northern Malawi, with 18 species and two endemics. Elsewhere, diversity is markedly lower. The eastern Zimbabwe highlands, which includes a small part of western Mozambique, has nine species, including three endemics. The nearby Shire Highlands of southern Malawi (also including a small part of adjacent Mozambique) has nine species, of which *G. bellus* and *G. zambesiacus* occur nowhere else.

In West Africa, the highlands of Cameroon and central Nigeria have six species of *Gladiolus,* but there are seven species in a small portion of extreme West Africa, including the Fouta Djallon of western Guinea, Sierra Leone, and the Nimba Mountains of Guinea, Ivory Coast, and interior Liberia. There are three endemic species in the latter area, all members of section *Acidanthera, G. chevalieranus, G. leonensis,* and *G. praecostatus.* This section of subgenus *Ophiolyza* is the only clade that has radiated to any degree in West Africa.

Ethiopia can conveniently be treated as comprising two centers as far as *Gladiolus* is concerned. The northern half of Ethiopia plus Eritrea form one center, in which there are seven species of *Gladiolus.* Only one is endemic, the poorly understood *G. mensensis* (subgenus *Ophiolyza*). *Gladiolus abyssinicus* is centered there, but it also occurs locally across the Red Sea in the highlands of western Arabia south of Mecca, and in southeastern Ethiopia, which I do not regard as part of the northern Ethiopian center. Southern Ethiopia is richer; there are 10 species of *Gladiolus* in the area extending from Gamu Gofa in the west to Harerge in the east. No less than five endemics occur there. *Gladiolus* is not yet adequately known from the area, but at least *G. lithicola* and *G. calcicola* appear to be local endemics from the Harar District, and *G. negeliensis* and *G. balensis* have restricted distributions to the south. Only *G. longispathaceus,* closely related to *G. abyssinicus,* has a wide range in the area, from Mt. Guge in Gamu Gofa to the Bale Mountains to the east.

In East Africa, the mountains of northern Tanzania together with central Kenya form another weakly defined phytogeographical center for *Gladiolus* (Map 1). Among the seven species that occur there, *G. watsonioides* is endemic to areas above 2000 m, and *G. usambarensis* is limited to south of this area, including the Taita Hills, and the Usambara–Mweru–Ngorongoro mountain complex. There is one more center, albeit weakly defined, in western central Tanzania. Some nine species of *Gladiolus* occur there, four of which are endemic. Three of the endemics are known from just a single collection each, *G. canaliculatus* and *G. richardsiae* from the Itigi District, and *G. stenolobus* near Uvinza, to the west. The remaining *G. grantii* is better known and appears to be relatively common between Singida and Irangwe.

There are some additional locally endemic species that do not fall into convenient centers. Some of the most notable are *Gladiolus iroensis* (section *Acidanthera*), restricted to the seasonal marshes around Lake Iro in southern Chad, and *G. lundaensis* (subgenus *Gladiolus*), also a marsh plant, from northeastern Angola and nearby southern Zaire. *Gladiolus serenjensis* (subgenus *Ophiolyza*), from northern Zambia, is one of several species of *Gladio-*

lus that occur in the area, where an absence of a marked concentration precludes recognition of a center for the genus there. The patterns of endemism in general reflect past or present cycles of geographic isolation that have allowed local divergence to proceed, or they may reflect very diverse local ecological conditions, or combinations of both phenomena. These patterns are particularly evident in southern Shaba. There, a diversity of soil types as well as isolated high mountain ranges combine to form a greater diversity of biological niches than elsewhere in tropical Africa. The most striking of these niches is the localized outcrops of ore bodies containing high concentrations of heavy metal, especially copper, but also cobalt and uranium. A series of closely related species appears to have evolved there, each largely or exclusively restricted to a particular location.

Systematic Treatment

THE GENUS *GLADIOLUS* LINNAEUS

Species Plantarum 36 (1753). Baker, Handbook Irideae 198 (1892); Fl. Trop. Africa 7: 360 (1898). Hutchinson & Dalziel, Fl. West Trop. Africa 2: 378 (1936). Hepper, Fl. West Trop. Africa, Ed. 2, 3(1): 141 (1972). G. Lewis et al., J. S. African Bot., Suppl. 8 (1972). Geerinck, Bull. Jard. Bot. Nat. Belgique 42: 269 (1972). Marais, Kew Bull. 28: 311 (1973). Goldblatt & de Vos, Bull. Mus. Nat. Hist. Nat., Sér. 4, Sect. B, Adansonia 11: 417 (1989). Goldblatt, Fl. Zambesiaca 12(4): 66–103 (1993).

TYPE SPECIES. *Gladiolus communis* Linnaeus.

Synonymy

Antholyza Linnaeus, Species Plantarum 37 (1753). Baker, Handbook Irideae 229 (1892); Fl. Capensis 6: 165 (1896); Fl. Trop. Africa 7: 343 (1898). N. E. Brown, Trans. Roy. Soc. S. Africa 20: 265 (1932), excluding the type. Type: *A. cunonia* Linnaeus (lectotype designated by Hitchcock & Green, 1929) (= *Gladiolus cunonius* (Linnaeus) Gaertner).

Cunonia Miller, Figures Plants 1: 75 (1756), Gardeners' Dictionary, Ed. 8 (1768), name rejected in favor of *Cunonia* Linnaeus (1759), Cunoniaceae, conserved name. Type: *C. antholyza* Miller (= *Gladiolus cunonius* (Linnaeus) Gaertner).

Anisanthus Sweet, Hortus Britannicus, Ed. 2: 566 (1830). Type: *A. splendens* Sweet (= *Gladiolus splendens* (Sweet) Herbert).

Petamenes Salisbury ex J. W. Loudon, Ladies' Flower Garden Ornamental Bulbous Plants 42, pl. 8 (1841). N. E. Brown, Trans. Roy. Soc. S. Africa 20: 276 (1932). Phillips, Bothalia 4: 44 (1941). Type: *P. abbreviatus* (Andrews) N. E. Brown (= *Gladiolus abbreviatus* Andrews).

Sphaerospora Sweet ex Loudon, Ladies' Flower Garden Ornamental Bulbous Plants 66, pl. 14 (1841), not *Sphaerospora* Klatt, Linnaea 32: 725 (1863) (= *Geissorhiza*). Sweet, Hortus Britannicus 398 (1827), name without description. Type: *S. imbricata* (Jacquin) Sweet ex Loudon (= *Gladiolus italicus* Miller).

Acidanthera Hochstetter, Flora 27: 25 (1844). Baker, Handbook Irideae 185 (1892), in part; Fl. Capensis 6: 130 (1896), in part; Fl. Trop. Africa 7: 358 (1898). Hutchinson & Dalziel, Fl. West Trop. Africa 2: 376 (1936). Hepper, Fl. West Trop. Africa, Ed. 2, 3(1): 139 (1968). Type: *A. bicolor* Hochstetter (= *Gladiolus murielae* Kelway).

Ballosporum Salisbury, Genera Plants 142 (1866). Type: *Gladiolus segetum* Ker.

Homoglossum Salisbury, Genera Plants 143 (1866), neither type nor constituent species indicated; Baker, J. Linn. Soc., Bot. 16: 161 (1878). De Vos, J. S. African Bot. 42: 301 (1976). Type: *H. watsonium* (Thunberg) N. E. Brown, according to de Vos (1976) (= *Gladiolus watsonius* Thunberg).

Hyptissa Salisbury, Genera Plants 142 (1866). Type: *H. rosea* Salisbury, unpublished name, identity uncertain.

Ophiolyza Salisbury, Genera Plants 142 (1866). Type: *Gladiolus alatus* Linnaeus (lectotype designated here).

Ranisia Salisbury, Genera Plants 143 (1866). Type: *Gladiolus tristis* Linnaeus.

Symphydolon Salisbury, Genera Plants 142 (1866). Type: *Gladiolus floribundus* Jacquin.

Oenostachys Bullock, Kew Bull. 465 (1930). Goldblatt, J. S. African Bot. 37: 411, 443 (1971). Type: *O. dichroa* Bullock (= *Gladiolus dichrous* (Bullock) Goldblatt).

Anomalesia N. E. Brown, Trans. Roy. Soc. S. Africa 20: 270 (1932). Goldblatt, J. S. African Bot. 37: 412 (1971). Type: *A. cunonia* (Linnaeus) N. E. Brown (= *Gladiolus cunonius* (Linnaeus) Gaertner).

Kentrosiphon N. E. Brown, Trans. Roy. Soc. S. Africa 20: 271 (1932). Type: *K. saccatus* (Klatt) N.E. Brown (= *Gladiolus saccatus* (Klatt) Goldblatt & de Vos).

Dortania A. Chevalier, Bull. Mus. Hist. Nat. Paris, Sér. 2, 9: 402 (1937). Type: *D. amoena* A. Chevalier (= *Gladiolus chevalieranus* W. Marais).

Eponymy

Gladiolus, "little sword," referring to the sword-shaped leaves of the genus, and in fact, of most members of the *Iris* family.

Description

Perennial herbs with deciduous leaves and globose corms with membranous to fibrous, reticulate tunics; the base enclosed by two or three sheathing cataphylls. LEAVES usually contemporary with the flowers and borne on the same shoot, sometimes borne earlier or later than the flowers and on separate shoots, two to several, basal or some inserted above ground level, sometimes the blade reduced or lacking, thus the entire leaf partly to entirely sheathing, blades equitant, unifacial, linear to lanceolate, either plane and the margins, midrib, and sometimes other veins not or only lightly thickened and hyaline, or the margins or midrib strongly raised, sometimes even winged (thus H- or X-shaped in section), or the midribs and margins much thickened and the blade evidently terete but with four narrow longitudinal grooves. FLOWERING STEM aerial, terete, simple or branched, erect or flexed downward above, often above the sheath of the uppermost leaf.

INFLORESCENCE a spike (rarely reduced to a single flower), the flowers secund (distichous in a few species outside our area); FLORAL BRACTS two, usually green, sometimes dry above or entirely, usually relatively large, the inner (adaxial) usually slightly smaller than the outer, sometimes much smaller, or rarely slightly longer, usually notched apically for 1–2 mm or entire. FLOWERS medianly zygomorphic (actinomorphic in a few tropical African, Madagascan, and South African species), tepals united below in a tube, the lower tepals usually with contrasting markings constituting a nectar guide, the stamens arcuate and unilateral, ascending to horizontal, the style arched over them; PERIANTH TUBE well developed, usually obliquely funnel-shaped, often shorter than the bracts in subgenus *Gladiolus,* usually about as long as the bracts in subgenus *Ophiolyza,* or sometimes much longer than the bracts; TEPALS usually unequal, the uppermost (dorsal) broader and arched to hooded over the stamens, the lower three narrower, in subgenus *Gladiolus* usually clawed below and united for a short distance, as long, shorter than, or exceeding the upper tepals. FILAMENTS filiform, inserted at the base of the upper part of the perianth tube, usually exserted but occasionally extending only to the mouth of the tube; ANTHERS symmetrically disposed around the style in species with actinomorphic flowers, usually unilateral, parallel, ascending to horizontal, lying below the dorsal tepal, dehiscing longitudinally, subbasifixed to centrifixed, rarely with sterile tails, the connective obtusely mucronate above or prolonged into an prominent acute to apiculate appendage. OVARY ovoid to oblong; STYLE filiform, dividing opposite to or beyond the anthers, the branches simple, filiform below, expanded gradually to abruptly above and channeled to bilobed. CAPSULES large, usually slightly inflated, ovoid to ellipsoid or globose, rarely elongate and nearly cylindric; SEEDS usually many per locule, platelike with a broad, membranous, circumferential wing or, rarely, few per locule and the seeds wingless and more or less globose to angled by pressure. BASIC CHROMOSOME NUMBER x = 15.

Distribution

Gladiolus is a genus of more than 250 species, centered in southern Africa and extending throughout tropical Africa and Madagascar and into the Arabian Peninsula, the Mediterranean basin, Europe, and Asia as far east as Iran and Afghanistan.

KEY TO THE SPECIES OF *GLADIOLUS* IN TROPICAL AFRICA

The key deals with *Gladiolus* in all countries of sub-Saharan Africa except South Africa, Botswana, Lesotho, Swaziland, the southern half of Namibia, and Mozambique south of the Save River.

Notes

Measure perianth length from top of ovary to apex of dorsal tepal of mature flower.

Measure perianth tube length from top of ovary to first point of separation of any tepal, usually the dorsal.

Measure lower tepals from apices to point of fusion with each other.

For leaf number, *exclude* basal sheathing cataphylls but *include* both fully developed laminate foliage leaves and partly and entirely sheathing (bladeless) leaves.

Terete leaves always have four longitudinal grooves, thus are four-grooved.

For bract length, measure outer bract from the middle of the spike.

For leaf width, choose a well-developed, preferably basal leaf, and measure near midline of blade.

1 Perianth tube at least twice as long as the dorsal tepal (and at least 5 cm long and ± cylindric throughout); flowers white to cream, often with red or purple markings on the lower tepals 2

1′ Perianth tube shorter to as long as the dorsal tepal (sometimes longer but never more than twice as long), and less than 6 cm long and not cylindric throughout; flowers variously colored, including white 13

2(1) Bracts 2.5–4 cm long; anthers obtuse to acute but lacking apiculate appendages; tepals unmarked 3

2′ Bracts (3–)4–10 cm long; anthers with or without apiculate appendages; tepals unmarked or with pale to deep red to purple marks on the lower 3 tepals 5

3(2) Perianth tube 80–120 mm long; leaves with long blades; leaf sheaths glabrous 4

3′ Perianth tube 60–75 mm long; leaf blades short or lacking (leaves thus entirely sheathing); sheaths of the lower or all leaves pubescent . . . . 20. ***G. curtifolius***

4(3) Dorsal tepal 40–45 mm long; filaments c. 25 mm long . . . . 61. ***G. richardsiae***

4′ Dorsal tepal 25–30 mm long; filaments c. 9 mm long . . . . 60. ***G. gunnisii***

5(2′) Perianth tube 60–75 mm long; anthers with or without apiculate appendages 6

5′ Perianth tube (90–)120–150 mm long; anthers with acute apiculate appendages (1–)2–5 mm long 7

6(5) Tepals usually uniformly white; anthers with apiculate appendages c. 1 mm long . . . . 79. ***G. chevalieranus***

6′ Tepals usually with red markings on the lower tepals; anthers obtuse to acute but without apiculate appendages . . . . 42. ***G. bellus***

7(5′) Leaf blades rigid, with margins, midribs and often other veins heavily thickened and raised; filaments included or exserted up to 5 mm 8

7′ Leaf blades relatively soft-textured, the margins and midribs hardly or not at all thickened; filaments included or exserted up to 15 mm 9

8(7) Leaves 2–3 mm wide and only the midrib and margins thickened; leaf blades 2–5(–12) cm long, rarely reaching the middle of the stem . . . . 78. ***G. leonensis***

8′ Leaves (4–)8–12 mm wide and secondary veins as well as the midrib and margins thickened; leaf blades at least 15 cm long, usually reaching to at least the base of the spike . . . . . . 77. ***G. praecostatus***

9(7′) Filaments included or exserted up to 5 mm; perianth not or hardly marked with purple on lower tepals 10

9′ Filaments exserted 5–15 mm; perianth with conspicuous purple marks on the lower tepals 11

10(9) Tepals 20–30 mm long; plants 2- to 4-flowered (plants of eastern Africa and southern Arabia) . . . . 82. ***G. candidus***

10′ Tepals 35–40 mm long; plants (4–)5- to 8-flowered (plants of western Africa) . . . . 76. ***G. aequinoctialis***

11(9′) Spike 1- to 2-flowered; leaf blades short, not reaching the base of the spike . . . . 80. ***G. iroensis***

11′ Spike normally with at least 3 and up to 8 flowers; leaves with long blades, usually reaching to at least the base of the spike 12

12(11′) Filaments exserted 10–15 mm; anther appendages 2–4 mm long (plants of eastern Africa) . . . . 81. ***G. murielae***

12′ Filaments exserted 5–8(–10) mm; anther appendages 0.5–1.5 mm long (plants of western Africa) . . . . 76. ***G. aequinoctialis***

13(1′) Dorsal tepal 2–3 times as long as the upper lateral tepals; lower tepals shorter than the upper laterals and sometimes scalelike 14

13′ Dorsal tepal about as long, to 1.5 times as long as the upper lateral tepals; lower tepals as long, longer, or shorter than the upper lateral tepals 18

14(13) Perianth tube not abruptly expanded, widening gradually to the apex; dorsal tepal 35–42 mm long . . . . 52. ***G. magnificus***

14′ Perianth tube abruptly expanded into a ± cylindric upper part; upper tepal 10–35(–40) mm long 15

15(14′) Foliage leaves with short blades evenly spaced on the stem; leaf sheaths lightly pubescent . . . . 53. ***G. huillensis***

15′ Foliage leaves with long blades and mostly ± basal; leaf sheaths glabrous 16

16(15′) Bracts large and bright red (or white), nearly concealing the flowers; perianth including the tube 45–60 mm long . . . . . 66. ***G. dichrous***

16′ Bracts moderate to large and reddish or green, never concealing more than half the flower; perianth including the tube 25–66 mm long 17

17(16′) Bracts 35–70 mm long, usually red or purple; dorsal tepal 20–40 mm long . . . . 64. ***G. abyssinicus***

17′ Bracts 18–24(–30) mm long, usually dull purple; dorsal tepal 12–18 (–22) mm long . . . . 65. ***G. schweinfurthii***

18(13′) Foliage leaves well developed, the blades plane and fairly densely pubescent 19

18′ Foliage leaves with blades either vestigial or developed and then plane, ribbed, or terete but glabrous or sparsely villous (to ciliate) only 20

19(18) Flowers large with tube 15–25 mm and dorsal tepal 25–35 mm long; bracts (20–)25–40 mm long . . . . 6. ***G. puberulus***

19′ Flowers small with tube c. 10 mm and dorsal tepal c. 20 mm long; bracts 16–20(–25) mm long . . . . 5. ***G. intonsus***

20(18′) Flowers large, (40–)45–85 mm long (from tube base to dorsal tepal apex); perianth tube at least 11 mm and usually 15–50 mm long 21

20′ Flowers small to medium-sized, 11–36(–40) mm long (tube base to dorsal tepal apex); perianth tube 3.5–12(–18) mm long 52

21(20) Perianth tube 30–50 mm long 22

21′ Perianth tube 11–29 mm long 37

22(21) Anther apices with acute apiculate appendages 1–2 mm long 23

22′ Anther apices obtuse or acute but without apiculate appendages 26

23(22) Filaments included or barely exserted from the upper part of the tube (less than 2 mm); flower white or cream . . . . 71. ***G. grantii***

23′ Filaments barely to well exserted from the upper part of the tube (2 mm or more); flower pink or red with white to cream markings on the lower tepals 24

24(23′) Dorsal tepal strongly arched and exceeding the perianth tube by c. 10 mm; filaments exserted 20–25 mm . . . . 67. ***G. decoratus***

24′ Dorsal tepal arched weakly or not at all, and exceeding the tube by c. 5 mm or slightly shorter than the tube; filaments exserted 4–10 (–15) mm 25

25(24′) Anthers 8–9 mm long; dorsal tepal (25–)30–35 mm long; tube slightly shorter than tepals . . . . 69. ***G. mirus***

25′ Anthers (10–)12–15 mm long; dorsal tepal 38–45 mm long; tube slightly exceeding the tepals . . . . 68. ***G. oligophlebius***

26(22′) Lower tepals half to a third as long as the dorsal, arching toward the ground or directed forward 27

26′ Lower tepals about as long (barely longer to no more than one-third shorter) than the dorsal tepal and straight, not recurving downward throughout 30

27(26) Foliage leaves linear with margins and midribs heavily thickened and raised; leaves 2.5–3.5 mm wide . . . . 50. ***G. pallidus***

27′ Foliage leaves either lacking (produced on separate shoots) or present and plane or somewhat ribbed; leaves (5–)8–35 mm wide 28

28(27′) Flowers predominantly red; perianth tube comprising a slender lower part 20–25 mm long, abruptly expanded into a wide horizontal cylindrical upper part c. 15 mm long; dorsal tepal 20–25 mm long . . . . 63. ***G. longispathaceus***

28′ Flowers predominantly red, orange, yellow, or greenish, the tepals often speckled or streaked with color; perianth tube 30–35 mm long, curving gradually outward and slightly wider above, not abruptly expanded above into a cylindrical upper part; dorsal tepal (25–)30–60 mm long, hooded over the stamens and concealing them 29

29(28′) Plants with or without foliage leaves at flowering time, the leaves if present (5–)10–20(–30) mm wide and smooth (microscopically papillose), with the margins and midribs moderately thickened; perianth various shades of red, orange, yellow, or greenish . . . . 48. ***G. dalenii***

29′ Plants always with foliage leaves at flowering time, the blades (25–)30–35 mm wide and puberulous or velutinous, and the margins and midribs and at least one other pair of veins moderately thickened; perianth greenish yellow . . . . 49. ***G. velutinus***

30(26′) Bracts clasping the spike axis and imbricate, usually 2, sometimes 3, internodes long . . . . 24. ***G. harmsianus***

30′ Bracts not normally clasping the spike axis, either not imbricate or if so not reaching beyond the middle of the next bract, thus less than 1.5 internodes long 31

31(30′) Bracts about as long or longer than the perianth tube 32

31′ Bracts shorter than the perianth tube 33

32(31) Dorsal tepal 45–50 mm long, longer than the perianth tube; flowers whitish to pale yellow, the tepals usually clearly veined in red to purple . . . . 44. ***G. verdickii***

32′ Dorsal tepal 30–35 mm long, about as long or shorter than the perianth tube; flowers white to pale yellow, rarely pink or red, the lower tepals sometimes with purple nectar guides but tepal venation never darkly colored . . . . 59. ***G. pauciflorus***

33(31′) Dorsal tepal longer than the perianth tube; flower predominantly red, the lower tepals with large white nectar guides . . . . 40. ***G. rupicola***

33′ Dorsal tepal shorter than the perianth tube; flowers either predominantly red, then without white nectar guides, or predominantly white 34

34(33′) Flowers brick-red . . . . 62. ***G. watsonioides***

34′ Perianth white, with or without contrasting nectar guides on the lower tepals 35

35(34′) Plants with only 2 or 3 foliage leaves (plants of southern Ethiopia) . . . . 37. ***G. negeliensis***

35′ Plants with 4–10 foliage leaves 36

36(35′) Perianth tube 45–50 mm long; dorsal tepal 25–28 mm long (plants of western Angola) . . . . 45. ***G. stenosiphon***

36′ Perianth tube 35–50(–60) mm long; dorsal tepal 30–32 mm long (plants of northern Tanzania) . . . . 41. ***G. usambarensis***

37(21′) Perianth tube 11–15 mm long 38

37′ Perianth tube 16–29 mm long 43

38(37′) Filaments short, 5–8 mm long, and included in the perianth tube 40

38′ Filaments 10–20 mm long, exserted at least 5 mm from the upper part of the perianth tube 40

39(38) Filaments c. 5 mm long; lower 2–3 mm of the anthers included in the perianth tube; anther apices with acute apiculate appendages 0.5–1 mm long . . . . 73. ***G. stenolobus***

39′ Filaments 7–8 mm long; anther exserted from the tube; anther apices obtuse . . . . 55. ***G. manikaensis***

40(38′) Leaves with heavily thickened margins and midribs, exceeding the spike; flowers green minutely dotted with red to purple . . . . 47. ***G. sericeovillosus*** subsp. ***calvatus***

40′ Leaves with margins and midribs lightly to moderately thickened, and shorter than the spike; flowers red, pink, blue, or purple, never minutely dotted 41

41(40′) Leaves sparsely pubescent, at least on the sheaths . . . 11. ***G. laxiflorus***

41′ Leaves glabrous 42

42(41′) Bracts 25–50 mm long, clasping the stem, and at least 2 or 3 internodes long . . . . 22. ***G. macrospathus***

42′ Bracts 25–30 mm long, not clasping the stem and rarely more than 1 internode long . . . . 39. ***G. boranensis***

43(37′) Leaves of the flowering stem with short, poorly developed blades, or bladeless; foliage leaves borne after flowering on separate shoots 44

43′ Leaves of the flowering stem normally with fully developed blades, sometimes the blades fairly short; separate leafy shoots not produced after flowering 45

44(43) Filaments included in the perianth tube; perianth orange to brick-red, or pink; style branches 5–6 mm long . . . . 57. ***G. melleri***

44′ Filaments exserted 4–5 mm from the perianth tube; perianth white, rarely pale pink; style branches 4–5 mm long . . . . . 58. ***G. roseolus***

45 Filaments included or exserted from the upper part of the tube for no more than 3 mm 46

45′ Filaments exserted at least 5 mm from the upper part of the perianth tube 47

46(45) Flowers yellow; anthers 10–12 mm long and with obtuse apices lacking apiculate appendages (plants of southwestern Tanzania) . . . . 56. ***G. oliganthus***

46′ Flowers red or pink, the lower tepals with white to pale yellow nectar guides; anthers c. 6.5 mm long and with acute apiculate appendages 0.5–1 mm long (plants of Sudan and Ethiopia) . . . . 70. ***G. sudanicus***

47(45′) Foliage leaves linear with margins and midribs heavily thickened and raised; leaves 2.5–3.5 mm wide . . . . 50. ***G. pallidus***

47′ Foliage leaves either lanceolate and more than 3 mm wide, or linear, less than 3 mm wide, but the margins and midribs lightly to moderately thickened 48

48(47′) Dorsal tepal horizontal and more or less hooded over the stamens, thus concealing them 49

48′ Dorsal tepal ascending, lightly inclined over the stamens but not concealing them 51

49(48) Dorsal tepal 35–40 mm long; bracts 40–60 mm long . . . . 59. ***G. pauciflorus***

49′ Dorsal tepal 20–28 mm long; bracts 20–30 mm long 50

50(49′) Perianth yellow (plants of Tanzania and Malawi) . . . . 51. ***G. nyasicus***

50′ Perianth white with yellow nectar guides on the lower tepals (plants of southern Ethiopia) . . . . 36. ***G. balensis***

51(48′) Lower tepals about as long as the dorsal tepal; perianth whitish to cream, or pale pink, the tepals darkly veined in red to purple; dorsal tepal 25–35 mm long . . . . 43. ***G. erectiflorus***

51′ Lower tepals shorter than the dorsal tepal; perianth pink or partly red, the lower tepals each with a white median stripe; dorsal tepal 25–55 mm long . . . . 40. ***G. rupicola***

52(20′) Foliage leaves well developed, with blades of the basal leaves longer than the sheaths, and leaf blades oval to terete in transverse section with the margins and midribs heavily thickened and arching over the laminar surface, thus with 4 narrow grooves 53

52′ Foliage leaves either well developed and with plane blades, sometimes with the midrib and margins strongly thickened, or poorly developed and largely to entirely sheathing (blades shorter than sheaths) 60

53(52) Flower actinomorphic (tube slightly curved); filaments included in the tube; tepals 6–7 mm long . . . . 27. ***G. actinomorphanthus***

53′ Flower zygomorphic; filaments exserted 4–9 mm; tepals 12–25 mm long 54

54(53′) Leaves, excluding the cataphylls, only 3, all with well-developed blades, or the upper largely sheathing; leaves much longer than the spike . . . . 28. ***G. robiliartianus***

54′ Laminate and sheathing leaves present, numbering 4–5(–7), shorter than or not much exceeding the spike, blades progressively reduced above; basal laminate leaves sometimes lacking due to fire 55

55(54′) Upper tepals narrow below, and flower in profile windowed (plants of southwestern Angola) . . . . 33. ***G. fenestratus***

55′ Upper tepals not especially narrow below, and flower in profile not windowed 56

56(55′) Dorsal tepal c. 20 mm long; edges of leaf grooves usually ciliate; flowering in July to September . . . . 4. ***G. juncifolius***

56′ Dorsal tepal 13–17 mm long; edges of leaf grooves not ciliate; flowering in November to May 57

57(56′) Bracts green, not becoming dry above at time of flowering; leaf sheaths lightly villous . . . . 3. ***G. zimbabweensis***

57′ Bracts becoming dry above or entirely dry at time of flowering; leaf sheaths glabrous 58

58(57′) Leaves 8 or 9, the lower 3 or 4 basal and with long blades . . . . 26. ***G. pungens***

58′ Leaves 3–6(–7), the lower 2 basal and with long blades 59

59(58′) Corm tunics of course thick fibers, these accumulating in a thick neck around the stem base . . . . 25. ***G. ledoctei***

59′ Corm tunics composed of fairly fine, soft-textured fibers, these not accumulating around the base . . . . 7. ***G. debeerstii***

60(52′) Upper cataphyll pubescent; flowering stem bearing only 1 leaf, this with a vestigial blade 0–3 cm long . . . . 19. ***G. curtilimbus***

60′ Upper cataphyll pubescent or glabrous; flowering stem bearing 2 or more leaves, these either with or without developed blades 61

61(60′) Filaments included in the perianth tube or barely exserted for less than 2 mm 62
61′ Filaments exserted at least 2 mm from the tube 68
62(61) Perianth tube 15–18 mm long; leaves of the flowering stem more than 3 and with well-developed plane blades 63
62′ Perianth tube 7–10 mm long; foliage leaves at least 3 and with well-developed plane blades 66
63(62) Anthers 7–10 mm long, and the apices obtuse, without appendages 64
63′ Anthers 5–6.5 mm long, and the apices acute with short appendages 0.3–0.5 mm long 65
64(63) Flowers orange-red with yellow nectar guides on the lower tepals or yellow; style branches 3–4 mm long . . . . 54. ***G. benguellensis***
64′ Flowers white to pale pink or lilac, without nectar guides; style branches . . . . 55. ***G. manikaensis***
65(63′) Perianth deep pink; perianth tube c. 15 mm long; anthers c. 6.5 mm long (plants of Zambia) . . . . 72. ***G. serenjensis***
65′ Perianth lilac; perianth tube c. 18 mm long; anthers c. 5 mm long (plants of eastern Ethiopia) . . . . 75. ***G. lithicola***
66(62′) Perianth evidently white; perianth tube 8–9 mm long; base of anthers included in the tube; leaves more than 3 and with well-developed blades, soft-textured, and margins and midrib not or hardly thickened . . . . 74. ***G. salmoneicolor***
66′ Leaves either all with vestigial blades or up to 2 leaves with long blades, these firm-textured 67
67(66′) Perianth pink or blue; anthers c. 6 mm long . . . . 18. ***G. canaliculatus***
67′ Perianth white to cream, somewhat flushed pink; anthers 7–9 mm long . . . . 17. ***G. muenzneri***
68(61′) Flowering stem bearing only 2 leaves, the lower sheathing the stem for at least half its length and with a blade shorter than the sheathing part (usually less than half as long); second leaf much smaller, inserted in the upper half of the stem 69
68′ Flowering stem bearing at least 3 leaves, these with or without well-developed blades and the lowermost not sheathing the lower half of the stem 71
69(68) Leaf blades terete and 4-grooved; corm tunics of tough brown fibers, usually accumulating in a neck around the base . . . . 29. ***G. tshombeanus***

69′ Leaf blades linear, more or less plane, but the margins and midribs moderately thickened; corm tunics of fine grayish fibers, not normally accumulating in a neck around the base 70

70(69′) Flowers white to yellow with deep yellow nectar guides; dorsal tepal 7–10 mm long . . . . 31. ***G. pusillus***

70′ Flowers pale blue-lilac to dark purple; dorsal tepal 10–15 mm long . . . . 30. ***G. gracillimus***

71(68′) Bracts firmly clasping the stem, imbricate and (1–)2–3 internodes long, concealing the spike axis 72

71′ Bracts weakly or not at all clasping, 0.5–1(–2) internodes long, only partly concealing the axis 73

72(71) Perianth tube 10–12 mm long; dorsal tepal 16–20 mm long; leaves linear, 1.5–3(–7) mm wide; corm tunics and cataphylls often accumulating in a dense fibrous mass and forming a neck around the base . . . . 23. ***G. microspicatus***

72′ Perianth tube 12–15 mm long; dorsal tepal 20–26 mm long; leaves narrowly to broadly lanceolate, 6–12(–16) mm wide; corm tunics seldom accumulating and base without a neck of fibers . . . . 21. ***G. gregarius***

73(71′) Leaves of the flowering stem poorly developed, reduced largely (or entirely) to sheaths (foliage leaves with long blades sometimes produced on separate shoots after flowering) 74

73′ Some or most of the leaves of the flowering stem with well-developed blades, the longest usually basal, and progressively decreasing in size above 77

74(73) Leaves (or sheaths) imbricate; tunics finely to coarsely fibrous and sometimes clawlike below, usually pale straw-colored 75

74′ Leaves (or sheaths) not all imbricate, at least the uppermost not overlapping; tunics membranous or irregularly fragmenting, sometimes ± fibrous, usually reddish brown 76

75(74) Perianth white entirely, partly shaded with purple, or uniformly dark purple to violet . . . . 15. ***G. atropurpureus***

75′ Perianth greenish yellow and brown . . . . 16. ***G. serapiiflorus***

76(74′) Leaves, or at least the lower sheaths, sparsely pubescent, usually microscopically; flowers large with tube 10–15 mm long and dorsal tepal 30–35 mm long . . . . 11. ***G. laxiflorus***

76′ Leaves blades and sheaths glabrous; flowers moderate sized, the tube 9–12 mm long and dorsal tepal 18–24 mm long . . . . 13. ***G. unguiculatus***

77(73′) Spike distichous and erect; tepals generally minutely and densely spotted throughout, most densely on the keels . . . . 47. ***G. sericeovillosus*** subsp. ***calvatus***

77′ Spike secund and usually inclined; tepals not minutely dotted throughout 78

78(77′) Flowers comparatively large, at least 35 mm long 79

78′ Flowers medium-sized, 16–34 mm long 87

79(78) Perianth tube 16–20 mm long 80

79′ Perianth tube 10–15 mm long 83

80(79) Tepals subequal; flowers cream to pink, conspicuously veined with pink to purple . . . . 43. ***G. erectiflorus***

80′ Tepals unequal, the lower 3 smaller then the dorsal; upper tepals unicolored, the lower often with contrasting nectar guides 81

81(80′) Flowers yellow . . . . 51. ***G. nyasicus***

81′ Flowers white, pink, or red 82

82(81′) Flowers white with yellow nectar guides . . . . 36. ***G. balensis***

82′ Flowers pink to red . . . . 70. ***G. sudanicus***

83(79′) Leaf sheaths (and often the leaves) sparsely pubescent; bracts obtuse and the inner often exceeding the outer; flowers uniformly purple . . . . 11. ***G. laxiflorus***

83′ Leaf sheaths and blades glabrous; bracts unequal, acute, and the inner smaller than the outer 84

84(83′) Leaf blades narrowly lanceolate to linear, but then the margins and midribs not or hardly thickened; leaves not exceeding the spike 85

84′ Leaf blades linear and with the margins and midribs heavily thickened; some leaves exceeding the spike 86

85(84) Perianth tube c. 15 mm long; flowers cream (plants of northern Eritrea) . . . . 38. ***G. mensensis***

85′ Perianth tube 10–12 mm long; flowers pale pink, the lower tepals marked with pale yellow . . . . 34. ***G. zambesiacus***

86(84′) Stem straight and erect; anthers c. 12 mm long . . . . 35. ***G. chelamontanus***

86′ Stem flexed near the base of the spike and spike inclined; anthers c. 6.5 mm long . . . . 10. ***G. sulcatus***

87(78′) Flowers 18–20 mm long and perianth tube 6–8 mm long; inner bracts, especially in the middle of the spike, longer than the outer . . . . 12. ***G. lundaensis***

87′ Flowers 24–34 mm long and perianth tube 9–15 mm long; inner bracts shorter than the outer 88

88(87′) Leaf sheaths and upper cataphyll lightly pubescent 89

88′ Leaf sheaths and upper cataphyll glabrous 90

89(88′) Dorsal tepal 22–26 mm long; flowers pale pink with yellow to cream nectar guides on the lower tepals . . . . 34. ***G. zambesiacus***

89′ Dorsal tepal 15–17 mm long; flowers yellow to greenish yellow . . . . 2. ***G. flavoviridis***

90(88′) Upper 3 tepals much narrowed toward the base, in profile the flower windowed (gaping between bases of dorsal and upper lateral tepals); tepal apices attenuate to caudate . . . . 32. ***G. permeabilis*** subsp. ***edulis***

90′ Upper 3 tepals not notably narrowed near the base and flowers not windowed in profile; tepal apices not attenuate to caudate 91

91(90′) Tepals more or less equal, the lower not united basally for a short distance 92

91′ Tepals unequal, the dorsal largest, and the lower 3 united basally for a short distance 93

92(91) Perianth tube c. 15 mm long; bracts 20–25(–30) mm long; anther apices obtuse, without appendages . . . . 46. ***G. calcicola***

92′ Perianth tube c. 38 mm long; bracts 14–20(–25) mm long; anther apices with short acute apiculate appendages . . . . . . 38. ***G. mensensis***

93(91′) Leaves linear, 2–4 mm wide 94

93′ Leaves narrowly to broadly lanceolate, at least 4 mm wide 95

94(93) Flower c. 30 mm long; dorsal tepal c. 21 mm long . . . . 9. ***G. linearifolius***

94′ Flower c. 26 mm long; dorsal tepal c. 16 mm long . . . . 8. ***G. camilae***

95(93′) Flowers shades of pink, the lower tepals with a dark pink to purple transverse band on the upper third of the tepals; stem often branched . . . . 1. ***G. crassifolius***

95′ Flowers white, white flushed with purple, or dark purple to violet, without transverse bands of color on the lower tepals; stem never branched 96

96(95′) Leaves 10–18 mm wide, leathery in texture; corm tunics composed of dark brown, papery, brittle layers . . . . 14. ***G. amplifolius***

96′ Leaves 2–6(–12) mm wide, and not leathery in texture; corm tunics composed of pale, straw-colored, fairly coarse fibers . . . . 15. ***G. atropurpureus***

SUBGENUS *GLADIOLUS*

Description

Plants small to medium, the stems usually slender, and branched or unbranched. LEAVES lanceolate to linear, or terete, the upper smaller and often entirely sheathing. SPIKES usually inclined toward the ground, rarely erect; BRACTS usually short, (6–)10–20(–30) mm long, usually slightly longer than the perianth tubes, sometimes shorter. FLOWERS small to medium, (12–) 20–40(–70) mm long, the lower tepals with nectar guides often on the upper third; PERIANTH TUBE (4–)10–15(–35) mm long, generally shorter than the bracts (exceeding them in a few derived species); TEPALS usually unequal, the dorsal about twice as long as the tube, except in the long-tubed species, lower tepals often united for some distance and narrowed below into claws, the limbs often abruptly expanded, the free parts usually shorter than the dorsal but in profile typically exceeding them. ANTHERS rounded apically. CAPSULES moderate in size, 7–15(–22) mm long; SEEDS 4–7(–8) × 2–5(–6) mm, with the wing well developed or smaller without wings.

Comprising over 120 species as circumscribed here, and with 33 species in tropical Africa, subgenus *Gladiolus* extends over virtually the entire range of the genus, including all of Africa, Madagascar, and Eurasia as far east as Afghanistan. The subgenus is especially diverse in the Cape region of southern Africa. Three sections are recognized here: section *Gladiolus*, the largest; section *Heterocolon;* and section *Hebea.* Section *Gladiolus* is widespread in tropical Africa and also occurs in southern Africa, Madagascar, and Eurasia. Section *Heterocolon,* restricted to southern tropical Africa and the Transvaal, South Africa, has its greatest diversity in Shaba Province, Zaire. Section *Hebea*, mainly southern African, has one species in southern Angola, and one, common in southern Africa, extends into Zimbabwe.

Despite extensive floral variation, members of subgenus *Gladiolus* are united by having comparatively small flowers and bracts, and short perianth tubes. Typically, the dorsal tepals are about twice as long as the tubes, and the tubes are shorter than the floral bracts so that they tend to emerge between them. Like the flowers, the capsules and seeds are also relatively small. The capsules are typically 10–15 mm long, and the seeds seldom more than 4–5 mm long. The spikes are borne on slender stems and are almost always flexed outward at the base, or the upper part of the stem is flexed so that the spike is inclined. Additionally, in most members of the subgenus, the lower tepals are united for a short distance and often divided into narrow claws and abruptly expanded limbs. The nectar guides are always on the distal part of the tepals, most often forming transverse bands of contrasting color across the middle of the limbs. A few exceptional species admitted to the subgenus do not fit all the above criteria. Both *G. curtifolius* and *G. harmsianus* have white flowers with weakly developed or no nectar guides, and the long perianth tubes exceed the dorsal tepals and the bracts. The dimensions of the tepals and capsules, however, do accord with those for the subgenus. Another exception, members of the *G. gregarius* complex have stiffly erect stems and spikes, and long floral bracts, usually two or three internodes long. Their flowers correspond exactly to those of the subgenus in every way, and despite their unusual bracts and spikes they fall comfortably within section *Gladiolus*. Perhaps the most remarkable development in subgenus *Gladiolus* is in section *Heterocolon*. There the leaves are rounded in section with four narrow longitudinal grooves, and the seeds usually lack wings and are more or less round to weakly angular. Based on floral criteria the species fall readily within the limits that define subgenus *Gladiolus*. Wingless seeds also characterize a few Eurasian species of subgenus *Gladiolus*, but this is undoubtedly a convergent development.

SECTION *GLADIOLUS*

Synonymy

Subgenus *Limonia* Klatt, Abh. Naturforsch. Ges. Halle 15: 339 (Ergänzungen 5) (1882). Type: *Gladiolus brevifolius* Jacquin (lectotype designated here).

Description

Plants medium or small, branched or unbranched. LEAVES lanceolate to linear, or terete, the upper smaller and sometimes entirely sheathing, usually contemporary with the flowers, but sometimes largely sheathing, or borne on separate shoots after flowering (*Gladiolus unguiculatus*). SPIKES usually inclined, or erect; *bracts* usually short, but in a few species two or three internodes long, (6–) 10–20 mm long, about as long as the perianth tubes or exceeding them. FLOWERS small to medium, (18–)25–40(–70) mm long, the lower tepals with median white to cream nectar guides; PERIANTH TUBE (6–)10–15(–75) mm long, usually included in the bracts but sometimes exceeding them; DORSAL TEPALS usually about twice as long as the tubes, occasionally slightly shorter when tubes elongate (*G. curtilimbus, G. gunnisii, G. harmsianus*); LOWER TEPALS much narrower and usually shorter than the dorsal, often clawed below.

As circumscribed here section *Gladiolus* includes 24 species in tropical Africa, mostly in eastern and southern tropical Africa, and some 40 more in Eurasia, Madagascar, and southern Africa. The section thus extends over virtually the entire range of the genus. Ideally, the section should be subdivided, but there are no clear species groups or major species clusters among the tropical African species in the section. The tropical and southern African members of the section need to be studied together before a satisfactory subdivision of the species can be achieved.

1. *Gladiolus crassifolius* Baker

PLATES 1, 2, FIGURE 8, MAP 2. Baker, J. Bot. (London) 14: 334 (1876); Handbook Irideae 215 (1892); Fl. Capensis 6: 150 (1896). G. Lewis et al., J. S. African Bot., Suppl. 10: 60 (1972). Goldblatt, Fl. Zambesiaca 12(4): 68 (1993). Type: South Africa, Eastern Cape, 1861, *Cooper 3185* (K, lectotype designated by Lewis et al., 1972: 60; G, PRE, isolectotypes).

SYNONYMY

Gladiolus thomsonii Baker, Handbook Irideae 223 (1892); Fl. Trop. Africa 7: 372 (1898). Type: Tanzania, high plateau north of Lake Nyasa, 1880, *Thomson s.n.* (K, holotype).

Gladiolus masukuensis Baker, Kew Bull. 1897: 283 (1897); Fl. Trop. Africa 7: 365 (1898). Type: Malawi, "Mesuku Mts. [type label], Masuku Plateau [protologue]," cultivated in Zomba, *Whyte s.n.* (K, lectotype designated by Goldblatt, 1993: 68, the sheet with a branched plant and well-preserved flowers; K, isolectotype).

Gladiolus mosambicensis Baker, Fl. Trop. Africa 7: 576 (1898). Type: Mozambique, Beira, without date, *Braga 117* (B, holotype, one of the two plants is 20 cm tall and has the leaves charred to the base, and the other, incomplete, has a branched stem).

Gladiolus gazensis Rendle, J. Linn. Soc., Bot. 40: 210 (1911). Plowes & Drummond, Wild Fl. Rhodesia, pl. 37 (1976). Type: Zimbabwe, Chimanimani (Melsetter), 1800 m, 23 Sept. 1907, *Swynnerton 779* (BM, lectotype designated by Goldblatt, 1993: 70; K, isolectotype); Zimbabwe, Chirinda outskirts, 1200–1350 m, 1 Sept. 1907, and Mt. Pene, 2100–2200 m, 28 Sept. 1907, both labeled *Swynnerton 779* and mounted on the same sheet with the lectotype (BM, syntypes).

Synonyms based on South African plants, cited by Lewis et al. (1972), are not repeated here.

EPONYMY

crassifolius, "thick-leaved," in reference to the thick, leathery, strongly ribbed leaves.

DESCRIPTION

Plants (25–)35–90(–120) cm high, sometimes the leaves emergent or charred by fire, or the leaves reaching to at least the middle of the spikes and even shortly exceeding them, sometimes producing 2–3(–5) stems per corm. CORM 2–3 cm in diameter, tunics coriaceous to more or less coarsely fibrous. CATAPHYLLS coriaceous, red-brown. LEAVES four to eight, either few and emergent at flowering time, then the stem bearing only one or two cauline and sheathing leaves and two laminate basal leaves, or with several basal laminate leaves and two or three cauline leaves, these with progressively shorter blades and the upper sometimes entirely sheathing, the blades narrowly lanceolate to linear, 5–12 mm wide, exceeding the spike in short-stemmed plants or reaching to about the base of the spike in long-stemmed plants, the midrib and margins hyaline and moderately thickened. STEM erect below, usually flexed outward above the sheath of the uppermost leaf, simple or 1- to 3(–5)-branched, c. 2 mm in diameter at the base of the spike.

SPIKE inclined, (6–)12- to 22-flowered, the branches with fewer flowers than the main axis; BRACTS usually pale green or flushed purple, soft-textured initially, becoming dry and membranous above (in fruit, typically translucent light red-brown), more or less obtuse to subacute, usually apiculate, 12–18 mm long, the inner slightly shorter than the outer and two-apiculate. FLOWERS pale to deep pink or purple, the lower lateral, and sometimes the lowermost tepals each with a dark band of color across the lower half of the limbs, the dark band often edged with cream; PERIANTH TUBE curved and obliquely funnel-shaped, cylindric below, widening and curving outward near the apex, 8–12 mm long; TEPALS unequal, the dorsal largest and arched to hooded over the stamens, 20–22 mm long, c. 8 mm at the widest (often less when dry), the upper laterals directed forward and curving outward above, 18–22 mm long, the lower three tepals more or less straight

FIGURE 8. *Gladiolus crassifolius.* Corm, leaves, and flowering spike, × 0.5; single flower, full size (leaves and flowering stem, *Chapman 8364*; capsules, *Lovett et al. 3276*).

and directed downward, united below for 2–3 mm, narrowed into claws below, the limbs channeled, the lowermost less so, the laterals 12–15 mm, the lowermost 16–20 mm long. FILAMENTS arcuate, c. 12 mm long, inserted at the base of the upper part of the tube, exserted for 6 mm; ANTHERS 7–8 mm long, purple, pollen whitish. OVARY obovoid, 3–4 mm long; STYLE arching over the filaments, usually dividing opposite the lower half of the anthers, the branches c. 2 mm long, seldom reaching past the middle of the anthers. CAPSULES obovoid, 9–12(–14) mm long; SEEDS elliptic, broadly winged, 5–6 × 2.5–4 mm, the wing usually more or less transparent. CHROMOSOME NUMBER $2n = 30$.

FLOWERING TIME. Either (August and) September to November, especially after fires, or toward the end of the rainy season, March to June.

DISTRIBUTION & HABITAT

Gladiolus crassifolius is a widespread eastern southern African species, common in South Africa, Swaziland, and Lesotho, and extending well into tropical Africa. The tropical African populations are centered in eastern Zimbabwe, the adjacent parts of western Mozambique, and on the Mt. Mulanje massif in southern Malawi. The species is particularly common in the highlands of eastern Zimbabwe, but scattered populations occur almost throughout the country, and it has been recorded at Great Zimbabwe and Mt. Buhwa in the south, and in the Mvurwi (Umvukwes) Mountains in the north. Isolated populations also occur in southwestern Tanzania where *G. crassifolius* has been recorded at a few localities, and in western Angola where it is known from two sites. If the type of the conspecific *G. masukuensis* is really from the Misuku Hills in northern Malawi (the type information indicates "Mesuku Mts."), then the series of northern populations includes northern Malawi as well as adjacent Tanzania. Angolan *G. crassifolius* differs from plants in eastern tropical Africa in having comparatively large flowers and narrow lower tepal claws, an indication of morphological differentiation in populations that are now isolated from the main range of the species by more than 1500 km. These differences do not appear to merit taxonomic recognition.

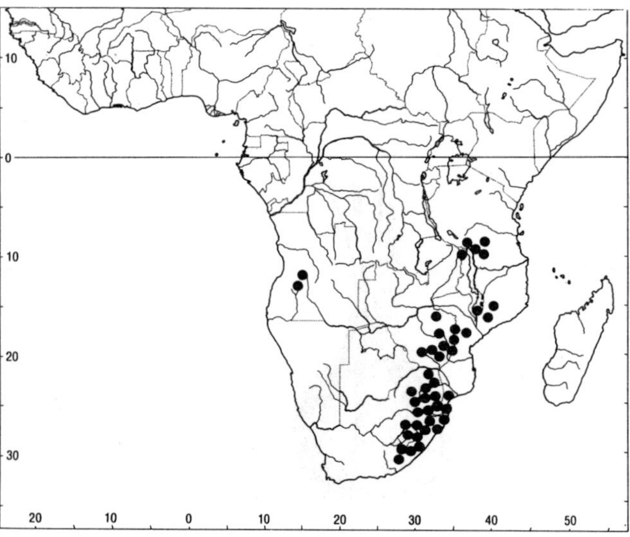

MAP 2. *Gladiolus crassifolius.*

Most collections of *Gladiolus crassifolius* in tropical Africa are from relatively high elevations, usually above 1800 m, but as low as 1100 m at Bikita and Chipinga in southeastern Zimbabwe, and even lower in Mozambique, at 1000 m, on the Gorongoza Massif. The type collection of *G. mosambicensis,* a synonym, is from "Beira," but this probably does not literally mean this coastal city but the higher country in the interior. The species grows in open grassland, sometimes in rocky situations, in both well-watered and relatively dry sites. Evidently fairly adaptable ecologically, it occurs in areas of relatively low rainfall as well as the high-rainfall montane grasslands of the southeastern African escarpment.

Flowering is concentrated in the late summer, toward the end of the wet season, over most of its range. In populations in eastern Zimbabwe and central Mozambique, however, flowering may occur at almost any time of the year. Fire seems to promote precocious flowering, and plants may bloom a few weeks after the fires, which are common there in October and November, when dry grassland is set ablaze before the rains to stimulate growth of new pasture. Flowering twice in a year is unknown in other species of *Gladiolus* and is not reported in the species elsewhere across its range.

In South Africa, Vogel (1954) has suggested that *Gladiolus crassifolius* is pollinated by bees because of the form and color of the flowers. My own observations in Zimbabwe suggest the contrary. There, I have seen only long-proboscid flies of the genus *Prosoeca* (Nemastrinidae) visiting the species. These flies may be important in pollinating other species of *Gladiolus* with similarly colored flowers of the same shape and size.

DIAGNOSIS & RELATIONSHIPS

The long spikes of many relatively small and densely arranged pale pink flowers, and the fan of several basal leaves, are characteristic of *Gladiolus crassifolius* and serve to distinguish the species in tropical Africa. It is a comparatively unspecialized plant. The structure of the flowers, with a large dorsal tepal and narrow lower tepals, united for some distance at the base, is typical of section *Gladiolus,* although *G. crassifolius* often has unusually narrow lower tepal claws. The several broad leaves, typical of several species of the section, are plesiomorphic and provide no indication of its relationships. The tropical African populations stand out in having comparatively soft-textured and apiculate bracts that turn tawny brown on drying, and dark purple markings on the lower tepals. Both these features are also found in some Transvaal plants, either separately or in combination, and plants from the northern Transvaal, in particular, are virtually identical with those from eastern Zimbabwe. Plants from the southern part of the range of *G. crassifolius,* especially those from coastal Natal and southern Mozambique, are notable for their firm and fairly narrow bracts, but apart from this there seem to be no significant differences between the populations of southern and tropical Africa. The closest relatives of *G. crassifolius* are probably the southern African *G. densiflorus* Baker and *G. invenustus* G. Lewis.

VARIATION

The unusual phenomenon of plants blooming at the end of the rainy season, from March to June, in populations in eastern Zimbabwe and central Mozambique, has been mentioned above. Plants that flower in the summer have bright pink flowers, well-developed foliage leaves, and tall stems with only one or two branches (Plate 1). Those that flower in the spring, however, have dark pink flowers, fewer and shorter basal leaves, and stems with up to five branches (Plate 2). The leaves sometimes appear to be emergent or are charred by fire. This pattern is remarkable and without parallel in *Gladiolus.* Whether the phenotype of a plant can vary due to its exposure to fire and other differences in growing conditions is uncertain. If field and greenhouse experiments indicate that early- or late-season flowering is genetically controlled, it will become necessary to recognize the two taxonomically. In the absence of such information, the early flowering and normally branched plants, represented by the types of *G. gazensis* and *G. mosambicensis,* are included here in *G. crassifolius.*

HISTORY

The first record of *Gladiolus crassifolius* in tropical Africa appears to be the two flowering spikes collected in 1880 by the naturalist Joseph Thomson in the highlands north of Lake Malawi. Honoring their collector, these specimens were described as *G. thomsonii* by J. G. Baker in 1892. A later gathering, made from plants grown in Zomba and said to have been collected in the "Mesuku Mts." ("Masuku Plateau" in the protologue), was named *G.*

masukuensis by Baker in 1898. This locality may be what we now know as the Misuku Hills of northern Malawi, but the species has not been re-collected there. A later collection from Mozambique, representing the early flowering and branched form of the species, was described as *G. mosambicensis* by Baker in 1898, whereas plants from eastern Zimbabwe, virtually identical with *G. mosambicensis,* formed the basis for A. B. Rendle's (1911) *G. gazensis.* While it is easy to see that the early- and late-season-flowering individuals of the species may have been regarded as separate species, the proliferation of names here is surprising. Lewis et al. (1972) appear to have been the first to regard *G. crassifolius* as encompassing tropical African as well as southern African plants, a treatment I follow without hesitation.

SELECTED SPECIMENS

Angola. Benguela: Between Ganda and Caconda, c. 1700 m, Dec. 1933, *Hundt 736* (BM, P). Huambo: Mt. Moco, low grassland, 1850 m, 19 Dec. 1973, *Huntley, Roberts, & Ward 91* (PRE).

Tanzania. Iringa: Njombe District, Upper Ruhudje, May 1931, *Schlieben 940* (B, LISC); Njombe District, 5 km north of Lupalilo, c. 2000 m, 27 Sept. 1970, *Thulin & Mhoro 1222* (UPS); Mufindi District, between Uhafiwa and Ihangana, 1850 m, 10 June 1989, *Lovett, Congdon, & Kayombo 3276* (MO). Mbeya: Rungwe District, Station Kyimbila, 1913, *Stoltz 2166* (B, C, G, LD, S, WAG, Z).

Malawi. Southern: Mt. Mulanje, without date, before 1896, *Adamson 336* (BR, BM, K, P); Mulanje, Sombani, halfway up Namisili, in seep, 2400 m, 16 May 1981, *Chapman & Patel 5704* (BR, K, MAL); Mulanje, Chambe Plateau, 23 Apr. 1958, *Jackson 2172* (K, MAL, SRGH); west slope on Mt. Mlanje, 1800 m, 21 June 1946, *Brass 16401* (NY).

Zimbabwe. Mashonaland East: Makoni, near Headlands, 15 Oct. 1964, *Plowes 2118* (K, SRGH). Marondera (Marandellas), 6 Nov. 1941, *Dehn 445* (SRGH). Mashonaland North: Mvurwi (Umvukwes), Horseshoe Mine, 16 Apr. 1960, *Leach & Brunton 9864* (K, SRGH). Manicaland: Mtare, Vumba Mountains, on granite, 1900 m, 16 Feb. 1964, *Chase 6024* (BM, BR, K, LISC, S, SRGH); Nyanga District, Mt. Inyangani, south end of Nyazengu Ridge, 14 Nov. 1980, *Pope & Muller 1702* (MO, SRGH); Pungwe Scenic Drive at Mtarazi Falls turnoff, 29 Mar. 1991, *Goldblatt 9077* (MO); Chimanimani District, near Martin's Falls, 11 Apr. 1967, *Plowes 2853* (K, LISC, SRGH); Pork Pie Hill, 4.5 km north of Chimanimani (Melsetter), 28 Mar. 1951, *Crook 390* (K, SRGH). Mtare, hill near Mozambique border, 8 Oct. 1959, *Chase 7178* (BM, EA, K, LISC, SRGH). Midlands: Bikita District, summit of Mt. Buhwa, 30 Oct. 1973, *Biegel, Pope, & Gosden 4333* (C, MO, SRGH, WAG); bank of Turgwe River, 3 km south of Cherne School, 4 May 1969, *Pope 55* (BR, K, SRGH). Masvingo: Zimbabwe District, Great Zimbabwe, 19 Oct. 1930, *Fries, Norlindh, & Weimarck 2066* (BM, BR, LD, S).

Mozambique. Manica e Sofala: Chimanimani Mountains, 8 June 1949, *Wild 2916* (K, LISC, S, SRGH); Gorongoza Mountains, 1000 m, 20 Oct. 1965, *Torre & Perreira 12476* (LISC); Gorongoza Mountains near Morombodzi Falls, 10 Aug. 1944, *Mendonça 2455* (BM, C, LISC, MO); Choa Mountains, 26 km from Vila Gouveia, 1400 m, 23 Mar. 1946, *Torre & Correia 15399* (LISC); Tsetsera, 2140 m, 7 Feb. 1955, *Exell, Mendonça, & Wild 247* (LISC). generally longer, or Zambezia: Mt. Ile, 26 Apr. 1943, *Torre 5587* (BM, LISC); Gurué, toward Namuli, 12 Aug. 1949, *Andrada 1860* (COI, LISC). Maputo: between Zitundo and Ponto da Ouro, 10 Dec. 1962, *de Lemos & Balsinhas 282* (BM, COI, K). Namaacha, Mt. Ponduine, 13 Apr. 1968, *Balsinhas 1219* (COI).

2. *Gladiolus flavoviridis* Goldblatt

MAP 3. Goldblatt, Fl. Zambesiaca 12(4): 71 (1993). Type: Zimbabwe, Manicaland, Mtare, north of Christmas Pass (Buru's grounds), 15 Feb. 1949, *Chase 1364* (SRGH, holotype; BM, K, isotypes).

EPONYMY

flavoviridis, "greenish yellow," alluding to the color of the flowers.

DESCRIPTION

Plants 55–85 cm high. CORM unknown. CATAPHYLLS pale and membranous, the upper longest, green, and pubescent above the ground. LEAVES six, usually the lower two more or less basal and largest, reaching to the base or middle of the spike, nearly linear, 4–5 mm wide, the midrib and margins lightly thickened, upper leaves cauline and progressively shorter above, the uppermost usually entirely sheathing, the sheaths and lower parts of the blades lightly to heavily pubescent. STEM simple or one branched, c. 2 mm in diameter at the base of the spike.

SPIKE 10- to 18-flowered, inclined from the base; BRACTS evidently membranous and dry and brown above at anthesis, 12–16 mm long, the inner about two-thirds as long as the outer. FLOWERS pale yellow or greenish, tepal markings unknown; PERIANTH TUBE obliquely funnel-shaped, widening and curving outward near the apex, c. 10 mm long; TEPALS unequal, the upper three largest, broadly lanceolate, 15–17 mm long or often less when dry, the dorsal inclined over the stamens, c. 12 mm wide, the lower three united below for c. 2 mm, 9–10 mm long, clawed and straight below, abruptly expanded above, the apices inclined toward the ground. FILAMENTS c. 10 mm long, arched below the dorsal tepal, exserted 6 mm from the tube; ANTHERS c. 5.5 mm long, apices obtuse, without appendages, evidently deep yellow. OVARY oblong, 3–4 mm long; STYLE arching over the filaments, dividing between the base and middle of the anthers, the branches 2–3 mm long, never reaching the anther apices. CAPSULES obovoid, c. 10 mm long; SEEDS unknown.

FLOWERING TIME. January to early March.

DISTRIBUTION & HABITAT

Restricted to eastern Zimbabwe, and evidently to Mtare and its immediate surroundings, *Gladiolus flavoviridis* grows in light woodland or open grassland. Occurring as it does near Mtare, it is all but certain that *G. flavoviridis* also be found across the nearby frontier in Mozambique. One collection, *Chase 6821*, is actually from close to the border between the two countries.

DIAGNOSIS & RELATIONSHIPS

The relatively small, pale yellow or greenish flowers on an inflexed secund spike, and bracts that are dry and rust-colored above by anthesis, readily distinguish *Gladiolus flavoviridis*. It is surprising that the plant has remained undescribed since its discovery, apparently by N. C. Chase in 1949. The paucity of collections probably reflects the very local distribution rather than rarity. The relationships of *G. flavoviridis* probably lie with the widespread eastern African *G. crassifolius* and with the eastern Zimbabwean *G. zimbabweensis*, both of which have pink flowers, and the latter also has pubescent upper cataphylls and leaf sheaths.

SELECTED SPECIMENS

Zimbabwe. Manicaland: Mtare, Umtali Heights, 1600 m, 1 Mar. 1959, *Chase 7073* (SRGH); Mtare, East Ridge, Nyamaganu Peak Commonage, 3 Jan. 1960, *Chase 7245* (SRGH); Mtare, The Kloof, Feb.

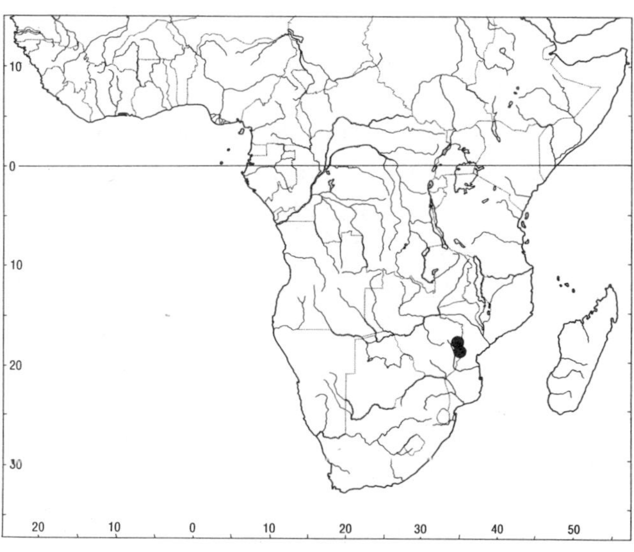

MAP 3. *Gladiolus flavoviridis*.

1965, *Pole Evans 6761* (PRE); Premier Estates, Old Umtali, 29 Jan. 1957, *Pole Evans 5080* (SRGH); Mtare, ridge south of Customs' Border Station, 20 Feb. 1958, *Chase 6821* (K, SRGH).

3. *Gladiolus zimbabweensis* Goldblatt

PLATE 3, FIGURE 9, MAP 4. Goldblatt, Fl. Zambesiaca 12(4): 72 (1993). Type: Zimbabwe, Manicaland, Nyanga District, slopes of Mt. Inyangani, 2100 m, 16 Feb. 1964, *Chase 8123* (SRGH, holotype; K, UPS, WAG, isotypes).

EPONYMY

zimbabweensis, "from Zimbabwe," the southern tropical African country where the species is most common.

DESCRIPTION

Plants 20–35 cm high. CORM 9–12 mm in diameter, tunics of fine pale, netted fibers, sometimes extending upward in a neck. CATAPHYLLS pale and membranous, the upper reaching 2–4 cm above the ground and then green or purplish and pubescent. LEAVES four or five, the lower two basal, reaching to about the base of the spike, the blades more or less linear, 1–1.5 mm wide, the margins and midrib strongly thickened and raised, thus arching over the surface and in transverse section more or less oval to terete with two narrow grooves on each face, edges of the raised parts ciliate to pubescent, the upper leaves inserted on the stem and progressively shorter, the uppermost sheathing for half its length; sheaths, especially the lower, lightly pubescent, lightly ribbed when dry. STEM erect, rarely branched, flexed at about the middle, c. 2 mm in diameter at the base of the spike.

SPIKE 4- to 8(–12)-flowered, inclined 30–45° to the ground; BRACTS green to purplish especially above, 10–15 mm long, the inner somewhat shorter than the outer. FLOWERS pale gray-blue or purple to pink, the lower lateral tepals each

FIGURE 9. *Gladiolus zimbabweensis.* Habit and leaves, full size; whole flower and vertical section, × 1.3 (*Goldblatt 9078, 9080*).

with a pale transverse band outlined in purple across the lower part of the limb; PERIANTH TUBE 7–10 mm long, curving outward between the bracts, widening gradually toward the mouth; TEPALS unequal, the dorsal arched to hooded, 14–19 mm long, c. 10 mm wide, the upper laterals smaller, directed forward and curving outward near the apices, the lower three straight, inclined, joined to the upper laterals for 2–4 mm, in profile usually slightly exceeding the dorsal tepal, c. 10 mm long, narrowed below into claws, the limbs fairly abruptly expanded, c. 4 mm at the widest, lower laterals channeled below. FILAMENTS 12–14 mm long, exserted for 8–9 mm; ANTHERS c. 6 mm long, with an apiculus 0.3–0.5 mm long, dark purple, pollen cream. OVARY c. 3 mm long; STYLE dividing opposite the base or lower third of the anthers, the branches c. 2 mm long, entwined with the anthers. CAPSULES obovoid, three-lobed above, 11–14 mm long; SEEDS oval, 4.5–5 × 3–3.5 mm, broadly winged. CHROMOSOME NUMBER $2n = 30$.

FLOWERING TIME. Mid January to early April.

DISTRIBUTION & HABITAT

Known from only a few collections from the highlands of eastern Zimbabwe and neighboring Mozambique, *Gladiolus zimbabweensis* is a rare endemic of the highest country in southern tropical Africa. Although it flowers in well-watered grassland in the wet season, from late January to March, it has unusual xeromorphic features such as heavily thickened margins and midribs, and light pubescence on the leaf sheaths. The species has been recorded from elevations above 1800 m on Mt. Inyangani, in the Nyanga Highlands of Zimbabwe, and from the Chimanimani Mountains that border Zimbabwe and Mozambique. It was first recorded by the Zimbabwean botanist, Hiram Wild, in 1954.

DIAGNOSIS & RELATIONSHIPS

The slender habit, short stature, spikes of four to eight small pink flowers, and narrow leaves with pubescent sheaths distinguish *Gladiolus zimbabweensis* from a group of fairly similar small-flowered species of *Gladiolus*. Its linear leaves, sometimes with strongly thickened margins, suggest a relationship with another eastern Zimbabwe species, the late-winter- and spring-flowering *G. juncifolius*. That species has dimorphic leaves, the lower with long blades and the upper entirely sheathing. The leaf sheaths of *G. juncifolius* are glabrous, but the edges of the leaf blades are often ciliate, at least below, and its flowers (Plate 4) are slightly larger than those of *G. zimbabweensis* (Plate 3) and have a slightly different form.

The affinities of *Gladiolus zimbabweensis* within section *Gladiolus* are uncertain, but it may be most closely related to the eastern Zimbabwe endemic, *G. flavoviridis*. Although more robust, this greenish yellow-flowered species also has lightly pubescent cataphylls and lower leaf sheaths.

SELECTED SPECIMENS

Zimbabwe. Manicaland: Nyanga, Mt. Inyangani, 2100 m, 16 Feb. 1964, *Chase 8122* (EA, K, LISC, SRGH), 30 Mar. 1991, *Goldblatt 9078* (MO, SRGH); base of Mt. Inyangani, open grassland, 2000 m, 16 Feb. 1964, *Plowes 2426* (K, SRGH); Mt. Inyangani,

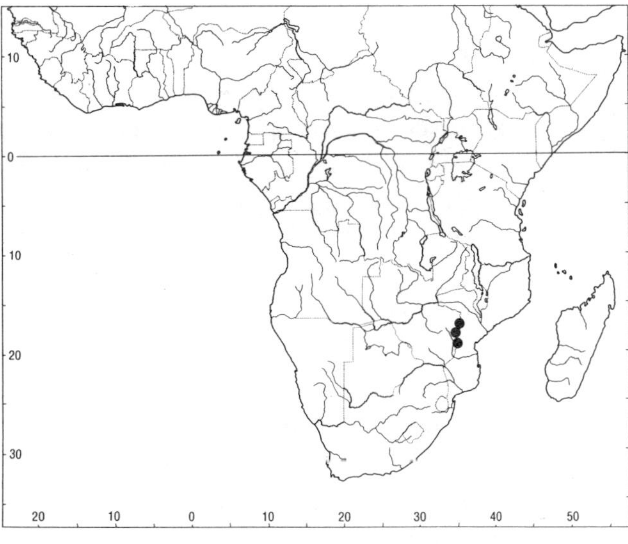

MAP 4. *Gladiolus zimbabweensis.*

2050 m, 7 Mar. 1981, *Philcox et al. 8919A* (K, SRGH); Nyanga National Park, Inyangani to Troutbeck road at Gairezi turnoff, 30 Mar. 1991, *Goldblatt 9080* (MO); Mtare District, Mt. Engwa, 1980 m, 8 Feb. 1955, *Exell, Mendonça, & Wild 294* (BM, LISC, SRGH); Mt. Engwa, open grassland, 2200 m, 2 Mar. 1954, *Wild 4458* (K, LISC, SRGH); Himalaya, Tambara, local in vlei, 2 Feb. 1960, *Drewe 37* (SRGH); Chimanimani Mountains, Stonehenge Plateau, 1900 m, 1 Feb. 1957, *Phipps 394* (PRE, SRGH).

Mozambique. Manica & Sofala: Tsetsera, 2140 m, grassland, 7 Feb. 1955, *Exell, Mendonça, & Wild 246* (BM, LISC, SRGH).

4. *Gladiolus juncifolius* Goldblatt

PLATE 4, FIGURE 10, MAP 5. Goldblatt, Fl. Zambesiaca 12(4): 74 (1993). Type: Zimbabwe, Manicaland, Chimanimani Mountains, damp ground in crevices in flat sandstone, 2200 m, 8 Oct. 1960, *Plowes 2129* (SRGH, holotype; K, isotype).

EPONYMY

juncifolius, with leaves rounded in transverse section, like those of some rushes *(Juncus)*.

DESCRIPTION

Plants 20–35 cm high. CORM unknown. CATAPHYLLS pale and membranous, the inner reaching c. 2 cm above the ground and then green or dry and light brown. LEAVES four or five, the lower two or three basal and with long blades reaching at least to the base of the spike (occasionally charred and then not evident), sometimes exceeding it, sheaths, especially the lower, conspicuously ribbed, blades more or less terete, 1–1.5 mm wide, the margins and midrib heavily thickened, the blades thus narrowly four-grooved in section, edges of the raised parts sometimes ciliate, the upper two or three leaves inserted above ground and

FIGURE 10. *Gladiolus juncifolius.* Habit and leaves, full size; single flower, × 1.3 (habit, *Grosvenor 177*; single flower, *Plowes 2129*).

entirely sheathing, decreasing in size above, widely spaced, the uppermost 15–30 mm long. STEM straight and erect, unbranched, c. 3 mm in diameter at the base of the spike.

SPIKE (2–)3- to 6-flowered; BRACTS entirely green or dry and brownish above, 9–14 mm long, the inner somewhat shorter than the outer. FLOWERS pale pink or purplish, the two lower lateral tepals each with yellow transverse band across the lower part of the limb and darker pink distally; PERIANTH TUBE c. 9 mm long, curving outward between the bracts, widening gradually toward the mouth; TEPALS unequal, the dorsal inclined to arched, c. 20 × c. 9 mm, the upper laterals smaller, directed forward and ultimately curving outward, the three lower tepals horizontal to downcurved, joined together for c. 3 mm, narrowed below into claws, the limbs fairly abruptly expanded, the lowermost c. 18 mm long, c. 7 mm wide, the lower laterals c. 15 mm long, c. 3 mm wide. FILAMENTS c. 15 mm long, exserted for 5 mm; ANTHERS c. 6.5 mm long, with an apiculus c. 0.5 mm long, yellow. OVARY c. 3 mm long; STYLE dividing toward the apex of the anthers, the branches c. 3 mm long. CAPSULES and SEEDS unknown.

FLOWERING TIME. Mid July to October.

DISTRIBUTION & HABITAT

Known from only a few collections from eastern Zimbabwe and adjacent Mozambique, *Gladiolus juncifolius* is a rare endemic of these southern tropical African highlands. Most of the records of the species are from the Chimanimani Mountains, but there is also a collection from Mt. Inyangani to the north. The species is unusual in flowering in the winter and spring, July to October. This is the dry season in most of southern and central Africa, but *G. juncifolius* is restricted to sites that retain some moisture throughout the year. The leaves are produced long before the flowers and are sometimes dry or broken at flowering time. The four-grooved terete leaf is a typical xeromorphic adaptation in *Gladiolus* and in many other Iridaceae. The first collection of the species seems to have been made by the South African biologist, Sheila Thompson, in July 1950.

DIAGNOSIS & RELATIONSHIPS

The dimorphic leaves, either basal and long-bladed or cauline and entirely sheathing, the short stature, and a spike of relatively small flowers distinguish *Gladiolus juncifolius* among tropical and eastern southern African *Gladiolus*. Its terete and four-grooved leaf blades suggest a relationship with the southern African *G. pretoriensis*, which flowers in the summer months, January to March, and which grows in relatively dry, rocky habitats. *Gladiolus juncifolius* may also be confused with the eastern Zimbabwean *G. zimbabweensis*, which flowers in the late summer. That species has narrow but not terete leaves that have lightly pubescent leaf sheaths and, sometimes, lower leaf blades, and the cauline leaves have short blades.

Until the capsules and seeds are known, the affinities of *Gladiolus juncifolius* must remain uncertain. *Gladiolus pretoriensis* stands out among the

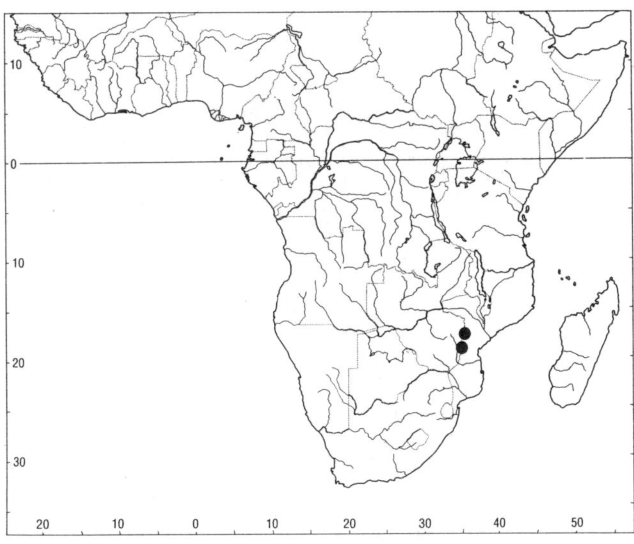

MAP 5. *Gladiolus juncifolius*.

southern African species in having angular instead of winged seeds typical of the genus. Other possible relatives, including *G. zimbabweensis,* have broadly winged seeds. *Gladiolus juncifolius* is notable in having the long-bladed basal leaves produced before flowering and sharply different from the entirely sheathing cauline leaves. In most species of *Gladiolus* the basal and cauline leaves intergrade, with generally only the uppermost being entirely sheathing.

SELECTED SPECIMENS

Zimbabwe. Manicaland: Nyanga District, Mt. Inyangani, 2500 m, damp ground, 7 Oct. 1962, *Plowes 2267* (SRGH). Chimanimani (Melsetter) District, level scarp en route to Martin Falls c. 1.5 km beyond Skeleton Pass, 1500 m, 18 Sept. 1956, *Taylor 1790* (SRGH); Chimanimani National Park, 18 July 1950, *Thompson 37* (PRE, SRGH); Chimanimani Mountains, Mt. Peza, 2200 m, 15 Oct. 1950, *Munch 277* (K, SRGH); Chimanimani Mountains near mountain hut, 24 Sept. 1966, *Grosvenor 177* (K, LISC, SRGH).

5. *Gladiolus intonsus* Goldblatt

FIGURE 11, MAP 6. Goldblatt, Fl. Zambesiaca 12(4): 72 (1993). Type: Malawi, Northern, Nyika Plateau, 15 km north of M1, 1830 m, 11 Mar. 1978, *Pawek 14066* (MAL, holotype; BR, K, MO, PRE, SRGH, isotypes).

EPONYMY

intonsus, "unshaven," alluding to the densely villous leaf surface.

DESCRIPTION

Plants 50–75 cm high. CORM globose, c. 15 mm in diameter, tunics coriaceous, the outer layers breaking into medium to coarse vertical fibers, brown. CATAPHYLLS pale below, the upper reaching to 6

FIGURE 11. *Gladiolus intonsus,* × 0.67 (*Pawek 14066*).

cm above the ground and then green to purple and densely pubescent. LEAVES six to eight, the lower two or three basal and largest and reaching shortly above the base of the spike, the upper leaves cauline and progressively reduced in size, narrowly lanceolate to sublinear, 7–12 mm at the widest, usually densely pubescent but sometimes only lightly so, the hairs arising from the margins and veins, the margins and midrib comparatively lightly thickened and hyaline. STEM unbranched, glabrous, to 4 mm in diameter at the base of the spike.

SPIKE 9- to 14-flowered; BRACTS evidently green below, becoming dry and light brown above, 16–20(–25) cm long, narrowly lanceolate, attenuate, often dry and membranous above, the inner about two-thirds as long as the outer. FLOWERS white to cream (sometimes flushed with pink) or pink to light purple, the lower lateral or all three lower tepals each with a broad yellow median streak, the lower part of the tube dark purple; PERIANTH TUBE c. 10 mm long, slender and cylindric below, widening and funnel-shaped in the upper half; TEPALS unequal, the dorsal largest, obovate, arched over the stamens, c. 20 mm long, c. 12 mm wide, the lower three spreading more or less horizontally, lanceolate, c. 20 mm long, c. 8 mm wide, narrowed into claws below. FILAMENTS c. 12 mm long, exserted for c. 7 mm; ANTHERS c. 9 mm long. OVARY fusiform, c. 5 mm long; STYLE dividing opposite the apex of the anthers, the branches c. 2.5 mm long. CAPSULES and SEEDS unknown.

FLOWERING TIME. Mainly March and April, occasionally in February or into early May.

DISTRIBUTION & HABITAT

Although it is known from relatively few collections, *Gladiolus intonsus* has a fairly wide distribution. It extends across central Africa from the mountains of eastern Shaba Province in Zaire through northern Zambia and into northern Malawi and western and central Tanzania. Plants occur in woodland in hilly country and flower in late summer, mainly March and April. The first collection of *G. intonsus* was evidently made in 1918 near "Tabora" (*sic*) in western central Tanzania (*Autriche s.n.*). It is still poorly known from Tanzania, but *G. intonsus* has subsequently become well known from northern Malawi, largely as a result of the work of the American botanist, Jean Pawek, who collected intensively in northern Malawi during the years she worked in the country. Only one gathering is known from Zaire. The relative paucity of records for the species suggests that it is fairly rare and truly has a scattered distribution.

DIAGNOSIS & RELATIONSHIPS

The vegetatively rather uniform *Gladiolus intonsus* can immediately be recognized by its normally heavily pubescent leaves and cataphylls (sometimes lightly pubescent, e.g., *Moors K23*), and small, white to cream or pink to light purple flowers. It is presumably related to the only other tropical African pubescent-leafed *Gladiolus, G. puberulus* of southeastern Zaire. *Gladiolus puberulus* has

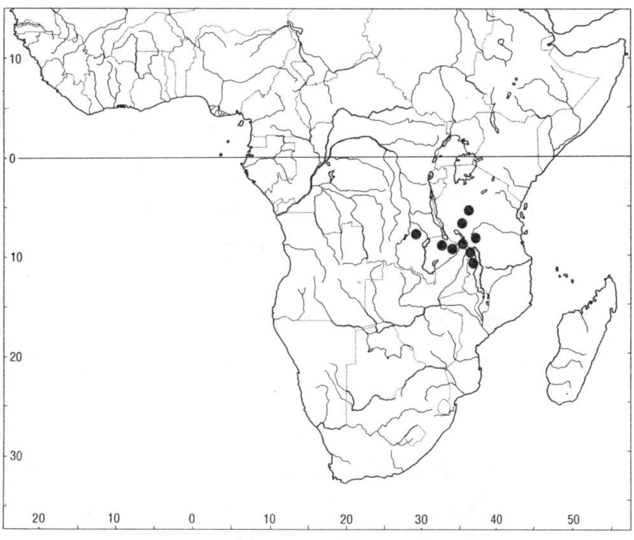

MAP 6. *Gladiolus intonsus.*

larger pink or purple or rarely white flowers with a tube (15–)24–35 mm long, a dorsal tepal 28–35 mm long, and leaves that are generally more heavily pubescent and with veins more heavily thickened than those of *G. intonsus*. That these are not merely regional variants of a single species is evidenced by the nature of the stamens. In *G. intonsus* the filaments are exserted c. 7 mm from the tube and the anthers are about 9 mm long. In the larger-flowered *G. puberulus*, however, the filaments are included in the tube and the anthers are c. 11–14 mm long.

SELECTED SPECIMENS

Zaire. Shaba: Mitwaba, Feb. 1956, *Vanden Brande M193* (BRLU).

Zambia. Northern: East of Mbala (Abercorn), Mbosi road, c. 1650 m, 31 Mar. 1932, *Thompson 1103A* (K); Mporokoso, 6 Jan. 1944, *Bredo 5918* (BR).

Tanzania. Tabora: Tabora (District?), in 1918, *Autrique 12* (BR); Urambo (5°04′ S, 32°03′ E), 1950, *Moors K23* (K). Rungwe: Ulambya, Ulanda, 10 Mar. 1970, *Leedal 434* (EA). Mbeya: Mbosi, Feb. 1935, *Horsbrugh-Porter s.n.* (K); Mbosi vicinity, Unyamwanga, 1500 m, 3 Apr. 1932, *Davies 153* (K); Namanyeri, Karungu, 12 Dec. 1921, *Swynnerton 1556* (BM); Chimala, 1220 m, 31 Dec. 1970, *Leedal 566* (EA).

Malawi. Northern: 8 km northeast of Mpora, 1800 m, 29 Mar. 1976, *Phillips 1604* (K, MO); Nyika Plateau, 40 km south of Thazima Gate, 1600 m, 2 Mar. 1987, *la Croix 995* (MO); Nyika road, mile 5, woodland, Mar. 1953, *Chapman 116* (BM); 50 km from Njakwa to Fort Hill, 5 Apr. 1951, *Hodge 8563* (K); Kaseye Mission, c. 16 km east of Chitipa, 1200 m, 5 Apr. 1969, *Pawek 1938* (K); Chitipa District, 15 km east of crossroads to Karonga, Songea stream, 19 Apr. 1969, *Pawek 2253* (MAL, K); Karonga District, Chiteka, 1300 m, 4 May 1947, *Benson 1271* (K); Kaẓaronti Hill near Mussissi, 29 Apr. 1947, *Benson 1282* (BM).

6. *Gladiolus puberulus* Vaupel

FIGURE 12, MAP 7. Vaupel, Bot. Jahrb. Syst. 48: 539 (1913). Geerinck, Bull. Jard. Bot. Nat. Belgique 42: 286 (1972). Type: Zaire, Mt. Senga, 1 June 1908, *Kassner 2927* (B, holotype; BM, HBG, K, isotypes).

SYNONYMY

Gladiolus decipiens Vaupel, Bot. Jahrb. Syst. 48: 538 (1913). Type: Zaire, Shaba, banks of the Lufuka River, 25 May 1908, *Kassner 2866* (B, holotype; E, HBG—the specimen with this number is apparently mislabeled Kundelungu Lopoi, Mar. 1908; J, K, P, Z, isotypes).

EPONYMY

puberulus, "downy, softly pubescent," referring to the soft, almost silky hairs that cover the leaf surface.

DESCRIPTION

Plants (30–)70–130 cm high. CORM depressed globose, 15–25 mm in diameter, tunics firm, the layers fragmenting into coarse vertical fibers below and to a lesser extent above, ultimately breaking up in irregular pieces, light brown. CATAPHYLLS pale, the upper reaching to 15 cm above the ground and then green and fairly densely and shortly pubescent. LEAVES four to six, the lower two or three basal and largest and reaching almost to the base of the spike, the upper leaves cauline and progressively reduced in size, narrowly lanceolate to sublinear, 8–12 mm at the widest, densely pubescent, the hairs arising from the margins and veins, the margins, midvein, and usually two other veins moderately thickened and hyaline. STEM unbranched, glabrous, to 4 mm in diameter at the base of the spike.

SPIKE 7- to 14-flowered; BRACTS evidently green and soft-textured, becoming dry toward the apex, (20–)25–40 mm long, narrowly lanceolate,

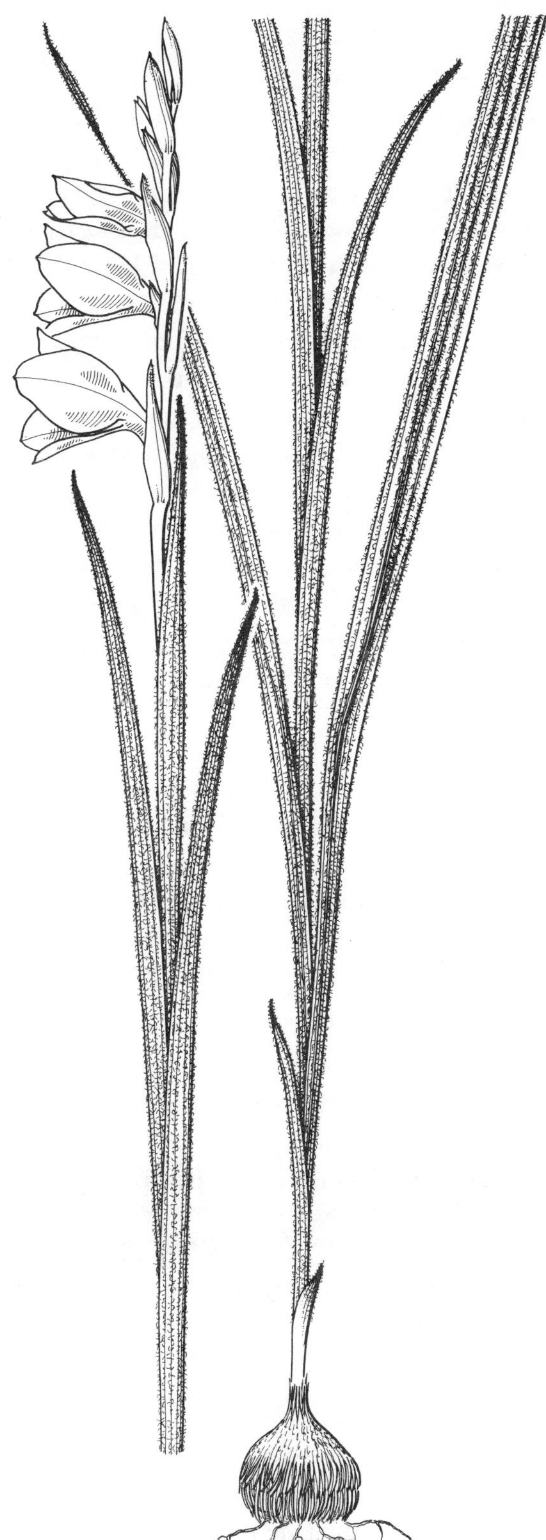

FIGURE 12. *Gladiolus puberulus,* × 0.67 (*Dubois 1358*).

attenuate above, the inner about two-thirds as long as the outer. FLOWERS cream to ivory, sometimes flushed with pink, or pink to purple with the lower three tepals cream with a central yellow band; PERIANTH TUBE 15–20(–35) long, slender and cylindric below, widening and funnel-shaped in the upper 7–8 mm; TEPALS broadly lanceolate, unequal, the dorsal largest, erect to inclined over the stamens, (25–)30–35 mm long, c. 18 mm wide, the lower three more or less spreading more or less horizontally, joined for 2–3 mm, narrowed into claws below. FILAMENTS c. 10 mm long, usually reaching only the mouth of the tube or exserted 1–4 mm; ANTHERS 10–12(–14) mm long. OVARY fusiform, c. 5 mm long; STYLE dividing close to the apex of the anthers, the branches c. 4 mm long. CAPSULES ellipsoid, 3–3.8 cm long; SEEDS oval, 7–8 × c. 5 mm, broadly winged.

FLOWERING TIME. March to May, occasionally later in July.

DISTRIBUTION & HABITAT

Known from just a handful of collections, *Gladiolus puberulus* is restricted to Shaba Province, Zaire. It occurs in the highlands of eastern Shaba, where

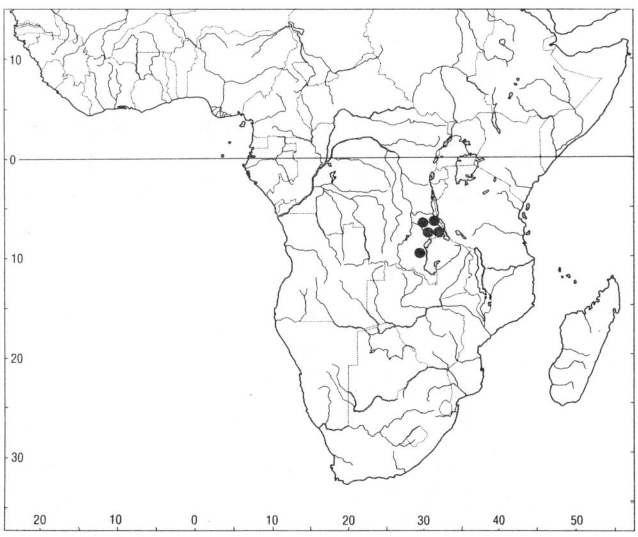

MAP 7. *Gladiolus puberulus.*

it grows in woodland on rocky sites at elevations above 1450 m. Most records of the species are from the Mugila (Muhila) Mountains that lie just to the west of Lake Tanganyika. The type collection of *G. decipiens,* regarded as conspecific with *G. puberulus,* is unusual in having somewhat longer-tubed and white flowers. It is from high elevations in the Kundelungu Mountains farther to the west than the main range of the species.

DIAGNOSIS & RELATIONSHIPS

Gladiolus puberulus can easily be distinguished by its strongly pubescent leaves and cataphylls and comparatively large pink to red-purple, or sometimes white, flowers (Figure 12). The size of the flowers is somewhat variable, with the tube reaching 30 mm and tepals 30–35 mm in some individuals of the type collection of the white-flowered *G. decipiens,* but collections such as *Kassner 2996* and *3324* have a tube only c. 15 mm long and tepals about 25 mm long. Some other floral parts are also smaller, with the anthers 10–12 mm. Some of the smaller-flowered plants are comparatively short for the species and reach only about 30 cm high (e.g., *Kassner 3324, Dubois 1358*).

The apparently closely related *Gladiolus intonsus* of eastern Zambia, Malawi, and southwestern Tanzania also has pubescent leaves and cataphylls, but it has considerably smaller flowers and bracts. The flowers have a short perianth tube 10 mm long, and the shorter anthers are exserted 7 mm from the mouth of the tube. The differences in the size of the flowers and bracts and in the exsertion of the filaments, in addition to the separate ranges, make confusion between these two species unlikely. *Gladiolus puberulus* is included in section *Gladiolus* with some misgiving. It has fairly large flowers borne on erect spikes, both features unusual for the section. Its close resemblance vegetatively to *G. intonsus,* which accords well with the critical features of the section, however, provides the main reason for placing this large-flowered species here.

HISTORY

Gladiolus puberulus was discovered by the German botanist and traveller, Theo Kassner. In the early 20th century, when the flora of interior central Africa was poorly known, Kassner undertook a collecting expedition (1906–1908) from Cape Town through central Africa to Egypt. As a consequence, Kassner discovered numerous new plants, including *G. puberulus* and the conspecific *G. decipiens.* A few other collections of *G. puberulus* have since been made, but the record of this species is still inadequate. It remains to be established whether the white-flowered and longer-tubed *G. decipiens* is really no more than a color form of typical *G. puberulus,* as maintained by Geerinck (1972), whose decision is followed here in the absence of additional information.

ADDITIONAL SPECIMENS

Zaire. Shaba: Mugila Mountains, west slopes, 23 May 1908, *Kassner 2996* (B, BM, E, HBG, K, P); west of ?Karsi Lake, July 1908, *Kassner 3324* (P); Kitendwe, 2000 m, Apr. 1945, *Dubois 1358* (BR); Marungu, in 1940, *Jurion 358* (BR); between Kampela and Kayabala on the Pweto–Moba (Baudoinville) road, 1450 m, Apr. 1926, *Robyns 2173* (BR); Mugila Plateau, Mt. Kapona, 1680 m, 13 May 1971, *Lisowski 81622* (POZG).

7. *Gladiolus debeerstii* de Wildeman

MAP 8. De Wildeman, Feddes Repert. Spec. Nov. Regni Veg. 12: 296 (1913). Type: Zaire, Upper Katanga, Tanganyika, without date but probably 1894 or 1895, *Debeerst s.n.* (BR, holotype; K, photo).

SYNONYMY

Gladiolus mitwabaensis Duvigneaud & van Bockstal ex Còrdova, Bull. Jard. Bot. Nat. Belgique 60: 327 (1990). Type: Zaire, Shaba, 15 km north of Mitwaba, Jan. 1960, *Duvigneaud 5081* (BRLU, holotype; BRLU, isotype).

EPONYMY

debeerstii, named to honor the first collector of the species, Gustave Debeerst.

DESCRIPTION

Plants 40–70 cm high. CORM 15–20 mm in diameter, TUNICS of fairly fine, pale, reticulate fibers. CATAPHYLLS pale or green above the ground. LEAVES three to five, the lower basal, reaching to about base of the spike or sometimes shortly exceeding it, 1–2 mm wide, terete and four-grooved to linear with the midribs and margins thickened and raised, 1–2 mm wide, upper leaves progressively reduced in size and imbricate, leaving only the upper part of the stem exposed. STEM unbranched, c. 2 mm in diameter at the base of the spike.

SPIKE 3- to 5(–10)-flowered, fairly lax, inclined below the first flower; BRACTS green or membranous, evidently becoming dry from the apex toward the end of flowering, 12–18(–20) mm long, the inner slightly shorter than the outer. FLOWERS pale pink, the lower lateral tepals each with a dark purple and yellow nectar guide in the distal half; PERIANTH TUBE obliquely funnel-shaped, c. 10 mm long; TEPALS unequal, the dorsal c. 16 mm long, c. 10 wide, the lower three united for c. 2 mm, c. 12 mm long, evidently extended horizontally, in profile shortly exceeding the dorsal. FILAMENTS arcuate, c. 10 mm long, exserted 5 mm from the tube; ANTHERS c. 5.5 mm long, yellow. OVARY ovoid, c. 2.5 mm long; STYLE arcuate, dividing opposite the upper half of the anthers, the branches c. 2 mm long. CAPSULES and SEEDS unknown (described as flat, broadly winged, and 3–7 mm in diameter by Còrdova, 1990, evidently from plants of another species).

FLOWERING TIME. Late November to January.

DISTRIBUTION & HABITAT

Gladiolus debeerstii is a local endemic of northeastern Shaba Province, Zaire, and judging from the meager record, it is uncommon. It grows in wet, seasonally waterlogged grassland or marshy places, normally at elevations above 1800 m. Included here in *G. debeerstii* are plants from the Upemba National Park (*de Witte 3238*) and the Marungu Plateau (*Lisowski et al. 8240*) with leaves shorter than those of other collections but that otherwise appear to accord with the type specimen.

DIAGNOSIS & RELATIONSHIPS

Gladiolus debeerstii can be recognized by its slender form, lax spike of relatively few small pale pink flowers, and three to five fairly short leaves, the margins and midribs of which are thickened and raised. The relationships of *G. debeerstii* are uncertain. The small size and shape of the flower as well as its narrow lower tepals suggest section *Gladiolus,* although the lower tepals are not extended horizontally as far as in most species of that section. It is perhaps most easily confused with *G. erectiflorus* (a small-flowered species of subgenus *Ophiolyza*) and *G. linearifolius.* Both these species have larger flowers with the lower tepals as large as the upper or not much smaller. In addition, *G. erectiflorus* has leaves with the margins very little raised and pale flowers very rarely without conspicuous dark venation. Although poorly known,

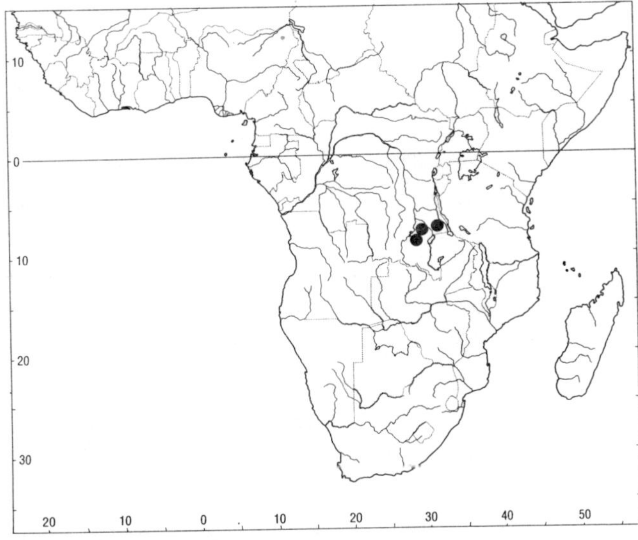

MAP 8. *Gladiolus debeerstii.*

specimens of *G. linearifolius* have leaves much longer than the spikes, and that species can be distinguished from *G. debeerstii* using leaves as well as flower size. Until more collections of *Gladiolus* are made in eastern Shaba, the species of the region will remain imperfectly understood. *Gladiolus debeerstii* has also been confused with *G. gracillimus* in herbaria, although I see little similarity between the two—*G. gracillimus* is a smaller plant with a strongly inclined spike, and it always has only two foliage leaves, the lower one largely sheathing and the upper reduced in size.

HISTORY

Although the date of its discovery and initial collection is not recorded, *Gladiolus debeerstii* was described as early as 1913 from plants gathered by Gustave Debeerst (collected 1894–1895) in "Tanganyka" (*sic*). This is presumably the region of northeastern Shaba, so called on early maps, located to the west of Lake Tanganyika, and not Tanzania, which was known as Tanganyika only after 1919. *Gladiolus debeerstii* remained so poorly known that when new collections were made by the Belgian botanists, G. F. de Witte and P. Duvigneaud, after World War II, they were not associated with this species. Duvigneaud and his colleague, Liliane van Bockstal, used a manuscript name, *G. mitwabaensis,* for two collections made near Mitwaba, a name only formally published in 1990 (Còrdova, 1990). There seems no doubt that *G. mitwabaensis* and *G. debeerstii* are conspecific. Both have small flowers of similar proportions, about 25 mm long, unusual linear leaves with thickened margins and midribs, and comparatively soft-textured, apically dry floral bracts. In the original description, *G. mitwabaensis* was compared with *G. laxiflorus,* but there is nothing to suggest that the two are closely related. *Gladiolus laxiflorus* has much larger flowers that lack banded nectar guides on the lower tepals, and the lightly pubescent leaves have broad, more or less plane blades.

ADDITIONAL SPECIMENS

Zaire. Shaba: 14 km north of Mitwaba, marshy plateau grassland, 16 Jan. 1960, *Duvigneaud 5072* (BRLU); Upemba National Park, source of the Mumbale, 1880 m, 16 Jan. 1948, *de Witte 3238* (BR); Marungu Plateau, edge of Mufufu marsh, 1900 m, *Lisowski, Malaisse, & Symoens 8240* (LSHI).

8. *Gladiolus camilae* Còrdova

MAP 9. Còrdova, Bull. Jard. Bot. Nat. Belgique 60: 325 (1990). Type: Zaire, Kivu, Itombwe Plateau, Lake Lungwe, Mar. 1953, *Thirion 25* (BRLU, holotype).

EPONYMY

camilae, named by the author of the species, Susana Còrdova, after her daughter, Camila.

DESCRIPTION

Plants 40–70 cm high. CORM 15–20 mm in diameter, *tunics* of fairly fine pale reticulate fibers. CATAPHYLLS membranous and pale, or green above the ground. LEAVES three to five, the lower basal, reaching to about base of the spike or sometimes shortly exceeding it, 1–2 mm wide, terete and four-grooved to linear with the midribs and margins thickened and raised, the upper leaves progressively reduced in size and imbricate, leaving only the upper part of the stem exposed. STEM unbranched, c. 2 mm in diameter at the base of the spike.

SPIKE 3- to 5(–10)-flowered, fairly lax, inclined below the first flower; BRACTS green or membranous, evidently becoming dry from the apex toward the end of flowering, 12–18(–20) mm long, the inner slightly shorter than the outer. FLOWERS pale pink, the lower lateral tepals each with a dark purple and yellow nectar guide in the distal half; PERIANTH TUBE obliquely funnel-shaped, c. 10 mm long; TEPALS unequal, the dorsal c. 16 mm long, c. 10 wide, the lower three united for c. 2

mm, c. 12 mm long, evidently extended horizontally, in profile shortly exceeding the dorsal. FILAMENTS arcuate, c. 10 mm long, exserted 5 mm from the tube; ANTHERS c. 5.5 mm long, yellow. OVARY ovoid, c. 2.5 mm long; STYLE arcuate, dividing opposite the upper half of the anthers, the branches c. 2 mm long. CAPSULES and SEEDS unknown (described as flat, broadly winged, and 3–7 mm in diameter by Còrdova, 1990, evidently from plants of another species).

FLOWERING TIME. Late November to January.

DISTRIBUTION & HABITAT

Gladiolus camilae is a regional endemic of eastern Kivu Province in eastern Zaire, known from only one gathering. It appears to be restricted to the Itombwe Plateau, but additional collecting in the highlands west of Lake Tanganyika and Lake Kivu may well extend its range.

DIAGNOSIS & RELATIONSHIPS

Gladiolus camilae can be recognized by its lax spikes of 3–10 small, pale pink flowers with pale markings on the lower tepals and three to five fairly short leaves, the margins and midribs of which are lightly thickened. The lower tepals extend forward and in profile exceed the upper tepals, typical of the species of subgenus *Gladiolus*. Still incompletely known, *G. camilae* seems most closely related to *G. debeerstii*, a species found in eastern Shaba Province. The longer, apparently firm-textured floral bracts and the long lower tepals of *G. camilae* make confusion with *G. debeerstii* unlikely. The latter has shorter bracts that appear, at least from the available herbarium material, to be rather soft-textured and apically dry at flowering time.

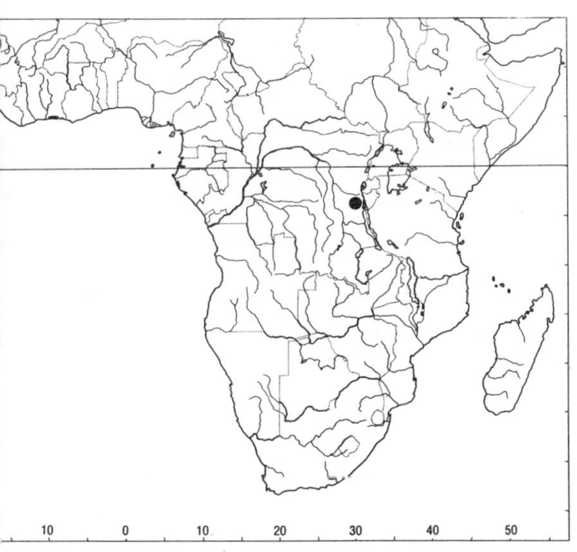

MAP 9. *Gladiolus camilae.*

SPECIMENS

Known only from the type, cited above.

9. *Gladiolus linearifolius* Vaupel

MAP 10. Vaupel, Bot. Jahrb. Syst. 48: 535 (1913). Type: Zaire, Shaba, Mt. Senga, rocky sites, 13 May 1908, *Kassner 2978* (B, holotype; BM, BR, E, HBG, P, Z, isotypes).

SYNONYMY

Gladiolus erectiflorus sensu Geerinck, Bull. Jard. Bot. Nat. Belgique 42: 274 (1972), not sensu Baker.

EPONYMY

linearifolius, "linear-leaved," alluding to the very narrow leaves.

DESCRIPTION

Plants 25–30 cm high, the leaves much exceeding the stems. CORM c. 12 mm in diameter, tunics coriaceous to membranous. CATAPHYLLS membranous, usually dull purple or brownish, the inner longest and lightly puberulent above ground level. LEAVES three to five, the lower two or three basal and longest (sometimes dry and broken at flowering time), the sheaths at least of the lowermost sometimes puberulent, blades linear, about twice as long as the stem and up to 30 cm long, 1.5–2 mm wide, the midrib and margins hyaline, moderately thickened, and raised, the upper one or two leaves sheathing the stem in the lower half, much shorter than the basal, the blades reaching to about the base of the spike. STEM unbranched, flexed above the sheath of the uppermost leaf, c. 1.5 mm in diameter at the base of the spike.

SPIKE 2- to 4-flowered, inclined c. 30° toward the ground; BRACTS apparently green below, membranous, dry and light brown in the upper half, 13–15 mm long, the inner about as long as the outer. FLOWERS of unknown color, probably red; PERIANTH TUBE more or less cylindric, widening and curving outward near the apex, c. 10 mm long; TEPALS more or less equal, narrowly obovate, c. 21 mm long (often less when dry), the dorsal arched over the stamens, the lower three inclined toward the ground. FILAMENTS arcuate, 12–15 mm long, exserted 6–9 mm from the tube; ANTHERS c. 5 mm long. OVARY obovoid, c. 3 mm long; STYLE arching over the filaments, dividing at about mid anther level, the branches c. 3 mm long, reaching to about the anther apices. CAPSULES and SEEDS unknown.

FLOWERING TIME. May.

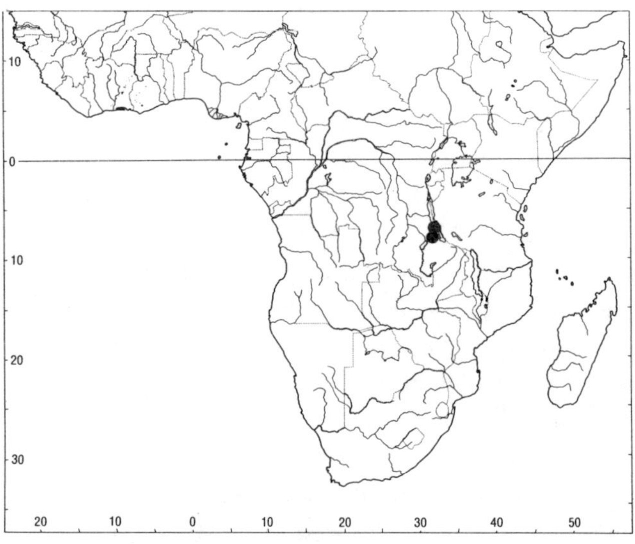

MAP 10. *Gladiolus linearifolius.*

DISTRIBUTION & HABITAT

First collected in May 1908 by German naturalist and explorer, Theo Kassner, *Gladiolus linearifolius* remains poorly documented. It is known from the type, fortunately widely distributed among the world's herbaria, and just one other collection. It is apparently restricted to the Marungu Mountains of southeastern Shaba Province, Zaire, and is recorded only from Mt. Senga (7°10′ S, 29°30′ E, according to Bamps, 1982) and Mt. Munda. According to collection information, it grows in rocky places, probably close to the summit. The higher ridges and peaks of the Marungus reach elevations of 2000 to 2400 m.

DIAGNOSIS & RELATIONSHIPS

Gladiolus linearifolius can be distinguished by its three to five relatively long, linear leaves, 1.5–2 mm wide, that are about twice as long as the spikes. The stems are fairly short, 20–40 cm high and inclined toward the ground in the upper half. The flowers are notable in having subequal oblanceolate tepals. Geerinck (1972) regarded *G. linearifolius* as conspecific with *G. erectiflorus* (Baker, 1876) and united them under the latter, earlier name. Except for a superficial general resemblance, the two probably have little in common. Vegetatively, *G. erectiflorus* differs in always being glabrous and in having lanceolate, rarely linear, leaves with the margins and midrib only weakly thickened. The leaves are also usually shorter than the stems, or occasionally exceed them by only 3 to 5 cm, unlike the much longer leaves of *G. linearifolius.* The stems of *G. erectiflorus* are erect, and minor floral differences include the length of the anthers, 6–7 mm long, the generally longer floral bracts, 20–40 mm long, and the lanceolate, acute tepals. In *G. linearifolius* the anthers are 5 mm long and the bracts 13–15 mm long.

More likely, *Gladiolus linearifolius* is immediately related to the southwestern Tanzanian endemic, *G. sulcatus,* which has a similar overall appearance despite its being a more robust species. It also has long linear leaves that much exceed the stems, inclined spikes, and leaf margins and midribs thickened, even more strongly than in *G. linearifolius.* The two also have subequal, oblanceolate tepals, and lightly pubescent upper cataphylls and lower leaf sheaths. Although I feel confident about the close relationship of *G. linearifolius* and *G. sulcatus,* their relationships to other species of subgenus *Gladiolus* are uncertain.

ADDITIONAL SPECIMENS

Zaire. Shaba: Marungu Plateau, Mt. Munda, above Mukonga, on rocks, 1600 m, 18 Feb. 1970, *Lisowski, Malaisse, & Symoens 10,003* (BRLU).

10. *Gladiolus sulcatus* Goldblatt, new species

FIGURE 13, MAP 11. Type: Tanzania, Matamba District, Kitulo Plateau, Ndumbi Stream near the Matamba–Kitulo road, grassland, 2500 m, 24 May 1986, *Lovett & Lovett 806* (DSM, holotype; K, MO, isotypes).

EPONYMY

sulcatus, "grooved," referring to the narrow grooves that run the length of the leaves between the thickened margins and midrib.

LATIN DIAGNOSIS

Plantae 20–40 cm longae, foliis 4–5 linearibus marginibus costistisque incrassatis hyalinisque vaginibus leviter pubescentibus vel ciliatis, spica 5–8 (–12) florum inclinatis, bracteis 20–25 mm longis, floribus atroroseis, tubo perianthii c. 13 mm longo, tepalis subequaelibus late obovatis c. 23 mm longis, filamentis c. 16 mm longis c. 9 mm exsertis, antheris c. 6.5 mm longis.

FIGURE 13. *Gladiolus sulcatus.* Habit, leaves, and flowering spike, × 0.5; cross-section of leaf, much enlarged (*Lovett & Lovett 806*).

DESCRIPTION

Plants 20–40 cm high, excluding the leaves. CORM globose, c. 2 cm in diameter, the tunics of moderately fine, mostly vertical fibers. CATAPHYLLS pale, firm-textured, the uppermost extending 5–7 cm above the ground and similar to the leaves except pubescent to puberulous. LEAVES four or five, the basal two or three longest, the blades of at least one usually exceeding the stem, sometimes by up to 15 cm, linear, 2.5–4 mm wide, with heavily thickened margins and midribs, the margin edges extended at right angles to the blade, thus with two narrow grooves on each surface, edges of the midrib often ciliate, the upper two leaves cauline and progressively shorter, the uppermost sometimes entirely sheathing and bractlike, sheaths of the basal leaves lightly pubescent to ciliate least below. STEM simple, inclined almost from the base, usually flexed at the node below the spike, ultimately inclined at c. 45°, c. 2 mm in diameter at the base of the spike.

SPIKE 5- to 8(–12)-flowered, unbranched, inclined at c. 45°; BRACTS green, usually turning purplish toward the apices, 20–25 mm long, the inner shorter than the outer, the outer acute, the inner minutely forked apically. FLOWERS dark pink to scarlet, the lower lateral tepals sometimes partly white and each with a broad triangular yellow mark in the center edged in dark pink or red; PERIANTH TUBE widening and curving outward near the apex, c. 13 mm long, extended between the bracts; TEPALS nearly equal, broadly obovate, obtuse-apiculate, the dorsal slightly larger than the others, c. 23 mm long, often less when dry, c. 12 mm wide. FILAMENTS arcuate, c. 16 mm long, inserted in the middle of the tube and exserted for c. 9 mm; ANTHERS c. 6.5 mm long, the connective obtuse apically, not reaching the tips of the anther lobes, pollen yellow. OVARY narrowly obovoid, c. 5 mm long; STYLE arching over the filaments, dividing near the apex of the anthers, the branches c. 6 mm long. CAPSULES obovoid, c. 13 mm long; SEEDS ovate, 6 × c. 4.5 mm.

FLOWERING TIME. Late March, April, and May.

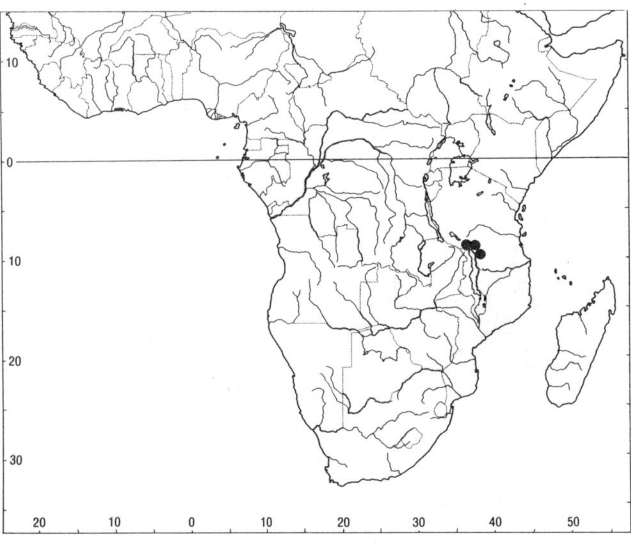

MAP 11. *Gladiolus sulcatus.*

DISTRIBUTION & HABITAT

Found only in southwestern Tanzania, *Gladiolus sulcatus* is known from just a few sites in the high mountains that lie to the north and east of Lake Malawi. Most of the known collections are from the Kitulo Plateau, possibly from only one or two places there. It also occurs in the Poroto Mountains, a short distance to the west of the Kitulo Plateau, and on the Lusitu Ridge west of Njombe, well to the south of the Kitulo. Plants typically grow in rocky situations in montane grassland, usually above 2300 m, but at somewhat lower elevations on the Lusitu Ridge.

DIAGNOSIS & RELATIONSHIPS

Gladiolus sulcatus is distinguished by its fairly large, deep pink or scarlet flowers, strongly inclined spike, and linear leaves with prominently thickened and raised margins and midribs. The margins and midribs are usually so enlarged that they are separated only by narrow grooves, two on each leaf surface, that run the length of the leaves. Occasionally, in specimens from less exposed woodland, presumably wetter sites (e.g., *Iverson et al. 87702*), the leaf margins may hardly be raised at all. The species appears to be most closely related to *G. linearifolius,* known only from two collections from eastern Shaba Province, Zaire. That is a smaller and less robust species with smaller flowers and linear leaves 1.5–2 mm wide, the margins and midribs less prominently thickened than in *G. sulcatus.* The two species share puberulous upper cataphylls and leaf sheaths, unusual in *Gladiolus,* and tepals that are obovate and obtuse, a feature that imparts a distinctive, somewhat bell-like appearance to the flowers. Unfortunately, *G. linearifolius* is so poorly known that it is impossible to assess its variability. In the absence of known intermediates, the Tanzanian *G. sulcatus* must be regarded as a separate species, since the morphological discontinuity between it and *G. linearifolius* is considerable in both leaf and flower characters.

ADDITIONAL SPECIMENS

Tanzania. Mbeya: Rungwe, Poroto Mts., rough pasture on volcanic soil, 2340 m, 16 May 1957, *Richards 9740* (BR, K); summit of Isalala, c. 2500 m, 3 June 1974, *Leedal 1878* (EA). Iringa: Makete, Kitulo Plateau, high-altitude grassland between Kitulo and Matamba, 2600 m, 30 Mar. 1987, *Lovett & Congdon 1870* (DSM, MO); Kitulo (Elton) Plateau, north of Ukinga, 3100 m, May 1953, *Eggeling 6595* (EA, K, PRE); Kitulo Plateau, above Salala waterfalls, Mt. Ishinga, 26 May 1987, *Iverson et al. 87702* (UPS); Njombe, Lusitu Ridge, Apr. 1992, *Spurrier 604* (DSM, MO).

11. *Gladiolus laxiflorus* Baker

PLATE 5, FIGURE 14, MAP 12. Baker, Trans. Linn. Soc. Bot., Ser. 2, 1: 268 (1878); Handbook Irideae 211 (1892); Fl. Trop. Africa 7: 362 (1898). Wild, Fl. Pl. Africa 33: pl. 1287 (1959). Geerinck, Bull. Jard. Bot. Nat. Belgique 42: 279 (1972). Goldblatt, Fl. Zambesiaca 12(4): 75 (1993). Type: Angola, Huila Mountains, grassy swamps, Dec. 1859, *Welwitsch 1536* (BM, lectotype designated here, very well preserved; B, BR, K, LISC, MO, PRE, SRGH, isolectotypes).

SYNONYMY

Gladiolus tritonioides Baker, Kew Bull. 1895: 74 (1892); Fl. Trop. Africa 7: 363 (1898). Type: Zambia, Fwambo [Lake Tanganyika], Sept. 1893, *Carson 37* (K, holotype).

Gladiolus antunesii Baker, Fl. Trop. Africa 7: 576 (1898). Type: Angola, Huila, without date, *Antunes 326* (B, holotype; COI, K—spike only with drawing of holotype, LISC, isotypes).

Gladiolus trichophyllus Diels, Notizbl. Bot. Gart. Mus. Berlin 13: 264 (1936). Type: Tanzania, upper Ruhudje, vicinity of Lupembe, 26 Aug. 1931, *Schlieben 1136A* (B, lectotype designated here, the specimen complete with corm; B, BR, K, LISC, MO, PRE, SRGH, isolectotypes).

EPONYMY

laxiflorus, "laxly flowered," in reference to the flowers that are fairly widely spaced on the spikes.

DESCRIPTION

Plants (20–)40–70 cm high. CORM 25–30 mm in diameter, depressed-globose, usually persisting for some years, usually dark red, the tunics coriaceous to somewhat membranous, fragmenting irregularly. CATAPHYLLS pale and membranous, the upper often flushed purple and lightly pubescent above the ground. LEAVES (of the flowering stem) three to six, the lower two to four basal and laminate, the blades sometimes very short, linear to narrowly lanceolate or falcate, 4–6 mm wide, often emergent and short at the beginning of flowering, ultimately the longest to 35 cm, usually lightly pubescent, especially on the sheaths, the margins and midribs moderately thickened, the upper one to three leaves largely to entirely sheathing and glabrous; longer leaves sometimes produced after flowering from separate shoots. STEM often one- to two-branched, occasionally simple, sometimes more than one stem produced from the same corm, the branching often subdivaricate but the main axis always thicker, c. 4 mm in diameter at the base of the spike.

SPIKE 6- to 9(–16)-flowered, the lateral spikes with fewer flowers; BRACTS herbaceous below, becoming more or less membranous above or dry at anthesis, 10–15(–20) mm long, the inner usually slightly longer than the outer. FLOWERS purple to pinkish, without markings or with one or all of the lower tepals with a small yellow mark in the distal half; PERIANTH TUBE expanding gradually from base to apex, curving outward above, 10–17 mm long; TEPALS nearly equal, or unequal with the upper three larger than the lower, the upper three ascending, the lower descending, lanceolate, narrowed below, 30–35 mm long, 13–15 mm at the widest (poorly pressed flowers may shrink up to 50% of their original size). FILAMENTS c. 20 mm long, arcuate, exserted 12–15 mm from the tube; ANTHERS 6–8 mm long, yellow. OVARY 4–5 mm long; STYLE arched over the stamens, dividing beyond the anther apices, the branches c. 3 mm long, much expanded apically. CAPSULES 15–22 mm long; SEEDS broadly winged, 8 × 6 mm, light brown. CHROMOSOME NUMBER $2n = 30$.

FLOWERING TIME. Mostly in October and November, sometimes in September, and into late December in years when the rainy season is late.

DISTRIBUTION & HABITAT

Widely distributed across southern tropical Africa, *Gladiolus laxiflorus* extends from across Angola, Zambia, and the provinces of Shaba and Kasai in Zaire, to western and southern Tanzania and northern Malawi. It is a species of wet habitats such as marshes, dambos (poorly drained grass-

FIGURE 14. *Gladiolus laxiflorus.* Corm, leaves, and spike, × 0.67; single flower, full size (*la Croix 737, 855*).

lands along watercourses), and stream banks, where it flowers toward the end of the dry season, mainly October and November, or at the beginning of the wet season, November and December. At this time the surrounding vegetation is usually fairly short and the flowers are visible above the low ground cover, later to become tall and dense.

DIAGNOSIS & RELATIONSHIPS

Unmistakable when seen alive, *Gladiolus laxiflorus* has large pink flowers subtended by unusually short floral bracts. The tepals are usually subequal and spread outward from close to the base. The stems are frequently branched, and the fairly short and broad leaves are pubescent, although sometimes sparsely so. Poorly preserved specimens with the flowers distorted and shrunk to half their size may be confused with *G. unguiculatus*. The latter is distinguished by the reduced and entirely glabrous leaves on the flowering stem, and the smaller flowers with dark spade-shaped markings on the lower lateral tepals. At the time of flowering, the leaves of *G. laxiflorus* are sometimes very short or barely developed, especially on stalks produced very early in the season. Possibly, these leaves enlarge later. Sometimes separate leafy shoots are also produced toward the end of the flowering season.

HISTORY

First recorded in the western part of its range, in Angola, *Gladiolus laxiflorus* was discovered by Friedrich Welwitsch during his botanical exploration of this country in the 1850s. It was described in 1878 by J. G. Baker. Baker seems to have had little understanding of the species for he recognized two other collections of the plant as distinct species. He described *G. tritonioides* in 1892 from northern Zambia, and in 1898 he named a second Angolan collection *G. antunesii*. Neither of these differs significantly from the type material of *G. laxiflorus*. The somewhat larger-flowered form of the species that grows in southwestern Tanzania was described as a separate species, *G. trichophyllus*, by Friedrich Diels in 1936. Available evidence suggests that although the Tanzanian form does differ consistently, it is connected by a series of intermediates to plants that occur to the west that exactly match the type of *G. laxiflorus* and does not merit taxonomic recognition.

SELECTED SPECIMENS

Zaire. Kasai: Near Mwene Ditu, 1952, *Dandoy 106* (BR, EA). Shaba: Katende, 50 km south of Dilolo, south of Dilungu village, 23 Aug. 1956, *Duvigneaud & Timperman 2465* (BR, BRLU, PRE); Upemba National Park, Mukana, 1810 m, 26 Aug. 1947, *de Witte 2770* (BR, EA, PRE); 13 km east of Mutshatsha, sandy dambo, 9 Nov. 1962, *Schmitz 7983* (BR, K).

Angola. Cuanza Sul: north of Kubango near Kassinga, 1400 m, 5 Oct. 1899, *Baum 231* (BM, BR, COI, E, K, S, Z). Malange: Malange, 1903, *Gossweiler 921* (K, P). Lunda: Saurimo, 31 Oct. 1932, *Young 1242* (BM). Moxico: Vila Luza, Luena River, 4 Nov. 1932, *Young 1346* (BM). Huambo: Huambo vicinity, near Chiva, 19 Sept. 1970, *Moreno 252* (BM,

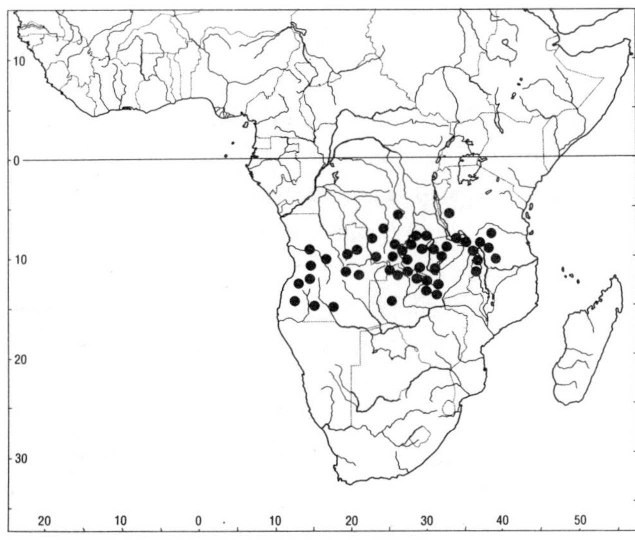

MAP 12. *Gladiolus laxiflorus*.

BRLU). Huila: Huila Plateau, Oct.–Nov. 1895, *Berthelot 326* (P); Ganguelas, Vila da Ponte, banks of Cubango River, 22 Oct. 1905, *Gossweiler 2125* (COI).

Zambia. Northwestern: 45 km east of Mwinilunga, vlei margin, 26 Oct. 1966, *Leach & Williamson 13484* (SRGH). Copperbelt: Kitwe, moist dambo, 17. Nov. 1967, *Fanshawe 2362* (K, NDO). Luapula: Kawambwa District, dambo near Mushota, Nov. 1950, *Lawton 1330* (K, NDO, P); 10 km from Kawambwa to Ntchilengi, 29 Nov. 1961, *Richards 15447* (BR, K, SRGH). Northern: Lake Chila–Mbalo, 1700 m, 27 Sept. 1968, *Sanane 277* (BR, K). Central: Mankoya, 76 km west of Kafue Hoek pontoon, 7 Nov. 1959, *Drummond & Cookson 6212* (E, K, LISC, MO, PRE, SRGH); Kabwe (Broken Hill), sandy soil, 3 Sept. 1938, *Pole Evans & Ehrens 1906* (BR, E, K, P, PRE). Southern: road to Mankoya, Barotse, 1 Oct. 1957, *West 3489* (K, SRGH).

Malawi. Northern: Nyika Plateau, boggy grassland at Chelinda Bridge, 2300 m, 24 Nov. 1986, *la Croix 855* (K, MAL, MO), 24 Sept. 1969, *Pawek 2799* (K, MAL). Central: Nkhata Bay, Mzenga dambo, 620 m, 21 Dec. 1985, *la Croix 737* (K, MAL, MO).

Tanzania. Kigoma: c. 55 km south of Uvinza, 31 Aug. 1950, *Bullock 3297* (BR, K, S). Rukwa: Ufipa, Chapota, c. 1700 m, 5 Dec. 1949, *Bullock 2058* (K, LISC, S, SRGH); dambo c. 24 km south of Sumbawanga, 17 Nov. 1986, *Goldblatt, Brummitt, & Lovett 8152* (MO). Iringa: Mufindi District, Ngwazi, marshy grassland, 21 Dec. 1960, *Paget-Wilkes 719* (EA, K, MO); Kitulo Plateau, Ndumbi, c. 2600 m, 20 Jan. 1978, *Leedal 4849* (EA, K); Mwakete–Njombe road, swamp, 2100 m, 17 Jan. 1957, *Richards 7875* (K). Mbeya: Kyimbila, north of Lake Nyasa, without date, *Stolz 2345* (BOK, BM, BR, K, P, PRE, UPS). Rovuma: Songea, Matengo Hills, Nyoni River, 18 Nov. 1956, *Semsei 2589* (K).

A doubtful record, *Surcouf 386* (P) from Mozambique, Vila Pery, right bank of the Zambezi, Nov. 1915, requires confirmation; the next nearest locality is in northern Malawi.

12. *Gladiolus lundaensis* Goldblatt, new species

MAP 13. Type: Angola, Lunda Sul, Saurimo, 24 Oct. 1932, *Young 1148* (BM, holotype; BR, BM, MO, isotypes).

EPONYMY

lundaensis, "from the country of the Lunda," the people in whose territory the species grows.

LATIN DIAGNOSIS

Plantae 38–60 cm altae, foliis 5–7 inferioribus laminis elongatis linearibus 3-4 mm latis, caulis eramosis, spicis erectis (3–)6–10 florum, bracteis viridis 6–10 mm longis, floribus purpureis vel albis, tubo perianthii 6–8 mm longo, tepalo dorsalis arcuato c. 12 mm longo, tepalis inferioribus anguste unguiculatis, filamentis 3–4 mm exsertis, antheris 5–6 mm longis.

DESCRIPTION

Plants 38–60 cm high. CORM 12–15 mm in diameter, the tunics either not accumulating and represented by the firmly membranous bases of the cataphylls, or composed of fine matted fibers. CATAPHYLLS more or less membranous, usually red-brown, the uppermost green above the ground. LEAVES five to seven, the lower three or four more or less basal and with long linear blades, reaching to about the base of the spike, 3–4 mm at the widest, the midrib lightly thickened, the margins not raised or thickened, the upper two or three leaves mostly to entirely sheathing and decreasing in length above. STEM erect, unbranched, c. 1.5 mm in diameter at the base of the spike.

SPIKE (3–)6- to 10-flowered; BRACTS green, 6–10 mm long, the inner about as long or slightly longer than the outer. FLOWERS purple (or white), the lower tepals each with dark purple markings on the limbs; PERIANTH TUBE 6–8 mm long, curving outward, widening near the mouth; TEPALS unequal, the dorsal arcuate, c. 12 mm long, c. 7 mm wide, the upper laterals smaller, joined to the lower three for c. 3 mm, directed forward and ultimately curving outward, the three lower smallest, united below for c. 1.5 mm, 9–10 mm long, narrowed below into claws, the limbs abruptly expanded and turned downward, c. 3 mm wide, in profile exceeding the dorsal. FILAMENTS c. 6 mm long, exserted 3–4 mm from the tube; ANTHERS 5–6 mm long, yellow. OVARY 2–3 mm long; STYLE dividing opposite the middle of the anthers, the branches c. 0.8 mm long. CAPSULES ellipsoid, c. 10 mm long; SEEDS not known.

FLOWERING TIME. October to mid November.

DISTRIBUTION & HABITAT

Known from just two collections, one from northeastern Angola in 1932 and the other from southern Zaire in 1954, *Gladiolus lundaensis* appears to be a local endemic of southwestern tropical Africa. Whether it is truly rare or simply restricted to a part of the continent poorly explored botanically remains to be established. *Gladiolus lundanensis* appears to favor permanently damp sites, where it flowers at the end of the dry season.

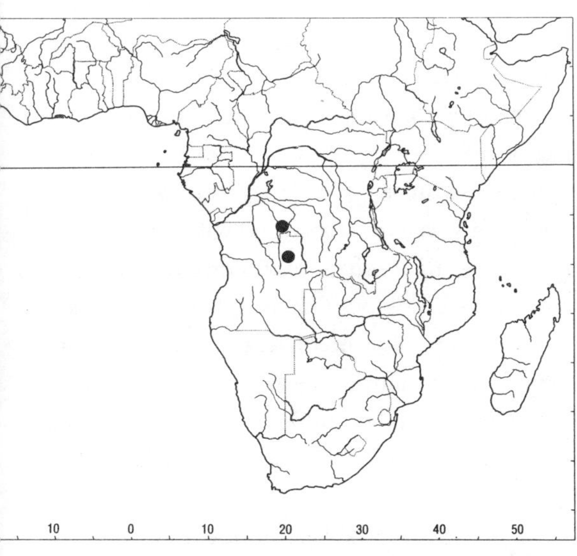

MAP 13. *Gladiolus lundaensis.*

DIAGNOSIS & RELATIONSHIPS

Gladiolus lundaensis is easily recognized by its nearly erect flowering spike of relatively small flowers, leaves of fairly coarse texture contemporaneous with the flowers, and the absence of accumulated corm tunics so that the corms are enclosed only by the bases of the current season's cataphylls. The fairly small purple or whitish flowers with enlarged and arcuate dorsal tepals and narrow, grooved lower tepals with dark nectar guides are typical of subgenus *Gladiolus.* The general impression of the plant is that it most closely resembles the widespread tropical African *G. unguiculatus.* This has flowering stems that bear only reduced sheathing leaves, occasionally with very short blades, and foliage leaves that are produced later in the flowering season on separate shoots. The leaves of *G. lundensis* are also distinctive in their narrowly lanceolate to linear shape, only lightly thickened midrib, and completely unthickened margins.

ADDITIONAL SPECIMENS

Angola. Lunda Sul: Saurimo, 24 Oct. 1932, *Young 1148* (BM, BR, MO).

Zaire. Bandundu: Gungu–Feshi road, Yembeshi marsh, 6 Nov. 1954, *Devred 1467* (BR, K).

13. *Gladiolus unguiculatus* Baker

PLATES 6, 7, FIGURE 15, MAP 14. Baker, J. Linn. Soc. Bot. 16: 178 ("1878," i.e., 1877) as a new name for *Gladiolus cochleatus* Baker, J. Bot., New Ser. 5: 333 (1876), an illegitimate homonym, not *G. cochleatus* Sweet (1832) (= *G. debilis* var. *cochleatus* (Sweet) G. Lewis); Handbook Irideae 233 (1892); Fl. Trop. Africa 7: 372 (1898). Mildbraed, Bot. Jahrb. Syst. 52: 232 (1923). Hepper, Fl. West Trop. Africa, Ed. 2, 3: 144 (1968). Lewis et al., J. S. African Bot., Suppl. 10: 292 (1972). Wickens, Fl. Jebel Marra 158 (1976). Adam, Fl. Descr. Monts Nimba, Part 5: 1643 (1981). Goldblatt, Fl. Zambesiaca 12(4): 76 (1993). Type: Sierra Leone, c. 1800, *Morson s.n.* (K, holotype).

SYNONYMY

Gladiolus brevicaulis Baker, Trans. Linn. Soc., Ser. 2, 1: 267 (1878). Type: Angola, Huila, damp sandy sites between Humpata and the Lopolo River, Dec. 1859–Jan. 1860, *Welwitsch 1534* (BM, lectotype designated here; C, COI, K, LD, P, isolectotypes).

Gladiolus oatesii Rolfe in Oates, Matabeleland, Ed. 2, 410 (1889). Baker, Handbook Irideae 226 (1892); Fl. Trop. Africa 7: 373 (1898). Type: Zimbabwe, Matabeleland, without date, *Oates s.n.* (K, holotype).

Antholyza labiata Pax, Bot. Jahrb. Syst. 15: 156, pl. 7(1–4) (1893). Baker, Fl. Trop. Africa 7: 374 (1898). *Gladiolus labiatus* (Pax) N. E. Brown, Trans. Roy. Soc. S. Africa 20: 267 (1932). Type: Togo, Bismarckburg, 12 Nov. 1889, *Kling 209* (B, holotype; K, photo).

Antholyza cabrae de Wildeman, Ann. Mus. Congo, Sér. 5, 15 (1902). *Gladiolus cabrae* (de Wildeman) N. E. Brown, Trans. Roy. Soc. S. Africa 20: 267 (1932). Type: Zaire, Mayumbe, "vallée de la Sangu," 1902, *Cabra s.n.* (BR, holotype; K, photo).

Antholyza thonneri de Wildeman, Études Fl. Bangala Ubangi 208 (1911). *Gladiolus thonneri* (de Wildeman) Vaupel in Mildbraed, Wiss. Ergeb. Deutsch. Zentr.-Afr.-Exped. 1910–1911, 2: 10, 67 (1922). Type: Zaire, near Yakoma (Ubangi), 26 Feb. 1909, *Thonner 235* (BR, lectotype designated by Geerinck, 1972).

Antholyza sudanica A. Chevalier, Études Fl. Afr. Centr. Franç. 1: 306 (1913), Explor. Bot. Afr. Occ. Franç. 1: 634 (1920), name without description (Guinea, *Brossart 15685*; Mali?, *Chevalier 939, 977*; Central African Republic, *Chevalier 8662*).

Antholyza djalonensis A. Chevalier, Explor. Bot. Afr. Occ. Franç. 1: 634 (1920), name without description (Guinea, *Chevalier 13268, 13590*).

Antholyza fleuryi A. Chevalier, Explor. Bot. Afr. Occ. Franç. 1: 634 (1920), name without description (Guinea, *Chevalier 21024, 21496*).

Gladiolus atropurpureus sensu Geerinck, Bull. Jard. Bot. Nat. Belgique 42: 271 (1972), in part. Obermeyer, Fl. Pl. Africa 44: pl. 1760 (1977), not sensu Baker (1876).

EPONYMY

unguiculatus, "clawed," referring to the lower tepals that are narrowed and clawlike in the lower half and abruptly expanded above.

DESCRIPTION

Plants 30–60 cm high. CORM 15–25(–35) mm in diameter, often dark red on the outside and sometimes internally, tunics membranous and irregularly broken to fibrous and reticulate, red-brown. CATAPHYLLS membranous, the lower two usually red-brown, the upper green and reaching shortly above the ground. LEAVES of the flowering stem usually short and entirely sheathing, then hardly distinguishable from the cataphylls, two or three, usually 6–9 cm long, somewhat longer below and exceeding the internodes, shorter above and shorter than the internodes, thus imbricate below but rarely so above, sometimes the lower with blades up to 4 cm long (rarely, two basal leaves fairly well developed and with limbs exceeding the sheaths); foliage leaves produced on separate shoots after blooming, (one? or) two or three, linear to narrowly lanceolate, ultimately 30–45 cm long, 4–8(–12) mm wide, the margins and midrib lightly thickened and hyaline). STEM erect, rarely branched, c. 2 mm in diameter at the base of the spike.

SPIKE 10- to 18-flowered; BRACTS green, 10–15 mm long, the inner somewhat shorter to nearly as

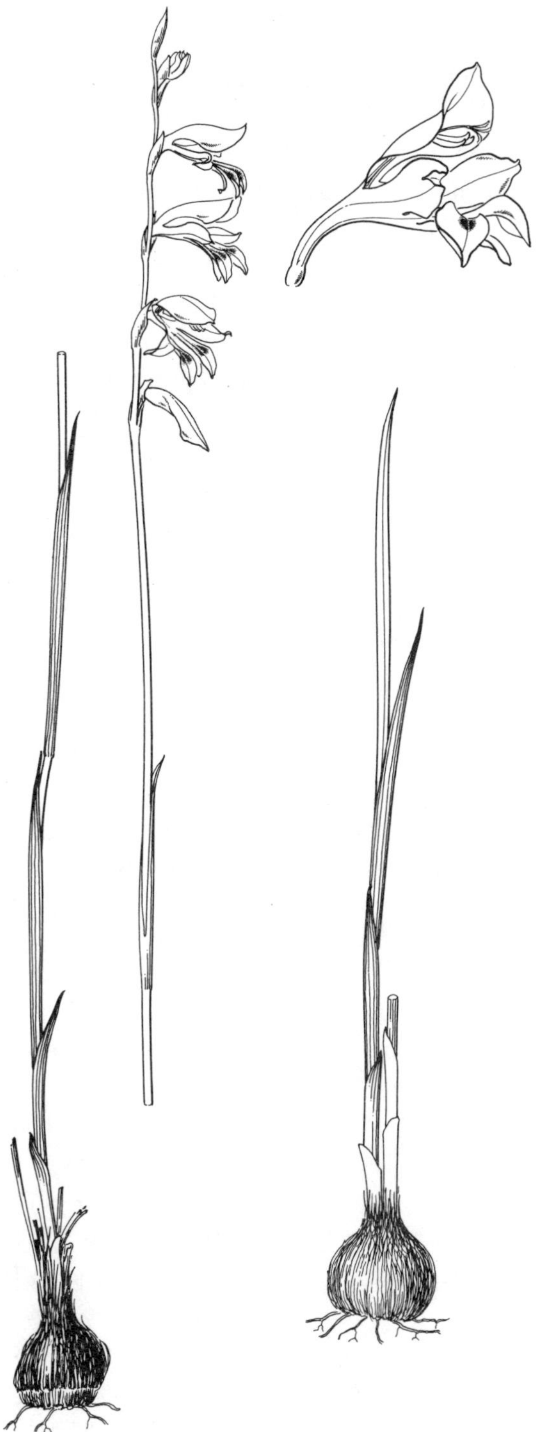

FIGURE 15. *Gladiolus unguiculatus.* Flowering spike, corm with flowering stem, and corm with separate leafy shoot, × 0.5; single flower, full size (flowering spike with corm and leafy shoot, *la Croix 941*; flower, *Schaijes 5123*).

long as the outer. FLOWERS cream to light purple, the upper tepals flushed light to deep purple, the lower three each with deep purple spear-shaped marking in the upper third, surrounded by a lighter area, sometimes with a dark spot at the base of the dorsal tepal, in profile windowed with a gap between the dorsal and upper lateral tepals; PERIANTH TUBE c. 10 mm long, curving outward between the bracts, widening near the mouth; TEPALS unequal, the dorsal arched over the stamens and style, 18–20(–24) mm long, (8–)10–12 mm wide, much narrower toward the base, the upper laterals smaller, joined to the lower three for 3–5 mm, directed forward and ultimately curving outward, the lower three tepals smallest, usually united for 1–2 mm, horizontal or directed downward distally, in profile usually exceeding the dorsal, 10–12 mm long, narrowed below into claws, the limbs abruptly expanded. FILAMENTS 10–12 mm long, exserted 4–5 mm from the tube; ANTHERS 6–8 mm long, yellow. OVARY 2–3 mm long; STYLE dividing opposite the lower half to the middle of the anthers, the branches 2–2.5 mm long, reaching to between the middle and apices of the anthers. CAPSULES ellipsoid-ovoid, 12–16 mm long; SEEDS more or less elliptic, c. 7 × 4 mm. CHROMOSOME NUMBER $2n = 24, 26, 26 + 2B$.

FLOWERING TIME. At the end of the dry season or early in the wet season, October to December in southern tropical and central Africa, mostly May to July in West Africa.

DISTRIBUTION & HABITAT

Gladiolus unguiculatus is the most widespread species of *Gladiolus* after the well-known *G. dalenii.* It occurs throughout West Africa, from Senegal and Gambia in the west to the Central African Republic and western Sudan in the east. Its range then extends southward through Gabon and Zaire to Angola and across central tropical Africa, including Zambia, Zimbabwe, and Botswana, to Mozambique and the Transvaal, South Africa. Surprisingly, it is absent from most of East Africa except for southwestern Tanzania. Judging from the records it is fairly common throughout its range and is a conspicuous element of the pre-rain or early wet-season flora. Plants generally favor seasonally wet sites with shallow soils and poor drainage and are sometimes found in flushes on rock outcrops. It also occurs in open savanna, along the upper margins of dambos, and in light deciduous woodland, where it completes its flowering before the surrounding trees have produced their seasonal flush of leaves.

DIAGNOSIS & RELATIONSHIPS

Perhaps the most distinctive feature of *Gladiolus unguiculatus* is the absence of foliage leaves on the flowering stem. Instead, the flowering stems bear three, sometimes only two, fairly short, nonoverlapping, entirely sheathing leaves. This feature combined with an erect spike of small, distinctly windowed flowers makes it relatively easy to recognize the species. Unlike other species of small-flowered tropical African *Gladiolus* that lack foliage leaves at flowering, leafy shoots are produced from separate buds on the same corm toward the end of the flowering cycle. *Gladiolus unguiculatus* is often confused with another common tropical African species, *G. atropurpureus,* that also flowers early in the season. The latter species has flowers of a similar size and coloring and also lacks foliage leaves on the flowering stem. Vegetative and floral details of the two species differ considerably, and despite their having been considered conspecific at times, they are probably not immediately related. The spikes of *G. atropurpureus* are always inclined toward the ground and the flowers have broader tepals so that they are not windowed in profile. The corms of *G. atropurpureus* also differ in being small and in having coarsely netted tunics, often thickened below into clawlike ridges. These are quite different from the rather large corms of *G. unguiculatus* with their reddish, membranous to finely fibrous tunics.

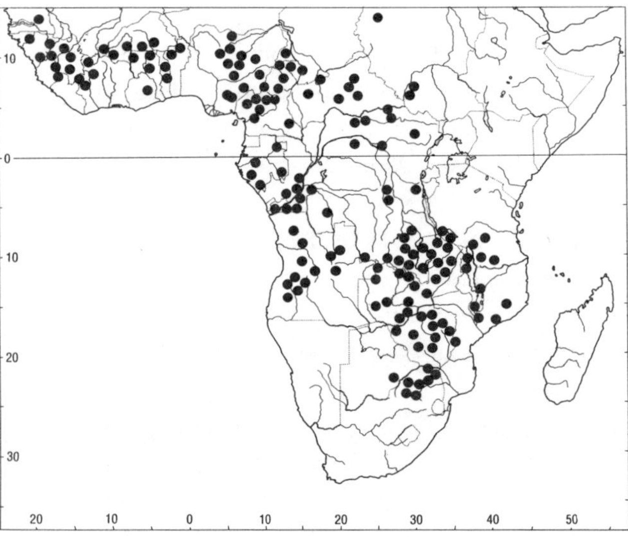

MAP 14. *Gladiolus unguiculatus.*

Plants from the southern part of its range appear to differ significantly from typical *Gladiolus unguiculatus.* Specimens from Botswana, the Transvaal, and western Zimbabwe have smaller corms, the lower leaves consistently have a short blade, and, as far as it is possible to tell from herbarium material, plants do not produce foliage leaves from separate shoots. The corm tunics are always fibrous, sometimes coarsely so, in these plants, and the corm tissue is white whereas in typical *G. unguiculatus* the corm tissue is most often reddish. The flowers do, however, conform to typical *G. unguiculatus.* A good argument can be made for separating these populations as a separate species or subspecies but they may equally be no more than a local race. There is not enough information available to make an informed decision. Plants matching this southern variant were described as *G. oatesii* (Rolfe, 1889), named in honor of the explorer and natural historian, Frank Oates, who died in Zimbabwe. The species was recognized by J. G. Baker in his *Flora of Tropical Africa* treatment of *Gladiolus* (1898), but the name *G. oatesii* is not in current use.

HISTORY

Few species of tropical African *Gladiolus* have had a history as confused as that of *G. unguiculatus*. Based on a collection from Sierra Leone made in the mid 19th century, it was first named *G. cochleatus* by J. G. Baker in 1876. Baker replaced this epithet, a later homonym for a southern African plant, with the name *G. unguiculatus* in 1877. Later collections of the species from other parts of western and central Africa were given new epithets, often being assigned to a different genus, *Antholyza*. Thus plants from interior Zaire were described as *A. thonneri*, others from Togo were named *A. labiata*, and collections from Sudan, Mali, and Guinea were assigned to *A. sudanica*. Specimens collected by Welwitsch in Angola were described as *G. brevicaulis* by Baker in 1878, despite his having originally described *G. unguiculatus*. All these additional species differed hardly at all from typical *G. unguiculatus*, and it is difficult to understand the reasons for the confusion, which was only resolved by the German botanist, G. W. J. Mildbraed (1923). Mildbraed also suggested that the Zimbabwean *G. oatesii* might be conspecific with *G. unguiculatus*, although he did not see the type.

The treatment of *Gladiolus unguiculatus* in the second edition of the *Flora of West Tropical Africa* (Hepper, 1968) follows Mildbraed's concept of the species. Geerinck (1972), revising *Gladiolus* in central Africa, however, included *G. unguiculatus* in the earlier *G. atropurpureus* (Baker, 1876). This taxonomy is not correct, for the latter is a strictly southern tropical African species that only superficially resembles *G. unguiculatus* in its reduced leaf blades and fairly small flowers. *Gladiolus atropurpureus* differs from *G. unguiculatus* in its small corms with coarsely fibrous, often clawlike tunics, absence of foliage leaves produced after flowering on separate shoots, flexed spike, and in some floral features as well, especially in not having a gap between upper and upper lateral tepals that in profile gives *G. unguiculatus* their windowed appearance.

SELECTED SPECIMENS

Senegal. Niokolo, 7 June 1958, *Adam 14201* (P).

Guinea-Bissau. Gabu, Canquelifá, 11 July 1950, *Espíritu Santo 2936* (LISC).

Guinea. Near Macenta, Apr. 1936, *Jacques-Felix 851* (P). Mt. Nimba, Nzérékoré, 30 Apr., *Adam 4885* (MO, PRE).

Mali. Sikasso District, Finkolo, 28 Apr. 1968, *Leferre 91* (K).

Sierra Leone. Loma Mountains, 1600 m, 16 Apr. 1966, *Jaeger 9853* (K, MO); Bintumani Peak, 1800 m, 2 May 1949, *Deighton 5091* (K).

Burkina Faso: Between Pô and Boni, 21 June 1984, *Le Joly 84/003* (BR).

Ivory Coast. Orodougou, between Fouénan and Sifié, 1 June 1909, *Chevalier 21807* (P).

Ghana: Halfway between Lawra and Wa, May 1952 (fruit), *Morton s.n.* (GC 7672, K); Kete Krachi, Apr. 1943, *Cox 111* (K, P).

Togo. Between Pangulam and Koumoniadé, 20 Apr. 1978, *Raadts 290* (B, K).

Benin. Atacora Mountains, Nioro–Kouandé, 400–600 m, *Chevalier 24001* (P).

Nigeria. Kano: Liruwen, Kano Hills, 1921, *Carpenter s.n.* (K). Kaduna: Kaduna to Kaciya, mile 23, 6 June 1957, *Summerhayes 97* (BR, K, P). Bauchi: Tangale–Waja, 3 km southwest of Tula, 20 May 1955, *Summerhayes 65* (K, P). Benue: Abinsi and Katsina Ala, June–July, *Dalziel 845* (C, E, K).

Sudan. Darfur: Jebel Marra, 45 km south of Zalinga, 1300 m, 17 July 1964, *Wickens 1967* (K). Bahr El Ghazal: Wau District, in 1932, *Macintosh 77* (K); "Seriba Ghattas, Lande der Djur," 10 June 1869, *Schweinfurth 1918* (E, K, PRE, Z).

Cameroon. Mbéré to Meiganga, June 1939, *Jacques-Felix 4282* (K, MO, P, PRE, Z); Bangwa, c. 15 km northwest of Baganté, 30 Apr. 1964, *de Wilde 2348* (B, EA, MO, P); near Nkolbisson, 7 km west of Yaoundé, 750 m, 15 Apr. 1965, *Leeuwenberg 5452* (C, EA, LISC, MO, P, WAG).

Chad. Fort Archambault, 10–11 June 1903, *Chevalier 8663* (P).

Central African Republic. between Yalinga and Bria, 24 Apr. 1921, *le Testu 2680* (BM, MO, P); Park Manovo-Gounda-St. Floris, 5 June 1985, *Fay 7077* (ALF, K, MO).

Gabon. c. 28 km from Fougamou to Mouila, 15 Oct. 1985, *Reitsma & Reitsma 1614* (CM, MO, NY); Tchibanga, savanna, 19 Aug. 1907, *le Testu 1117* (BM, P); Gamba, 23 Sept. 1968, *Breteler & van Raalte 5667* (MO, PRE, WAG).

Congo: Bafuru Plateau, near Outchouo village, 680 m, 18 Oct. 1991, *Thomas, Harris, & Moutsambothé 8696* (MO); Upper Logone, between Bouala and Mt. Kadé, 12 Jan. 1907, *Lenfant 1121* (P).

Zaire. Bas Zaire: Matadi District, Suadi, 10 Nov. 1932, *Dacremont 329* (BR, C, K, LD, LISC, MO, NY, SRGH, UPS). Haute Zaire: Yangambi, 2 June 1955, *Germain 8591* (BR). Kivu: Luemba, road to Itombwe, 29 Oct. 1954, *Christiaensen 724* (BR). Shaba: Dilolo, sandy marsh, Sept. 1931, *Overlaet 1253* (BR); Kapanga–Tshibalaka, Sept. 1933, *Overlaet 785* (BR, EA, K).

Angola. Uige: 12 km north of Uige, 1 Sept. 1958, *Stanton 69* (BM). Cuanza Sul: Mussende, Quibala, 1300 m, 8 Aug. 1963, *Murta 210* (COI, LISC). Malange: Capunda, Mulundo, 5 Aug. 1965, *Henriques 566* (K, SRGH). Huambo: between Cuemba and Manhango, 1100 m, *Texeira et al. 8906* (COI, LISC). Lunda Sul: Mona Quinbundo, 2 Sept. 1932, *Young 683* (BM). Moxico: Vila Luzo, Luena River, 7 Nov. 1932, *Young 1386* (BM). Benguella: Highlands between Ganda and Caconda, Nov. 1933, *Hundt 710* (BM, BR, G, Z).

Zambia. Luapula: 10 km from Kawambwa to Mansa (Fort Roseberry), 27 Nov. 1961, *Richards 15382* (K). Northern: Mbala District, 6 km from Kalolo village, 9 Dec. 1964, *Richards 19324* (B, BR, K). Northwestern: Solwezi, dambo, 12 Oct. 1953, *Fanshawe 387* (BR, K, NDO); Matonchi Farm, 17 Nov. 1937, *Milne-Redhead 3087* (BR, K, PRE). Central: 75 km east of Mankoya, 21 Nov. 1947, *Drummond & Cookson 6718* (LISC, PRE). Southern: Vlei near Kalomo, 16 Dec. 1961, *Whellan 1984* (K, LISC, SRGH).

Tanzania. Rukwa: Near the Sumbawanga–Mbala (Abercorn) road, 1800 m, 7 Dec. 1954, *Richards 3546* (K). Iringa: Njombe–Milo road, 28 Jan. 1961, *Richards 14030* (K). Rovuma: Songea district, 9.5 km west of Songea, 990 m, *Milne-Redhead & Taylor 8054* (B, BR, EA, K, SRGH).

Malawi. Northern: Katoto, 5 km west of Mzuzu, 7 Nov. 1970, *Pawek 3943* (K, MAL), 1 Dec. 1973, *Pawek 7555* (WAG). Central: Grassland just north of Kasungu, 1100 m, 27 Dec. 1986, *la Croix 4247* (K, MAL, MO). Southern: Zomba District, Mbidi Estate, 11 Jan. 1960, *Banda 365* (BM, MAL, SRGH).

Zimbabwe. Mashonaland West: Hurungwe (Urungwe) Nature Reserve, Dec. 1956, *Davies 2251* (BR, K, LISC, PRE, SRGH). Mashonaland Central: Harare (Salisbury), 26 Dec. 1926, *Eyles 4572* (K, SRGH). Matabeleland North: Shangani–Bubi, Gwampa Forest Reserve, c. 900 m, Jan. 1956, *Goldsmith 71/56* (K, LISC, SRGH). Matabeleland South: Bulawayo, Nov. 1902, *Eyles 1224* (BM, SRGH). Masvingo: 3 km east of Masvingo (Fort Victoria), Dec. 1957, *Miller 4891* (SRGH).

Mozambique. Nampula: Between Nametil and Nampula, 2 Dec. 1936, *Torre 1087* (COI, LISC). Zambézia: road to Pebane, 75 km from Namagoa, 23 Dec. 1947, *Faulkner K153* (COI, K, PRE, SRGH).

Botswana. Ngwaketse: 10 km north of Lobatse, rocky hill c. 1350 m, 23 Oct. 1977, *Hansen 3242* (K, SRGH, UPS).

14. *Gladiolus amplifolius* Goldblatt, new species

FIGURE 16, MAP 15. Type: Angola, Huambo, vicinity of Huambo (Nova Lisboa) near Caála (Belém), 28 Dec. 1970, *da Silva 3379* (COI, holotype; BM, BR, K, LISC, PRE, SRGH, isotypes).

EPONYMY

amplifolius, "broadly leaved," alluding to the broad lanceolate leaves, up to 18 mm wide, unusual in subgenus *Gladiolus.*

LATIN DIAGNOSIS

Plantae 50–70 cm longae, foliis 6 inferioribus basalibus, laminis lanceolatis 9–18 mm latis, marginibus incrassatis hyalinis costis obscuris, spicis 10–15 florum, bracteis 11–15 mm longis, floribus purpureis atropurpureis notatis, tepalis inaequalibus superiore c. 12 mm longis 9–10 mm latis inferioribus connatis c. 4 mm c. 8 mm longis, filamentis c. 12 mm longis 4 mm exsertis, antheris c. 6 mm longis.

DESCRIPTION

Plants 50–70 cm high. CORM 12–18 mm in diameter, tunics coriaceous to papery, brown or straw-colored, evidently not becoming fibrous with age. CATAPHYLLS membranous, red-brown, the inner reaching above the ground and then green. LEAVES six or seven, the lower four or five basal, reaching to about the base of the spike, lanceolate, 9–18 mm wide, the margins lightly thickened and hyaline, the midribs obscure, the upper one or two leaves reduced in size, cauline, largely sheathing. STEM erect, simple, 2.5–3 mm in diameter at the base of the spike.

SPIKE 10- to 15-flowered, nearly straight; BRACTS green, fairly soft-textured, half to nearly as long as the internodes, 11–15 mm long, the inner somewhat shorter to nearly as long as the outer. *Flowers* dark to light purple fading to cream in the throat, the lower laterals each with a deep purple spade-shaped mark in the distal third, surrounded by a lighter area; PERIANTH TUBE c. 14 mm long, curving outward between the bracts, widening near the mouth; TEPALS unequal, the dorsal arched to hooded, c. 12 mm long, 9–10 mm wide, the upper laterals smaller, directed forward and ultimately curving outward, the three lower joined with the upper laterals for c. 4 mm and to themselves for a further 1 mm, curving downward distally, usually slightly exceeding the dorsal in profile, c. 8 mm long, narrowed below into claws, the limbs abruptly expanded, c. 4 mm at the widest. FILAMENTS c. 12 mm long, exserted 4 mm from the tube; ANTHERS c. 6 mm long, reaching to

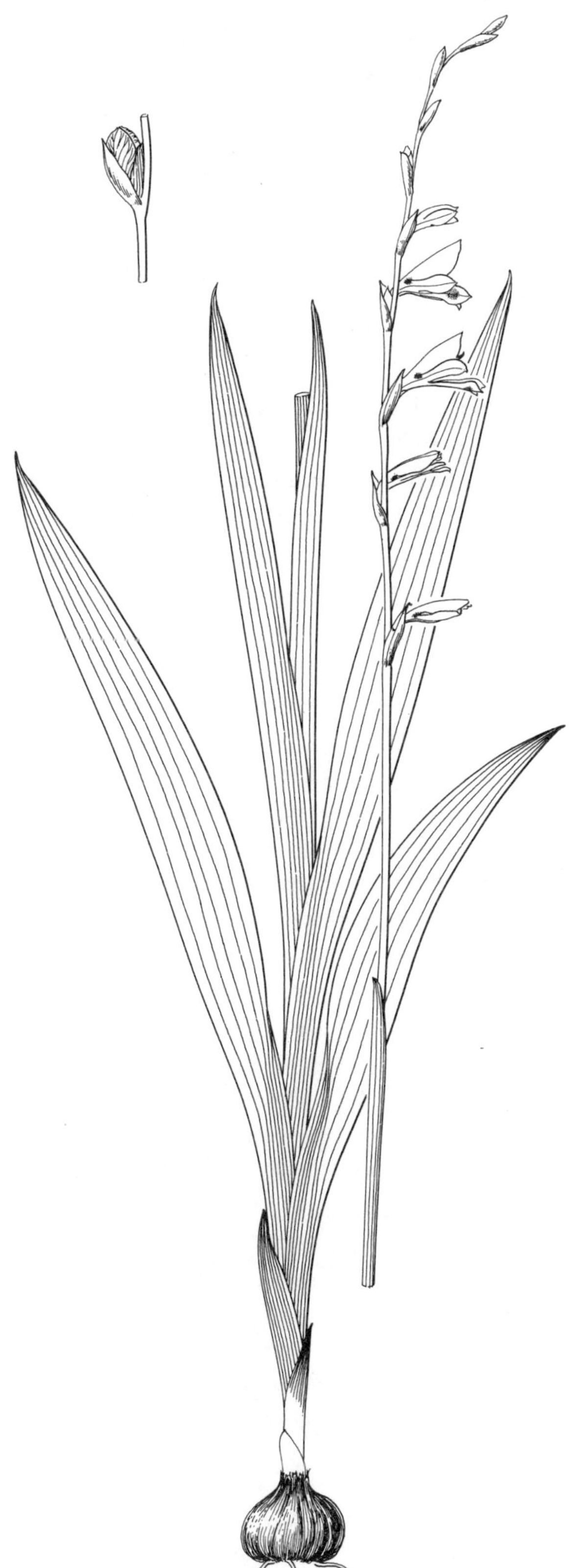

FIGURE 16. *Gladiolus amplifolius*, × 0.5 (*da Silva 3379*).

within 2 mm of the dorsal tepal apex, yellow. OVARY c. 4 mm long; STYLE dividing opposite the lower half to middle of the anthers, the branches c. 2.2 mm long, broader in the upper half, reaching almost to the anther apices when fully extended. CAPSULES ellipsoid-ovoid, c. 12 mm long (only immature capsules known); SEEDS unknown.

FLOWERING TIME. December and January.

DISTRIBUTION & HABITAT

Known from only a single gathering, *Gladiolus amplifolius* is apparently restricted to the highlands of western Angola, in the vicinity of the city of Huambo. Its habitat is unknown but the appearance of the medium-sized corms with largely unbroken rather than fibrous tunics suggests to me a preference for wet sites. The species was apparently discovered by the Portuguese botanist, Manuel da Silva, as late as 1970.

DIAGNOSIS & RELATIONSHIPS

The relationships of *Gladiolus amplifolius* are clearly with the smaller-flowered members of section *Gladiolus,* perhaps most closely allied to the southeastern African *G. crassifolius.* The flowers resemble particularly those of some forms of the widespread tropical African *G. atropurpureus* in size, general shape, and coloration. That species has a more slender stem, the leaves are usually largely sheathing and, when the blades are developed, are more or less linear, seldom exceeding 4 mm in width. The corms of *G. atropurpureus* also differ in their globose shape and coarsely to finely fibrous tunics. The similarity of the flowers of *G. amplifolius* and *G. atropurpureus* may be coincidental rather than a reflection of close relationship.

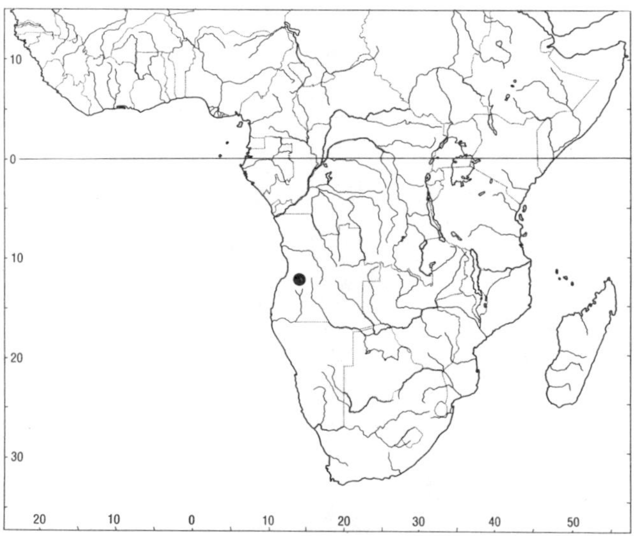

MAP 15. *Gladiolus amplifolius.*

SPECIMENS

Known only from the type, cited above.

15. *Gladiolus atropurpureus* Baker

PLATE 8, FIGURE 17, MAP 16. Baker, J. Bot., New Ser. 5: 335 (1876); Handbook Irideae 211 (1892); Fl. Trop. Africa 7: 364 (1898). Geerinck, Bull. Jard. Bot. Nat. Belgique 42: 271 (1972), in part. Goldblatt, Fl. Zambesiaca 12(4): 80 (1993). Type: Mozambique, Morrumbala (Zambezi highlands), "3000 ft," 30 Dec. 1858, *Kirk s.n.* (K, lectotype designated here); without date, *Meller s.n.* (K, syntype). The lectotypification by Geerinck (1972), for plants collected on 18 Jan. 1863, mostly in fruit, "2000 ft" and "1800 ft," is not accepted.

SYNONYMY

Gladiolus caerulescens Baker, Trans. Linn. Soc. Bot. 1: 267 (1878); Handbook Irideae 211 (1892); Fl. Trop. Africa 7: 364 (1898). Type: Angola, Huila, near Lopolo, Apr. 1860, *Welwitsch 1537* (BM, lectotype designated here; K, isolectotype, poor material).

Gladiolus luridus Welwitsch ex Baker, Trans. Linn. Soc. Bot. 1: 267 (1878); Handbook Irideae 211 (1892); Fl. Trop. Africa 7: 365 (1898). Type: Angola, Huila, maize fields near Lopolo, Dec. 1859,

Welwitsch 1533 (BM, lectotype designated here, specimens well preserved and with corms; BM, COI, K, isolectotypes; the specimen of *Welwitsch 1533* at G is *G. unguiculatus*).

Gladiolus gracilicaulis G. Lewis, J. S. African Bot. 7: 29 (1941), as a new name for *G. flexuosus* Baker, Kew Bull. 390 (1894); Fl. Trop. Africa 7: 372 (1898), an illegitimate homonym, not *G. flexuosus* Linnaeus f. (1782) (= *Tritoniopsis flexuosa* (Linnaeus f.) G. Lewis). Type: Zambia, Urungo, Fwambo (south of Lake Tanzania), Jan. 1893, *Carson 79* (K, holotype; white flowers with purple guides, leaf blades very short).

Gladiolus whytei Baker, Kew Bull. 282 (1897); Fl. Trop. Africa 7: 363 (1898). Type: Malawi, Mt. Malosa, Nov. & Dec. 1896, as "Mt. Zomba and Mt. Malosa, 4000–6000 ft," *Whyte s.n.* (K, lectotype designated by Goldblatt, 1993; B, isolectotype); Mt. Zomba, Dec. 1896, *Whyte s.n.* (G, K, syntypes); Mt. Zomba, "4000–6000 ft," Dec. 1896, *Whyte s.n.* (K, syntype, without corms = *G. unguiculatus*).

EPONYMY

atropurpureus, "dark purple," referring to the color of the flowers in the type specimens.

DESCRIPTION

Plants 30–60 cm high. CORM (10–)15–20(–30) mm in diameter, tunics fibrous, pale straw-colored, fibers mostly vertical and often thickened and clawlike below. CATAPHYLLS pale and membranous, the upper sometimes light brown above the ground. LEAVES of the flowering stem three or four, sometimes five, imbricate, entirely sheathing or with blades to 5(–10) cm long, narrowly lanceolate to linear, the margins and midribs hyaline and usually lightly thickened (long-bladed foliage leaves on separate shoots are not known and probably not produced); leaves of nonflowering plants solitary, as far as known, linear, 15–20 cm long, 4–6(–12) mm wide, usually with lightly

FIGURE 17. *Gladiolus atropurpureus.* Corm, stem, and spike, × 0.5; single flower, full size (*la Croix 519*).

thickened margins. STEM erect, unbranched or, rarely, branched, usually lightly flexed above the sheath of the uppermost leaf, 1–2 mm in diameter at the base of the spike.

SPIKE (3–)5- to 10(–15)-flowered, lightly flexuose, inclined toward the ground; BRACTS green, rather soft-textured, the margins often membranous, attenuate, 10–15(–20) mm long, the inner somewhat shorter to nearly as long as the outer. FLOWERS either white to cream or light purple, the upper tepals often flushed light to deep purple, the lower laterals each with a deep purple spade-shaped mark in the distal third, or partly to entirely dark purple, rarely dark red, sometimes with pale markings on the lower tepals; PERIANTH TUBE 10–12 mm long, curving outward between the bracts, widening near the mouth; TEPALS unequal, the dorsal arched to hooded, 15–20 mm long, 9–12 mm wide, the upper laterals smaller, directed forward below, distally curving outward, the lower three tepals united with the upper laterals for c. 5 mm, horizontal to downcurved, usually exceeding the dorsal in profile, (10–)12–15 mm long, narrowed below into claws, the limbs abruptly expanded, 6–9 mm at the widest. FILAMENTS 6–8 mm long, exserted c. 4 mm from the tube; ANTHERS 5–8 mm long, yellow. OVARY 2–3 mm long; STYLE dividing opposite the upper half of the anthers, the branches 2–2.5 mm long., usually reaching nearly to the apices of the anthers, sometimes exceeding them. CAPSULES ovoid to ellipsoid, (10–)12–18 mm long; SEEDS oval to elliptic, 5–7 × 3.5–4.5 mm. CHROMOSOME NUMBER $2n = 24 + 0\text{–}5B, 36$.

FLOWERING TIME. November and December (to early January), at the end of the dry season or early in the wet season.

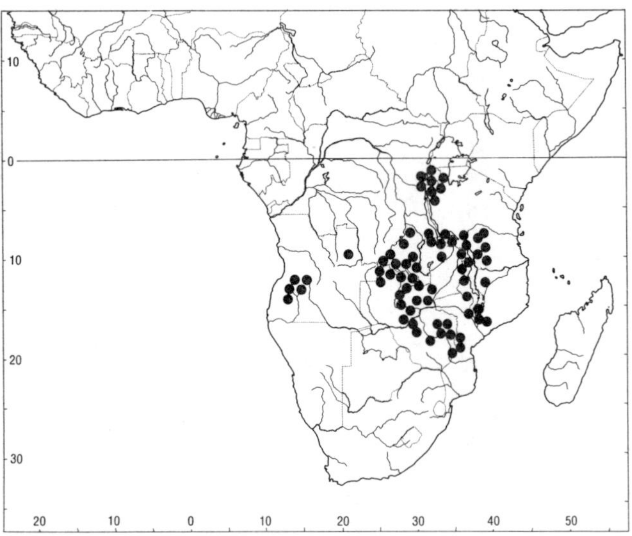

MAP 16. *Gladiolus atropurpureus.*

DISTRIBUTION & HABITAT

Most often found in light woodland or low montane grassland, sometimes in rocky soils and rock outcrops, *Gladiolus atropurpureus* occurs throughout southern tropical Africa. It extends from the western Angolan highlands through Zambia and Shaba Province of Zaire to the Southern Highlands of Tanzania, and south through Malawi to central Mozambique and Zimbabwe. Despite its fragile and delicate appearance, it is one of the more successful species of tropical African *Gladiolus* and is adapted to a variety of humid to dry situations. Mature plants flower at the beginning of the wet season, sometimes even before heavy rains have fallen, set fruit in late December and January, and then become dormant through the wet months of February to April. Whether juveniles and nonflowering individuals follow the same phenology is unknown. Probably at least each season's seed crop germinates during the wet months and seedlings become dormant only at the onset of the dry season.

DIAGNOSIS & RELATIONSHIPS

Because of its variability, particularly in the degree of development of the leaf blades, it is not always easy to recognize *Gladiolus atropurpureus.* The small plants have inclined spikes of either whitish to purple-flushed or intensely purple flowers, usually with conspicuous contrasting nectar guides on the lower tepals. The corms are unusual in their pale fibrous tunics, often thickened into clawlike ribs below. Leaf blades are normally vestigial or only 2–3 cm long, and photosynthesis presumably occurs in the imbricate leaf sheaths. Occasionally, especially in late flowering specimens, the foliage leaves may be fairly well developed, and the type of *G. luridus* seems to represent a particularly foliose population. In some specimens of this gathering the longest leaf blades range from 10 to 20 cm and may extend beyond the base of the spike. Despite the short leaves of most plants, leafy shoots are not produced after flowering. Seedlings and nonflowering individuals produce a single linear leaf that remains green long after the flowering shoots have died back.

Although flowers vary little in perianth color within populations, there seems no clear pattern to the distribution of the color forms. The type, from Morrumbala in interior Mozambique, has fairly dark purple flowers, but there are a number of collections made almost throughout its entire range that have even more intensely purple flowers, at least sometimes with white nectar guides. The form that corresponds to the types of both *G. caerulescens* and *G. whytei* has predominantly white to cream flowers, usually with dark purple nectar guides and the dorsal tepal lightly suffused with purple. A series of populations from the Mwinilunga District in western Zambia have long and quite broad leaf blades and always intensely deep purple flowers.

Originally included in *Gladiolus atropurpureus,* a series of populations with yellow or yellow-brown flowers has been segregated as *G. serapiiflorus* (Goldblatt, 1993). Apart from an obvious relationship between these two species, the affinities of *G. atropurpureus* are obscure. Both Geerinck (1972) and Lewis et al. (1972) considered *G. atropurpureus* conspecific with *G. unguiculatus,* but the similarities of the plants, although striking, are superficial. Like *G. atropurpureus, G. unguiculatus* has reduced leaf blades, but in the latter species, leafy shoots are produced on separate shoots of the same corm after flowering. The cauline leaves of *G. unguiculatus* are usually short and seldom imbricate as they consistently are in *G. atropurpureus.* The corms are also quite different; those of *G. unguiculatus* are usually large and reddish internally and the tunics are dark brown and coriaceous rather than fibrous. In addition, details of the flower structure of the two species differ so that they can usually be recognized in the absence leaves or corms. Reduction of the leaf blades in these two species is clearly convergent and does not indicate close relationship.

It is tempting to postulate that *Gladiolus atropurpureus* and the immediately related *G. serapiiflorus* are allied to southern African species such as *G. brevifolius, G. pubigerus,* and *G. woodii,* which also have reduced leaf blades on the flowering stems. These species do, however, produce leaves on separate shoots, and moreover, the leaves or leaf sheaths are always villous or at least pubescent. Perhaps equally likely, *G. atropurpureus* and *G. serapiiflorus* may be derived from tropical ancestors such as *G. amplifolius, G. crassifolius,* and the like, which have leaves and flowers produced on the same shoots. The reduction of the leaf blades would thus appear to be convergent in these tropical and southern African members of subgenus *Gladiolus.*

HISTORY

Described by J. G. Baker in 1876 and based on the first collections of *Gladiolus atropurpureus* to reach Europe, the dark purple-flowered form of the species is common in Mozambique and eastern Zimbabwe. These plants were collected at Morrumbala by the explorer John Kirk in 1858 at the beginning of David Livingstone's Zambezi Expedition. Then in 1878, Baker named the cream and light purple-flowered form, collected by Friedrich

Welwitsch in Angola, *G. caerulescens.* Baker also named a collection with longer leaves than normal for the species, *G. luridus,* also collected by Welwitsch in Angola. Additional cream and light purple specimens from northern Zambia and southern Malawi were described, respectively, as *G. flexuosus* in 1894 and *G. whytei* in 1897, both by Baker. Neither differs significantly from *G. caerulescens,* nor from *G. atropurpureus,* except in flower color. Baker's *G. flexuosus* was a later homonym for a South African plant, now *Tritoniopsis flexuosa,* and it was renamed *G. gracilicaulis* in 1941 by the South African botanist, G. J. Lewis, who at the time was unaware that valid earlier names were available for the tropical African species.

Gladiolus atropurpureus was united with the widespread tropical African *G. unguiculatus* by Geerinck (1972), but as explained above the two are separate species, almost certainly not immediately related to one another, and Geerinck's treatment is not followed here.

SELECTED SPECIMENS

Zaire. Kivu: Kalehi, Mt. Kahusi, 27 Nov. 1955, *Deube 55/21* (BR). Shaba: west of Kolwezi, 15 Dec. 1985, *Schaijes 2736* (BR, K); Luiswishi, 18 Nov. 1982, *Malaisse 12473* (BR, WAG); Fungurume, slope of copper hill, 22 Dec. 1959, *Duvigneaud 4727* (BRLU).

Burundi. Ruyigi: Murema, Kigamba, 1580 m, 29 Oct. 1965, *Lewalle 1173* (BR, K, MO); Bubanza, Cibitoke–Mabaye road, 1 Nov. 1969, *Lewalle 3969* (BR, MO).

Rwanda: Impara, 24 Nov. 1954, *Michel 4617* (BR).

Tanzania. Kagera: Ngara, Murganza–Mkisenye escarpment, 1 Dec. 1956, *Gane 89* (K). Kigoma: 15 km from Mangovu to Kasulu, 22 Nov. 1962, *Verdcourt 3424* (BR, K, PRE); Nyabibuye, 25 Nov. 1960, *Tanner 5660* (K, WAG). Rukwa: Ufipa, Chapota, 3 Dec. 1949, *Bullock 1992* (B, BR, K, S). Iringa: Lupembe District, upper Ruhudje, Jan. 1931, *Schlieben 106* (G). Mbeya: Station Kyimbila, 1900 m, 13 Dec. 1912, *Stolz 1019* (B, BM, C, G, K, MO, P, PRE, S, UPS). Rovuma: 9.5 km west of Songea, 1 Jan. 1962, *Milne-Redhead & Taylor 8022* (B, BR, K).

Angola. Huambo: Outskirts of Huambo (Nova Lisboa), Sacaála, 1700 m, 18 Nov. 1970, *da Silva 3348* (BM, BR, BRLU, COI, K, LISC, PRE, SRGH). Lunda Sul: Saurimo, Dala road, 1 Nov. 1932, *Young 1299* (BM).

Zambia. Northwestern: Mwinilunga, woodland and cleared ground, 23 Nov. 1937, *Milne-Redhead 3361* (BR, K, PRE). Copperbelt: Kitwe, 27 Dec. 1955, *Fanshawe 2680* (K, LISC, NDO, S). Eastern: 22 km south of Chitipa, Nyanyuni village, 1050 m, 27 Dec. 1972, *Richards 6161* (K). Northern: Mbala (Abercorn) Ndundu, 1600 m, 1 Dec. 1960, *Richards 13638* (K). Central: Lusaka, Chilanga, Jan. 1958, *Benson 226b* (BR, K). Southern: Choma, dambo at Siamambo, 1 Dec. 1962, *Lawton 1019* (K, NDO).

Malawi. Northern: Misuku Hills, Chisasu–Itera road, 27 Dec. 1977, *Pawek 13391* (K, MAL, MO); Livingstonia Escarpment, 1000 m, 31 Dec. 1973, *Pawek 7685* (K, MAL, MO). Central: Chongoni, Ciwau Hill, 19 Jan. 1959, *Robson 1259* (K, SRGH). Southern: Zomba Plateau, 27 Nov. 1976, *Goldblatt 4512* (MO, WAG).

Zimbabwe. Manicaland: Mtare (Umtali), 26 Nov. 1957, *Chase 6776* (COI, K, LISC, PRE, SRGH); Rusape, 5 Feb. 1949, *Munch 148* (K, SRGH). Mashonaland Central: Harare (Salisbury), 20 Dec. 1926, *Eyles 4581* (K, SRGH). Midlands: Gokwe, 20 Dec. 1963, *Bingham 972* (K, LMU, PRE, SRGH, WU). Mashonaland West: Lomagundi, Mangula, 3 Dec. 1961, *Jacobsen 1536* (PRE, SRGH).

Mozambique. Niassa: Massangulo, Jan. 1933, *Gomes e Sousa 1233* (COI). Zambézia: Morrumbala Mountain, 9 Dec. 1971, *Pope & Müller 569* (MO, PRE, SRGH). Manica & Sofala: Mavita, Serra Mocuta, 13 Dec. 1965, *Pereira & Marques 1057* (LMU, WAG); Chimoio, between Vila Pery and Garuzo, 14 Dec. 1943, *Torre 6299* (LISC).

16. *Gladiolus serapiiflorus* Goldblatt

PLATE 9, FIGURE 18, MAP 17. Goldblatt, Fl. Zambesiaca 12(4): 82 (1993). Type: Zambia, Southern Province, Machili, woodland, 17 Dec. 1960, *Fanshawe 5981* (K, holotype; NDO, SRGH, isotypes).

EPONYMY

serapiiflorus, named for the resemblance of the flowers to those of some species of the Eurasian orchid genus *Serapias* that have a similarly narrow, partly closed perianth.

DESCRIPTION

Plants 30–60 cm high. CORM 15–20 mm in diameter, tunics fibrous, pale straw-colored, fibers netted or mostly vertical and then often thickened and clawlike below. CATAPHYLLS pale and membranous, usually flushed purplish above the ground. LEAVES of the flowering stem four to six, imbricate, almost entirely sheathing or the lower laminate, blades linear, sometimes lanceolate, margins and midribs hyaline and lightly thickened, usually 3–8 cm long but occasionally to 30 cm and then reaching the top of the spike, upper leaves always predominantly sheathing (foliage leaves borne on separate shoots after flowering are not known and are probably not produced); leaves of nonflowering plants not known. STEM erect, unbranched, lightly flexed at the apex of sheath of the uppermost leaf.

SPIKE 4- to 8(–12)-flowered, inclined; BRACTS green, becoming dry above, 10–14 mm long, the inner somewhat shorter to nearly as long as the outer. FLOWERS dull to bright yellow, the dorsal and the limbs of the two lower lateral tepals sometimes brown or the lower laterals deeper yellow; PERIANTH TUBE 7–8 mm long, curving outward between the bracts, widening near the mouth; TEPALS unequal, the dorsal arched and hooded, c. 13

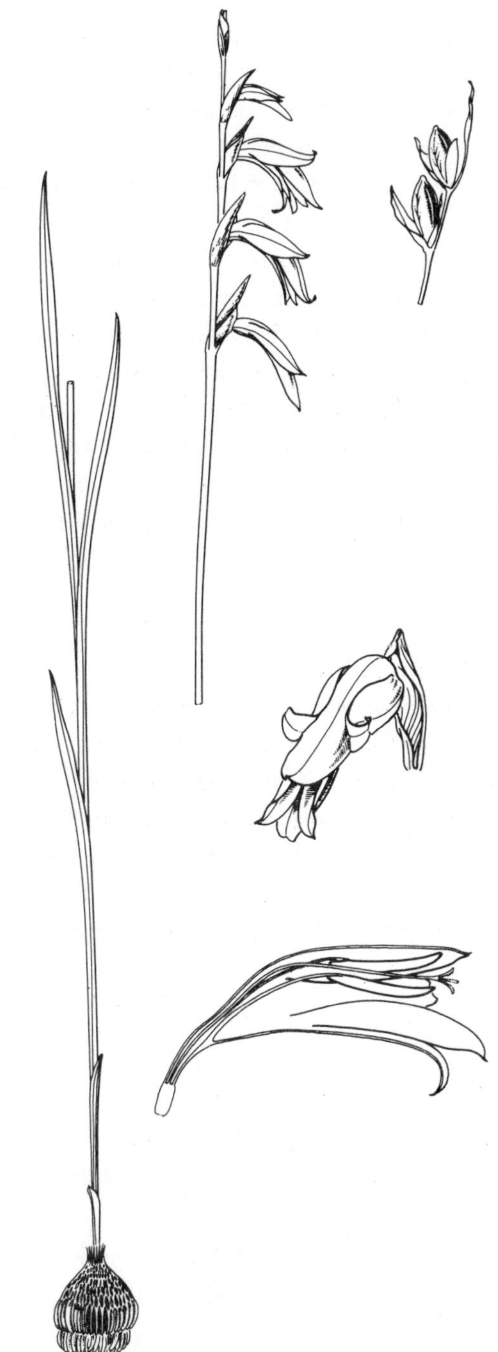

FIGURE 18. *Gladiolus serapiiflorus.* Corm, stem, flowering spike, and capsules, × 0.5; single flower and vertical section of flower, full size (stem and flowering spike, *Bamps & Malaisse 8086*; single flower, *Schaijes 2709, 2695*).

mm long, 7–10 mm wide, upper laterals smaller, c. 12 mm long, directed forward and curving upward apically, the three lower united for 3 mm, downcurved, in profile usually exceeding the dorsal, c. 10 mm long, narrowed below into claws, the limbs channeled, c. 3 mm at the widest. FILAMENTS c. 8 mm long, exserted 3–4 mm from the tube; ANTHERS c. 6 mm long, yellow. OVARY 2–3 mm long; STYLE dividing toward the anther apices, the branches c. 2.5 mm long. CAPSULES ellipsoid, 12–15 mm long; SEEDS elliptic, c. 7 × 3.3 mm, the wing well developed. CHROMOSOME NUMBER $2n = 22$.

FLOWERING TIME. November and December, at the end of the dry season or early in the wet season.

DISTRIBUTION & HABITAT

Gladiolus serapiiflorus is restricted to central and southern Shaba Province in Zaire and adjacent northwestern and central Zambia (Figure 18). It extends into Zambia into the Southern Province as far south as Machili, and it seems to be relatively common in the Kafue National Park. It is most often found in light woodland or open, low grassland, often in sandy soils that become waterlogged in the wet season.

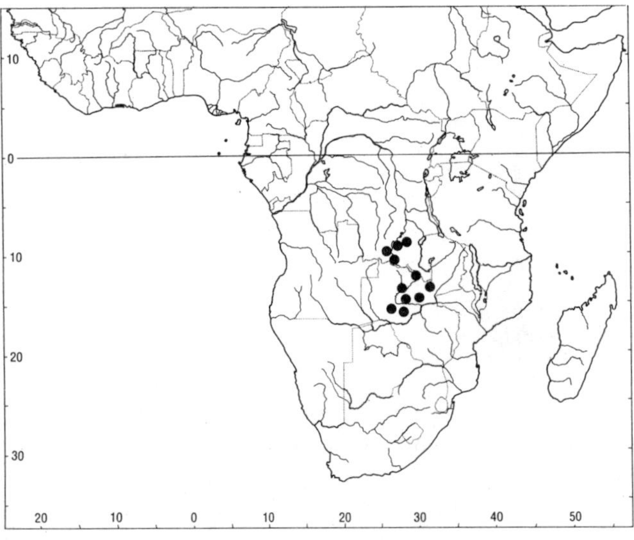

MAP 17. *Gladiolus serapiiflorus.*

DIAGNOSIS & RELATIONSHIPS

Although in general appearance *Gladiolus serapiiflorus* resembles the common and widespread southern tropical African *G. atropurpureus,* both in the coarsely fibrous corm tunics and weakly developed leaf blades on the flowering stems, it has strikingly distinctive flowers. These are yellow or yellow-brown and the long narrow tepals are extended forward, forming a tubelike structure with a narrow opening at the tepal apices (Figure 18). Both the shape and color of the flowers are so different from *G. atropurpureus* that there can be no doubt that it is a separate species. The flowers of *G. atropurpureus* are either purple or white with purple markings and the tepals are held fairly widely apart, making the flowers accessible to most moderate-sized bees, the presumed pollinators. The unusual flower of *G. serapiiflorus* suggests that it has an unusual pollination strategy.

HISTORY

Although first collected in Zambia in 1906 by C. E. F. Allen, Forester to the Rhodesia Railways, and recorded occasionally since then, with increasing frequency after World War II, *Gladiolus serapiiflorus* remained undescribed until 1993. Possibly the condition of flowers when pressed made it difficult to be sure that the species was distinct from *G. atropurpureus,* to which it is undoubtedly related, and thus botanists were reluctant to deal with it. Photographs provided by the Belgian mining engineer Michel Schaijes (Plate 9) have made it possible to describe this charming plant and to document the complex structure of the flowers.

SELECTED SPECIMENS

Zaire. Shaba: 41 km northwest of Kolwezi, 1100 m, 1 Dec. 1985, *Schaijes 2709* (BR); c. 33 km from Kolwezi to Musokatanda, 1408 m, 16 Nov. 1981, *Schaijes 1130* (BR); Kolwezi Hill, 12 Dec. 1959, *Duvigneaud 4523B* (BR); Manika Plateau, c. 12 km north of Lulumba, c. 1500 m, 23 Nov. 1969, *Lisowski, Malaisse, & Symoens 7839* (BR); Biano Plateau, Jan. 1912, *Homblé 889* (BR); Kolwezi–Musokatanda, km 36, Dilungu, 1490 m, 18 Jan. 1986, *Bamps & Malaisse 8086* (MO); 10.5 km southwest of Kolwezi, 11 Dec. 1991, *Schaijes 5136–5124A* (MO).

Zambia. Copperbelt: Luanshya, open ground, 19 Dec. 1954, *Fanshawe 1734* (NDO, K). Central: 60 km east of Lusaka, Chipata (Fort Jameson) road, 2 Jan. 1958, *Benson 225* (K); Kabwe District, Mpunde, 53 km northwest of Kabwe, on laterite in woodland, 7 Dec. 1972, *Kornas 2761* (K); Mkushi District, Fiwila, 4 Jan. 1958, *Robinson 2610* (K). Southern: Batoka Plateau near Kalomo, 5 Jan. 1906, *Allen 203* (K); 6 km from Namwala to Ngoma, woodland, 8 Dec. 1962, *van Rensburg 1033* (K, SRGH); Namwala, Ngoma, Kafue National Park, 8 Dec. 1961, *Mitchell 11/81* (SRGH).

17. *Gladiolus muenzneri* Vaupel

PLATE 10, FIGURE 19, MAP 18. Vaupel, Bot. Jahrb. Syst. 48: 542 (1913). Goldblatt, Fl. Zambesiaca 12(4): 83 (1993). Type: Tanzania, vicinity of Urungu on Lake Tanganyika, in moist places, 1200 m, 24 Dec. 1908, *Münzner sub Exped. Fromm 95* (B, holotype; K, photo).

EPONYMY

muenzneri, named to honor Max Münzner, botanist on the Fromm Expedition of 1908 to western Tanzania.

DESCRIPTION

Plants 30–60 cm high. CORM globose, (10–)20–35 mm in diameter, tunics more or less coriaceous or fragmenting into coarse to medium-textured fibers, sometimes entirely fibrous, brownish. CATA-

FIGURE 19. *Gladiolus muenzneri.* Habit, stem, and flowering spike, × 1.3; single flower and vertical section of flower, × 1.3 (habit, *Pawek 10432*; single flowers, *la Croix 746*).

PHYLLS pale and membranous, often green or purple to brown above the ground. LEAVES two or three, often imbricate, entirely sheathing or with short blades 1–8(–12) cm long, blades lanceolate, usually narrowly so, margins and midribs hyaline and usually lightly thickened (foliage leaves on borne on separate shoots after flowering are not known and probably not produced); leaves of nonflowering plants not known. STEM erect, unbranched, usually straight, c. 3 mm in diameter at the base of the spike.

SPIKE 2- to 8-flowered; BRACTS green or sometimes membranous or dry above, 18–25 mm long, the inner somewhat shorter than the outer and apically bifid. FLOWERS white to cream or yellow, sometimes flushed pinkish or light mauve, especially on the upper three tepals, or rarely uniformly mauve, pink, or light orange, lacking contrasting markings on the lower tepals; PERIANTH TUBE 7–8 mm long, curving outward between the bracts, widening near the mouth; TEPALS nearly equal, directed forward, nearly parallel to the ground, weakly curving outward distally, 20–25 mm long, hardly or not at all narrowed below into claws, (5.5–)7–8 mm at the widest. FILAMENTS 4–5 mm long, usually included in the tube but sometimes exserted for up to 1 mm; ANTHERS 7–9 mm long, sometimes the bases included for up to 2 mm, light purple, minutely apiculate. OVARY 2–3 mm long; STYLE dividing toward the anther apices, the branches 3.5–4 mm long. CAPSULES globose, 10–12 mm long; SEEDS more or less oval, c. 5 × 4 mm.

FLOWERING TIME. December to mid January.

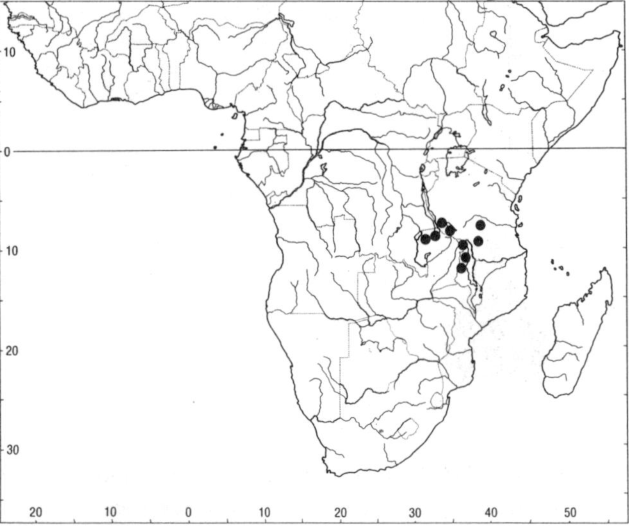

MAP 18. *Gladiolus muenzneri.*

DISTRIBUTION & HABITAT

Restricted to southwestern Tanzania, northern Zambia, adjacent southeastern Shaba Province of Zaire, and the northern half of Malawi, *Gladiolus muenzneri* grows in relatively dry woodland, occasionally in grassland, usually above 1200 m. Although first collected in 1908 and described in 1913, it remained poorly known until after World War II, when Mary Richards and A. A. Bullock established its presence in northern Zambia and western Tanzania. Only in the 1970s was its range in Malawi firmly documented. The only collection from Zaire, made in 1969, is from the Marungu Plateau, in the extreme southeast of Shaba Province. The species is evidently fairly rare and rather inconspicuous, which accounts for its being so poorly known.

DIAGNOSIS & RELATIONSHIPS

Gladiolus muenzneri is readily recognized by its pale, usually cream flower with subequal elongate tepals and short filaments, almost to entirely included in the wide upper half of the comparatively short perianth tube. As in the presumably related *G. atropurpureus,* the leaf blades are reduced to only a few centimeters long or are vestigial, so that the leaves consist only of a sheath. Like *G. atropurpureus,* the plants do not produce foliage leaves from separate shoots as they do in *G. unguiculatus,* the flowering stems of which also have partly to entirely sheathing leaves like those of *G. muenzneri.* Plants from Ilembula, Njombe District, Tanzania, *Brummitt & Polhill 13657,* are unusual in having leaves with fairly long blades, but the bracts and flowers accord exactly with *G. muenzneri.* That collection includes, coincidentally, the only known fruiting specimens.

Perianth color is variable in *Gladiolus muenzneri.* Collections from Tanzania and most of those from Malawi have cream flowers, but in a few the tepals are flushed with purple or apricot. Plants from Zambia show more pink color, and in one Zambian collection, *Bullock 1369,* the perianth was described as mauve and the flowers have dried a pink color. The only known collection from Zaire appears to have had pink flowers, somewhat darker in the throat, and tunics of unusually fine fibers. Its does, however, have the narrowly lanceolate subequal tepals and short stamens with filaments included in the tube that are typical of the species. Despite their vegetative similarities, the flowers of *G. muenzneri* are so different from those of *G. atropurpureus* and *G. unguiculatus* that they should never be confused. The two latter species have a strongly zygomorphic perianth with a larger dorsal tepal arched over the stamens, the filaments of which are well exserted from the tube, and the lower tepals are joined for a short distance, strongly unguiculate and, unless uniformly dark purple, with large, dark purple nectar guides on the two lower lateral tepals.

SELECTED SPECIMENS

Zaire. Shaba: Marungu Plateau, edges of Mufufu Pool, 2 km east of Luonde, 27 Nov. 1969, *Lisowski, Malaisse, & Symoens 8240* (BRLU).

Tanzania. Iringa: Njombe District, Ilembula, dambo, 1425 m, 9 Jan. 1975, *Brummitt & Polhill 13657* (EA, K); Mufindi District, 19 km northwest of Mafinga on Madibira road, grassy woodland, 1660 m, 23 Dec. 1988, *Gereau & Lovett 2681* (MO). Rukwa: Sumbawanga District, Ufipa, Malonje, 1800 m, marshy grassland, 3 Jan. 1962, *Richards 15889* (K); Sumbawanga, 2000 m, 29 Nov. 1954, *Richards 3414* (K); Nsangu Forest Reserve, 2 Jan. 1962, *Robinson 4867* (K, SRGH).

Zambia. Northern: Kasama–Mporokoso road, 25 Oct. 1949, *Bullock 1369* (K); Luapula: Kawambwa District, Luapula Leper Settlement, by Mbereshi River, 2 Dec. 1961, *Richards 15486* (K, SRGH); Mweru, 18 Nov. 1955, *Jones 25* (BM).

Malawi. Northern: Mzimba District, Champira, 18 km northwest of Katete over Lwanjati Pass, woodland, c. 1500 m, 25 Dec. 1975, *Pawek 10515* (K, MAL, MO); Mzimba District, South Vipya near Luwawa, 1690 m, 10 Dec. 1985, *la Croix 746* (K, MAL, MO); Vipya Plateau, 5 km south of Chikangawa, 1700 m, 1 Dec. 1978, *Phillips 4305* (K, MO, WAG).

18. *Gladiolus canaliculatus* Goldblatt, new species

FIGURE 20, MAP 19. Type: Tanzania, Singida, Manyoni District, Kazikazi, drainage valleys in hardpan short grassland, 17 Dec. 1933, *Burtt 4967* (K, holotype; BR, EA, K, MO, S, isotypes).

EPONYMY

canaliculatus, "channeled," alluding to the adaxially grooved leaf blades.

LATIN DIAGNOSIS

Plantae 15–25 cm longae, foliis 2, basalibus linearibus marginibus costistisque non incrassatis hyalinisque, spicis 8–12(–15) florum, bracteis (12–) 16–22 mm longis, floribus caeruleis, tubo perianthii 9–10 mm longis, tepalis subaequalibus lanceolato-attenuatis c. 15 mm longis, filamentis c. 5 mm longis in tubo inclusis, antheris c. 6 mm longis.

DESCRIPTION

Plants 15–20 cm high. CORM unknown. CATAPHYLLS membranous, the upper green and leaf-like above the ground. LEAVES two, both basal, reaching to about the base of the spike, linear, c. 2 mm wide, without obvious midribs, the sheath extending almost the entire length, the blade thus channeled, the abaxial margin lightly thickened. STEMS unbranched, 1.5–2 mm in diameter at the base of the spike.

SPIKE 8- to 12-flowered, fairly congested, internodes 6–10 mm long; BRACTS green, apices of the outer usually three-lobed, imbricate, about two internodes long or longer, (12–)16–22 mm long, the inner about two-thirds as long as the outer. FLOWERS blue, possibly the lower tepals with contrasting markings; PERIANTH TUBE narrowly funnel-shaped, curving outward between the bracts, 9–10 mm long; TEPALS subequal, lanceolate-attenuate, c. 15 mm long, c. 6 mm wide in the midline, all similarly oriented, directed forward and curving outward distally. FILAMENTS c. 5 mm long, included in the upper part of the tube; ANTHERS c. 6 mm long, symmetrically disposed?, blocking the mouth of the tube, the basal 1 mm included, evidently dark-colored (purple?). OVARY oblong, 2–3 mm long; STYLE orientation uncertain, dividing at mid anther level, the branches c. 2 mm long, extending between the anthers. CAPSULES and SEEDS unknown.

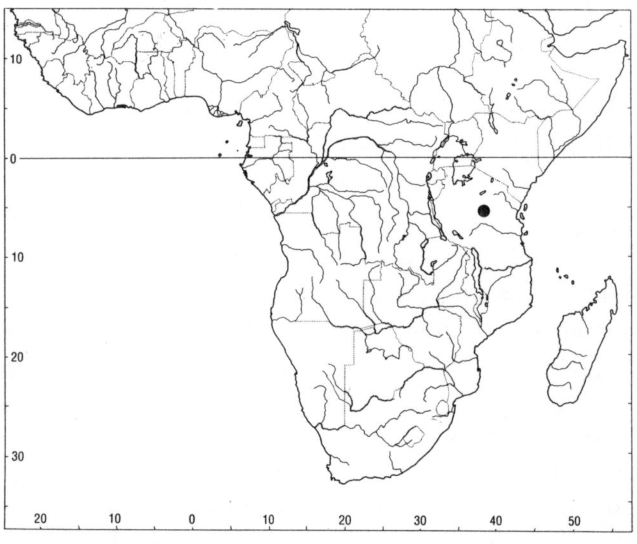

MAP 19. *Gladiolus canaliculatus.*

FIGURE 20. *Gladiolus canaliculatus.* Habit, full size; single flower, × 1.3 (*Burtt 4967*).

FLOWERING TIME. Recorded only in mid December, but probably flowering from late November in some years to early January in others.

DISTRIBUTION & HABITAT

The distinctive *Gladiolus canaliculatus* is known from a single gathering made in 1933 in the Manyoni District of central Tanzania, an area not well known botanically. The plants were recorded by the British botanist, B. D. Burtt, who discovered the species growing in hard ground along drainage lines in short grassland. The flowering time, mid December, suggests that the species flowers near the beginning of the rainy season, a common strategy among the smaller-flowered species of *Gladiolus*.

DIAGNOSIS & RELATIONSHIPS

Although known from just one gathering, the collection is ample and consists of small plants, 15–20 cm tall, that consistently have just two foliage leaves and a short spike of pale bluish flowers. Their general appearance is striking, and this combined with an unusual floral structure leaves no doubt that it is novel. The flowers are nearly actinomorphic and have more or less equal lanceolate-attenuate tepals and short stamens, c. 11 mm long, with the filaments entirely included in the perianth tube and the anthers partly included, unlike any other *Gladiolus*. The leaves, too, are unusual in being channeled on the adaxial surface almost to the apex. Evidently a member of subgenus *Gladiolus*, its immediate relationships are puzzling. The flowers resemble most closely those of *G. muenzneri*, a taller plant with reduced leaf blades that usually has white to cream or, occasionally, pink flowers. The long leaves of *G. canaliculatus* are, however, quite unlike the almost entirely sheathing leaves of *G. muenzneri*. Unfortunately, the corms of *G. canaliculatus* are unknown and thus cannot be used to help determine its relationships.

SPECIMENS

Known only from the type, cited above.

19. *Gladiolus curtilimbus* Duvigneaud & van Bockstal ex Còrdova

MAP 20. Còrdova, Bull. Jard. Bot. Nat. Belgique 60: 325 (1990). Type: Zaire, Shaba, Mulungwishi, woodland on schist, Nov. 1959, *Duvigneaud 4105G* (BRLU, holotype; BR, isotype).

EPONYMY

curtilimbus, "short-limbed," in reference to the comparatively short tepals of the small flowers.

DESCRIPTION

Plants 25–45 cm high. CORM 12–20 mm in diameter, tunics membranous, light brown, sometimes becoming matted and fibrous externally with age. CATAPHYLLS membranous, the uppermost extending above the ground, lightly pubescent, poorly differentiated from the leaf, pale or light brown above. LEAF of the flowering stem one or rarely two, inserted nearly at ground level, 4–6 (–16) cm long, reaching as far as the middle of the stem, entirely sheathing or with a blade 1–2(–3) cm long, this with lightly thickened hyaline midrib and margins (foliage leaves borne on separate shoots after flowering are not known and are probably not produced); leaves of nonflowering plants unknown. STEM erect, simple or with one branch, 1.5–2 mm in diameter at the base of the spike.

SPIKE (3–)5- to 9-flowered, flexuose, evidently erect; BRACTS green, firm-textured, becoming dry above in late bloom, the margins often membranous, attenuate, 12–16 mm long, the inner somewhat shorter to as long as the outer. FLOWERS apparently cream or pale pink to mauve at least on the upper tepals; PERIANTH TUBE 10–12 mm long, curving outward between the bracts, widening near the mouth; TEPALS unequal, the dorsal arched to hooded, c. 15 mm long, c. 8 mm wide, the upper laterals smaller, directed forward and ultimately curving outward, the three lower joined for c. 1 mm, more or less straight, horizontal or directed toward the ground, about as long as the dorsal in profile, c. 12 mm long, gradually narrowed below, c. 7 mm at the widest. FILAMENTS c. 10 mm long, exserted c. 4 mm from the tube; ANTHERS 5–6 mm long, yellow. OVARY 2–3 mm long; STYLE dividing opposite the upper third of the anthers, the branches c. 3.5 mm long, the apices extending beyond the anthers. CAPSULES broadly obovoid, c. 9 mm long; SEEDS oblong to elliptic, 4.5–5.5 × c. 3 mm.

FLOWERING TIME. October to December.

DISTRIBUTION & HABITAT

Gladiolus curtilimbus is restricted to southern and central Shaba, the known range extending from Lubumbashi west to Katentania and the Biano Plateau. According to collection notes, it usually occurs in shady situations in miombo (*Brachystegia boehmii*) woodland. Although first collected in 1911 by Rev. F. A. Rogers and given a provisional name in 1959 by Duvigneaud and van Bockstal, it has remained poorly known and was only formally named by Còrdova in 1990. All of the collections have poorly preserved flowers, and both floral and seeds details need to be more thoroughly investigated.

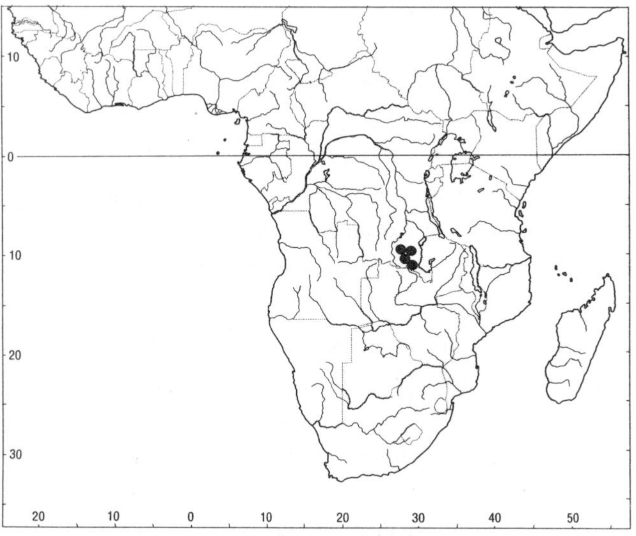

MAP 20. *Gladiolus curtilimbus.*

DIAGNOSIS & RELATIONSHIPS

The most striking feature of *Gladiolus curtilimbus* is its almost leafless habit. Apart from two or three cataphylls, the uppermost of which is lightly pubescent, it normally has just one leaf that hardly differs from the upper cataphyll except in its position and in being glabrous. The leaf is inserted close to ground level and is largely or entirely sheathing, sometimes with a short free blade 1–2 cm long. The spike of three to nine flowers and the short bracts seem typical of section *Gladiolus,* as are the pale, cream to mauve flowers. The immediate relationships of *G. curtilimbus* are uncertain, but it is possibly closely allied to the widespread *G. atropurpureus.* More and better preserved specimens are needed so that the relationships of this Shaban endemic can be properly assessed.

ADDITIONAL SPECIMENS

Zaire. Shaba: Kambove–Kamoia, gray schist hill, 13 Nov. 1959, *Duvigneaud 4102GC* (BRLU); Kambove, *Brachystegia* woodland, 13 Nov. 1959, *Duvigneaud 4104* (BRLU); Lubumbashi (Elisabethville), road to Sakania, *Brachystegia* woodland, 16 Nov. 1959, *Duvigneaud 4171* (BRLU); Biano Plateau, 1066 m, Oct. 1939, *Quarré 6082* (BR); vicinity of Katentania, Biano Plateau, Nov. 1912, *Homblé 781* (BR); Lubumbashi, 1500 m, 22 Oct. 1911, *Rogers 10125* (K).

20. *Gladiolus curtifolius* Marais

MAP 21. Marais, Kew Bull. 28: 313 (1973), as new name for *Acidanthera goetzei* Harms, Bot. Jahrb. Syst. 30: 278 (1901), not *Gladiolus goetzei* Harms (= *G. dalenii* Geel). Goldblatt, Fl. Zambesiaca 12(4): 102 (1993). Type: Tanzania, Mbeya Mountain (Unyika, Beyeberg), grassy slopes at 2000 m, Nov. 1899, *Goetze 1448* (B, holotype; K, photo).

EPONYMY

curtifolius, "short-leaved," alluding to the very short or virtually vestigial leaf blades.

DESCRIPTION

Plants 25–35 cm high. CORM 12–18 mm in diameter, tunics brittle-papery, usually broken into vertical fibers below. CATAPHYLLS membranous. LEAVES of the flowering stem two or three, almost to entirely sheathing, 4–8 cm long, sometimes with short nonsheathing blades 1–3 cm long, inserted on the lower half of the stem, sparsely pubescent on the sheaths and blades; leaves of nonflowering individuals evidently solitary, linear, c. 2 mm wide, sparsely pubescent. STEM erect, unbranched, the lower half sheathed by the leaves, the upper half occasionally with a short sheathing leaf bract 2–3 cm below the first flower, this with an open sheath, c. 1.5 mm in diameter at the base of the spike.

SPIKE (1–)2- to 4-flowered; BRACTS 15–20 mm long, green or membranous, usually dry and brownish above at anthesis, the inner about as long as the outer, or slightly longer, bifid. FLOWERS cream to pale lemon-yellow, the tepals symmetrically disposed, fragrant; PERIANTH TUBE 50–75 mm long, cylindric, very slightly expanded in the upper third; TEPALS subequal, lanceolate, (15–)18–22 mm long, the inner three slightly shorter than the outer. FILAMENTS unilateral, included or exserted 1–2 mm from the tube; ANTHERS c. 7 mm long. OVARY broadly ovoid, 3.5–4 mm long; STYLE dividing near the anther apices, the branches c. 4 mm long. CAPSULES ovoid-ellipsoid, 13–15 mm long; SEEDS broadly winged, 6–7 × c. 4 mm.

FLOWERING TIME. November to early January.

DISTRIBUTION & HABITAT

Gladiolus curtifolius is restricted to the Southern Highlands of Tanzania and adjacent northern Malawi. At present the distribution is incompletely known. There are only a few records scattered across Tanzania, where the species extends from Mufindi and Njombe westward to Mbeya and the Tukuyu highlands. In Malawi the two recordings are from the Mafinga Mountains and the Misuku hills, close to the northern end of Lake Malawi. *Gladiolus curtifolius* grows in grassland or open woodland or bush in highlands, generally above 1800 m.

DIAGNOSIS & RELATIONSHIPS

The cream-colored flowers with an elongate perianth tube and lightly pubescent short-bladed

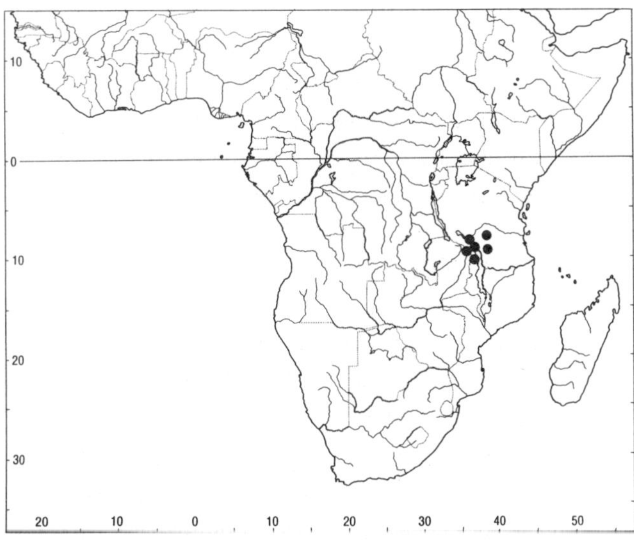

MAP 21. *Gladiolus curtifolius.*

leaves make *Gladiolus curtifolius* unmistakable. The perianth tube, 5–7.5 cm long, and pale flower suggest a relationship with section *Acidanthera*. This is, however, questionable. *Gladiolus curtifolius* lacks the long-apiculate anthers characteristic of section *Acidanthera* and also has long inner bracts sometimes exceeding the outer, another feature that does not accord with section *Acidanthera*. The relationships of *G. curtifolius* most likely lie with species of section *Gladiolus*, such as *G. atropurpureus*. The fairly small bracts and tepals accord with the section, and the reduced leaf blades are consistent with several species of the alliance. The long perianth tube is, however, a striking adaptation although not entirely unique for section *Gladiolus*.

HISTORY

The distinctive *Gladiolus curtifolius* was discovered by the German botanist, Walter Goetze, in the mountains that lie to the north of Mbeya in southwestern Tanzania in 1899. Two years later it was described by Hermann Harms, who referred the species to the genus *Acidanthera*, as *A. goetzei* in Goetze's honor. The species remains poorly known and has only been re-collected on a few occasions, although its range across western Tanzania and northern Malawi is fairly wide. It was renamed *G. curtifolius* when transferred to *Gladiolus* (Marais, 1973), because the epithet *goetzei* had already been used in the genus.

SELECTED SPECIMENS

Tanzania. Mbeya: Kyimbila, Kinga Mountains, Madahari, 1912, *Stolz 2330* (BM, K, P). Iringa: Njombe District, burnt grassland, Kitulo–Njombe road 50 km east of Bulongwa turnoff, 13 Nov. 1966, *Gillett 17827* (BR, K, P, SRGH); Dansland, Njombe, 2140 m, 2 Jan. 1992, *Spurrier 609* (DSM, MO).

Malawi. Northern: Karonga District, Mafinga Mountains, northeast slopes, 18 Jan. 1964, *Robinson 6294* (B, K, MAL); Chitipa District, 18 km southeast of Chisenga, Jembya Forest Reserve, 9 Dec. 1988, *Thompson & Rawlins 5449* (CM).

21. *Gladiolus gregarius* Welwitsch ex Baker

PLATES 11, 12, FIGURE 21, MAP 22. Baker, Trans. Linn. Soc., Ser. 2, 1: 268 (1878); Handbook Irideae 210 (1892); Fl. Trop. Africa 7: 365 (1898). Hepper, Kew Bull. 21: 493 (1968); Fl. West Trop. Africa, Ed. 2, 3(1): 144 (1968). Geerinck, Bull. Jard. Bot. Nat. Belgique 42: 276 (1972). Goldblatt, Fl. Zambesiaca 12(4): 79 (1993). Type: Angola, Malange, Pungo Andongo, Jan.–Mar. 1857, *Welwitsch 1528* (BM, lectotype designated here, the specimen complete and in good condition; B, C, COI, G, K, isolectotypes).

SYNONYMY

Gladiolus spicatus Klatt, Linnaea 35: 377 (1867), illegitimate homonym for *G. spicatus* Linnaeus (1753) (= *Thereianthus spicatus* (Linnaeus) G. Lewis). *Gladiolus klattianus* Hutchinson, Fl. West Trop. Africa 2: 379 (1936), as new name for the homonym *G. spicatus* Klatt. Aké Assi, Contrib. Fl. Côte d'Ivoire 233 (1963). Type: Niger, Nupe, near Jeba on the Kworra, 1858, *Barter s.n.* (S, holotype; G, K (designated the lectotype on insufficient grounds by Geerinck, 1972), P, probable isotypes, without detailed locality data).

Gladiolus multiflorus Welwitsch ex Baker, Trans. Linn. Soc., Ser. 2, 1: 269 (1878); Handbook Irideae 221 (1892); Fl. Trop. Africa 7: 369 (1898). Type: Angola, Huila, *Protea* woods at Monino, Dec. 1859–Feb. 1860, *Welwitsch 1538* (BM, lectotype, complete and well preserved; B, C, COI, G, K, LD, P, isolectotypes).

Gladiolus hanningtonii Baker, Handbook Irideae 212 (1892); Fl. Trop. Africa 7: 366 (1898). Type: Tanzania, "mountains of East Africa," possibly Iringa District, without date, *Hannington s.n.* (K, holotype).

Gladiolus karendensis Baker, Bull. Herb. Boiss., Sér. 2, 1: 867 (1901). Type: Tanzania, Karenda, 1882, *Bohm 22* (B, lectotype designated here; Z, isolectotype); Gonda, 1882, *Bohm 3* (Z, syntype).

Gladiolus uhehensis Harms, Bot. Jahrb. Syst. 28: 365 (1901). Type: Tanzania, Uhehe, near Bweni in light bush, 1700 m, 11 Mar. 1899, *Goetze 733* (B, holotype; BR, isotype; sturdy broad-leaved plants with short stems).

Antholyza gilletii de Wildeman, Ann. Mus. Congo, Sér. 4: 19 (1902). Type: Zaire, Bas-Congo, vicinity of Kimuenza, Jan. 1901, *Gillet s.n.* (BR, holotype).

Antholyza descampsii de Wildeman, Ann. Mus. Congo, Sér. 4: 18 (1902). Type: Zaire, Shaba (Katanga), 1891, *Descamps s.n.* (BR, lectotype designated by Geerinck, 1972); Shaba (Katanga), Lofoi, 1899, *Verdick s.n.* (BR, syntype). Both collections are the long-bracted form.

Gladiolus corbisieri de Wildeman, Feddes Repert. Spec. Nov. Regni Veg. 12: 296 (1913). Type: Zaire, Shaba, Welgelegen, 1912, *Homblé sub Corbisier 602* (BR, lectotype designated by Geerinck, 1972: 277; BR, isolectotype); Shinsenda, Mar. 1912, *Ringoet sub Homblé 435* (BR, syntypes); Shinsenda, Mar. 1912, *Ringoet sub Homblé 518* (BR, syntypes); Mugila Mountains, 24 May 1908, *Kassner 2996* (BR, J, Z, syntypes); Shiwele, 19 Feb. 1908, *Kassner 2477* (BM, BR, E, HBG, P, Z, syntypes).

Gladiolus elegans Vaupel, Bot. Jahrb. Syst. 48: 536 (1913). Type: Tanzania, Mfimbwo Mountain on Lake Tanganyika, 2300 m, grassland, 15 Mar. 1901, *Münzner sub Exped. Fromm 225* (B, holotype; K, photo; plants short, slender, and with cream flowers).

Gladiolus pseudogregarius Mildbraed ex Hutchinson, Fl. West Trop. Africa 2: 379 (1936), illegitimate name, without Latin description.

EPONYMY

gregarius, "gregarious," alluding to the habit of plants growing crowded together in clumps.

DESCRIPTION

Plants (15–)30–80 cm high. CORM (12–)18–30 (–40) mm in diameter, TUNICS membranous, fragmenting irregularly, occasionally subfibrous, the fibers than somewhat clawed. CATAPHYLLS usually dark brown, the upper green. LEAVES four to seven, narrowly lanceolate, the lower three to five more or less basal, lanceolate, half to two-thirds as long as the stem, (6–)9–16(–24) mm wide, the margins and midribs not or hardly thickened, upper leaves cauline and shorter. STEM erect and straight, unbranched, 2–3 mm in diameter at the base of the spike.

SPIKE (2–)8- to 20(–25)-flowered, often congested, evidently straight but regularly flexuose under the bracts; BRACTS green or becoming dry and brown at the end of flowering, firm, sometimes lightly striate (the veins raised and hyaline), imbricate, 2–2.5(–3) internodes long, 20–30(–50) mm long, 8–10 mm at the widest, the inner about two-thirds as long as the outer. FLOWERS usually light to dark purple, rarely dark red-brown, fading to cream in the throat or mostly white, the tube usually dark purple below, the lower tepals each with a with a dark purple diamond-shaped mark in the upper third; PERIANTH TUBE c. 12 mm long, curving outward and widening above; TEPALS unequal, the dorsal (16–)22–26 mm long, 12–18 mm wide, arched over the stamens, the lower three narrowed below into claws, the limbs flexed downward and channeled, joined to the upper laterals for 4–6 mm and to one another for c. 2 mm, c. 12 mm long, c. 5 mm at the widest, usually shortly exceeding the upper in profile or about as long. FILAMENTS 9–13 mm long, exserted 5–6 mm from the tube; ANTHERS 7–10 mm long, purple or yellow, pollen yellow. OVARY ellipsoid, 3–5 mm long; STYLE arched over the stamens, dividing opposite the base to middle of the anthers, branches c. 2 mm long. CAPSULES narrowly elliptic, 12–15 mm long, comparatively hard and woody; SEEDS elliptic, the wing sometimes poorly or not developed on the sides or occasionally even at the proximal end, 5–7 × 2.5–4 mm. CHROMOSOME NUMBER $2n = 22$.

FLOWERING TIME. In southern tropical Africa from January to March; in West Africa from June to October.

DISTRIBUTION & HABITAT

One of the most widespread species of *Gladiolus, G. gregarius* extends across tropical Africa, from Senegal in West Africa to Angola, Zaire, Zambia, Malawi, and southern Tanzania. It grows in open woodland, often in rocky habitats where it is sheltered from competition, and occasionally in wetter habitats such as dambo verges. It is particularly variable in size. Thus plants in a single population of *G. gregarius* may range from 25 to 70 cm in height, with a corresponding variation in the number of flowers on the spike. Some particularly small plants, less than 12 cm high, from parts of West Africa, have only two to five flowers per spike and narrow linear leaves. Such plants were treated as *G. pseudogregarius* by J. Hutchinson (in Hutchinson & Dalziel, 1936) based on a manuscript name of Mildbraed, but the name has never been validated. Some dwarf plants have seeds with a vestigial wing, restricted to the distal end rather than being circumferential. These plants may simply be depauperate due to poor growing conditions.

DIAGNOSIS & RELATIONSHIPS

Striking in its unusually long, overlapping floral bracts that clasp the spike and that are at least two internodes in length, *Gladiolus gregarius* is seldom mistaken for another species. Its fan of several broad basal leaves and its relatively small flowers with hooded dorsal tepals and narrow, darkly marked lower tepals are unremarkable and place the species squarely in section *Gladiolus*. The species stands out in the subgenus not only in the specialized floral bracts but in the rigidly erect

FIGURE 21. *Gladiolus gregarius*. Corm, leaves, and spike, × 0.67; single flowers, × 1.3 (habit and spike, *Richards 19040*; single flower, *Billiet & Jadin 4204*).

spikes and unbranched stems. Specialized in all these features, the relationships of *G. gregarius* lie with *G. harmsii, G. macrospathus,* and *G. microspicatus,* which also have long, clasping floral bracts and erect, unbranched stems. This small alliance is also unusual cytologically. The two species counted (Goldblatt et al., 1993) both have $x = 11$, the lowest base number in the genus.

Among the specimens generally referred to *Gladiolus gregarius,* a series of populations from Shaba and Kasai Provinces in Zaire stand out in their larger, more darkly marked flowers and longer, distinctively ridged bracts that tend not to discolor when dried. The status of these plants, first collected in 1891 in Shaba by Captain Descamps and described as *Antholyza descampsii* by Emile de Wildeman in 1902, has remained controversial. De Wildeman clearly distinguished them from *G. gregarius,* which he regarded as comprising two species, the West African *A. labiata* and the lowland Zairian *A. gilletii,* on the basis of their different bracts. Later, for no apparent reason, de Wildeman (1913) described plants exactly matching *A. descampsii* as *G. corbisieri,* based on several new collections. The differences between the bracts of the more common and widespread form of *G. gregarius* and the large-bracted populations is usually clear, but the differences in flower size are less so. Thus, except for poorly preserved specimens, the two can usually be distinguished. It is uncertain, however, how much their differences are correlated with variation in climate and soil. Until more is known about the two variants, I prefer to follow the example of Geerinck (1972) in treating them as a single species.

A more significant variant of the *Gladiolus gregarius* complex is *G. microspicatus,* which occurs at all heavy-metal-enriched sites in southern Shaba Province, Zaire, and in some montane habitats elsewhere in Shaba, northern Zambia, and Burundi. It is unusual in having narrow leaves, usually 2–3 mm wide, and a low stature. The comparatively short bracts are seldom more than 20 mm long and are one to two internodes in length. Flower color in these populations is light to dark purple, or violet, with a yellow throat and yellow nectar guides on the lower tepals.

A series of populations of particularly large-flowered and unusually robust plants from Burundi, until now included in *Gladiolus gregarius,* is treated here as a new species, *G. macrospathus.* They differ not only in the size of the flowers, but the perianth is blue-mauve, and the bracts, when dry, are generally straw-colored.

HISTORY

The complex history of *Gladiolus gregarius* begins with the almost contemporary discovery of the species in Angola by Friedrich Welwitsch, and in Nigeria by Charles Barter. The plants from Nigeria were described by F. W. Klatt in 1867 as *G. spicatus,* while it was not until 1878 that J. G. Baker described *G. gregarius,* based on white-flowered specimens from Angola. At the same time, Baker described *G. multiflorus* as a separate species, based on Angolan plants with darker-colored flowers and dark markings on the lower tepals. *Gladiolus*

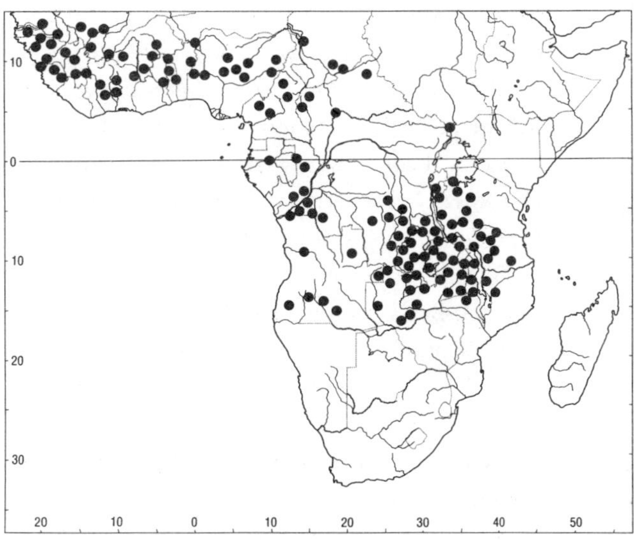

MAP 22. *Gladiolus gregarius.*

spicatus was a homonym for a southern African species, now *Thereianthus spicatus,* but the West African species was not renamed until 1936, when J. Hutchinson called it *G. klattianus,* not realizing that there were earlier valid names for the species. It was only in 1968 that F. N. Hepper first used the name *G. gregarius* for the species in West Africa. The first collections of *G. gregarius* from Tanzania also received a separate name, *G. hanningtonii* (Baker, 1892). Specimens collected later from that country were assigned by Vaupel, Harms, and Baker, respectively, to three different species, one of which, a short white-flowered plant from the mountains east of Lake Tanganyika, was called *G. elegans.* The series of names applied to the species in Zaire by Emile de Wildeman have been discussed above. All of the numerous synonyms of the species were united under *G. gregarius* only in 1972, when Geerinck revised *Gladiolus* for central Africa.

SELECTED SPECIMENS

Senegal. Kanéméré, Nov. 1965, *Fotius 485* (ALF, MO); Tambacounda, Niokolo Koba, 30 Oct. 1958, *Adam 15863* (MO).

Guinea Bissau. Gabú, Canquelifá, 20 Nov. 1951, *Espírito Santo 2962* (COI); between Piché and Canquelifá, 28 Oct. 1952, *Espírito Santo 3108* (COI, K).

Mali. Bamako, Oct. 1976, *Adam 26954* (MO); Kabiola, 15 Sept. 1958, *Adam 15063* (MO, P).

Sierra Leone. near Kambia (Magbema), 10 Oct. 1952, *Jordan 945* (K, P); Suribolomia (Samu Chiefdom), 15 Nov. 1950, *Jones 42* (K, P).

Guinea. Friguiagbé, 16 Oct. 1939, *Chillou 1755* (MO, P); Beyla, 26 July 1926, *Collenette 81* (K); Macenta, Sept. 1936, *Jacques-Felix 1859* (P).

Burkina Faso. Bobo–Dioulassa, cultivated savanna, 7 Oct. 1967, *Geerling & Bokdam 1179* (BR, C, K, MO); Deux Balé National Park, 21 July 1973, *Sihvonen 24* (MO).

Niger: Koulou, Sambera, 21 Sept. 1943, *Marechal s.n.* (P).

Chad. Kasselem, in 1973, *Garvigues 510* (MO); Bedjama, 28 Aug. 1964, *Audru 995* (P).

Ivory Coast. between Semien and Séguéla, 6 Oct. 1961, *Aké Assi 6054* (BR, K); Fésséléman, sandy savanna, 9 Nov. 1977, *Audru 6431* (P); 15 km west of Kotouba, 14 Sept. 1967, *Geerling & Bokdam 855* (BR, K).

Ghana. Tono, Navrongo, Aug. 1961, *Irvine 4663* (K, P); Maliata, Togo Plateau, c. 700 m, 14 Nov. 1958, *Morton A3498* (GC). Mole National Park, Konkori, 28 July 1976, *Hall & Lock s.n.* (GC46267, K).

Togo. Sokodé, 29 Sept. 1976, *Ern, Hein, & Pircher 804* (B).

Benin: Kalalé–Iou, 10°23′ N, 36°27′ E, 24 Sept. 1989, *Sinsin 1319* (BR).

Nigeria. Niger: northern Zungiru, 12 Sept. 1905, *Dalziel 252* (K). Kwara: Ilorin, Bongu, Nweri Grazing Reserve near Yashikira, *Latilo s.n.* (FHI 25384, K). Zaria: Gwari District, Mando Forest Reserve, west of Karu, 28 July 1950, *Keay 25987* (K, P); Kaciya–Zonkwa, mile 5, 20 July 1957, *Summerhayes 130* (K, P). Bauchi: Tangale Waja, junction of Gombe, Numan and Tula roads, 600 m, 16 Apr. 1955, *Summerhayes 71* (B, BR, K, P).

Cameroon. Mbam Menkoum Massif, 20 km northwest of Yaoundé, 18 Aug. 1972, *Letouzey 11642* (HBG, K, P); Mt. Neshele, 10 km east-southeast of Bamenda, 2621 m, 12 Aug. 1975, *Letouzey 14281* (K, P); Ngoundéré–Meiganga, June 1939, *Jacques-Felix 4169* (P).

Central African Republic. Ouham-Pendé: Bozoum–Uham, Janga Mountains, 900 m, *Mildbraed 9701* (K). Vakaga: summit of Mt. Toussoro (9°10′ N, 23°15′ E), 1330 m, 3 Nov. 1989, *Harris & Fay 2214* (MO). Ombela-Mpoko: 30 km north of Bangui, 3 Aug. 1921, *Tisserant 408* (P).

Uganda. West Nile: Mt. Otze, c. 1900 m, June 1954, *Jackson B9801* (EA); June 1954, *Bally 9801* (K).

Gabon: Ogooué, Booué, Jan. 1950 (fruit), *Hallé & le Thomas 233* (P).

Congo. Batéké Plateau, Maloukou road between Mandielé and Lake Ngatsou, in 1965, *Sita 1174* (P); km 45, Mbé road near Brazzaville, Jan. 1949 (fruit), *Koechlin 114* (P).

Zaire. Kinshasa: Mense Peak, 5 Apr. 1971, *Breyne 2052* (BR). Kasai: Gandajika, 19 Feb. 1957, *Liben 2658* (BR, K); Popokabaka, Kimbidi, 25 Aug. 1966, *Pauwels 2623* (BR, EA). Shaba: Kamina, Lovoi, Mar. 1932, *Quarré 2944* (BR, EA, G, Z); Kiala, near Manono, Feb. 1958, *Thiebaud 731/379* (BR, C, K, LISC, MO); Upemba National Park, 8 Mar. 1949, *de Witte 5821* (BR, K, PRE).

Burundi: Kitete, 65 km south of Tunda, 19 Feb. 1934, *Rossignol 107* (BR).

Angola. Huambo: Huambo–Sacaála, 7 Mar. 1962, *Murta 129* (COI, LISC). Bié: Chiangar, 1700 m, 15 Feb. 1962, *Texeira & Andrade 6523* (LISC); Chieque, 1400 m, Mar. 1973, *Raimundo, Matos, & Figuera 55* (BR, LISC, SRGH). Cuando Cubango: on the Longa above Lazingua, 1250 m, 20 Jan. 1900, *Harms 663* (G); Cuito Cuanavale, 19 Mar. 1960, *Mendes 2576* (LISC). Cuchi–Menongue (Serpa Pinto), km 15, 10 Mar. 1973, *Bamps, Raimundo, & Matos 4104* (BR, LISC) Lunda: Saurimo, Apr. 1937, *Gossweiler 11702* (COI, K).

Zambia. Copperbelt: Plateau woodland, Ndola, 27 Feb. 1954, *Fanshawe 867* (BR, K, NDO, SRGH); Mufulira, 12 Mar. 1961, *Linley 101* (G). Central: Mkushi District, 42 km from Kapiri Mposhi to Serenje, 3 Feb. 1973, *Strid 2893* (C, K, MO). Northern: Edge of Kalambo Falls Gorge, 1200 m, 8 Feb. 1965, *Richards 19603* (K, MO, P); Mbala Sand Pits, 4 Mar. 1969, *Sanane 475* (BR, EA). Eastern: Luangwa Valley near Kapimbi River, 3 Mar, 1970, *Astle 5793* (K, NDO, SRGH). Southern: Mongu, edge of woodland, 20 Jan. 1966, *Robinson 6808* (B, K); Mazabuka, 28 Feb. 1960, *Leach 9800* (K, MO, PRE).

Tanzania. Shinyanga: Old Shinyanga, 3 Apr. 1932, *Burtt 3721* (BR, K). Mpanda: ridge near Lake Katavi, 25 Jan. 1950, *Bullock 2330* (B, K, MO, S). Tabora: Tabora Game Reserve, 1200 m, 24 Jan. 1970, *Mangure 85* (EA). Rukwa: Ulanda, Ulambya, 10 Mar. 1970, *Leedal 435* (EA); Ufipa, Kawa River gorge, 15 Feb. 1959, *Richards 10892* (B, BR, K, S). Iringa: Ihemo, 30 km south of Iringa, 23 Feb. 1962, *Polhill & Paulo 1582* (B, BR, K, LISC, PRE); Ruaha National Park, Magangwe Air Strip, 1340 m, 8 Mar. 1972, *Bjornstad 1442* (UPS); Lupembe, upper Ruhudje, in 1931, *Schlieben 525* (BM, BR, K, P, S, Z). Mtwara: 150 km southwest of Lindi, 25 Apr. 1935, *Schlieben 6418* (BM, BR, P, Z).

Malawi: Northern: Karonga, Vintukhutu Forest Reserve, 3 km north of Chilumba, 26 Mar. 1975, *Pawek 9506* (K, MAL, MO). Central: Dzalanyama Forest Reserve near Chiungiza, 9 Feb. 1959, *Robson 1516* (BM, BR, K, MAL, PRE, SRGH).

Mozambique. Niassa: Massangulo, 24 Mar. 1933, *Gomes e Sousa 1351* (COI). Nampula: km 10, road from Mutuali to Malema, 15 Mar. 1953, *Gomes e Sousa 4058* (PRE). Tete: Angónia, near Domué, 22 Feb. 1980, *Macuacua & Mateus 1129* (WAG). Zambézia: Gurué, 26 km from Mutuali to Lioma, 10 Feb. 1964, *Torre & Paiva 10506* (LISC).

22. *Gladiolus macrospathus* Goldblatt, new species

PLATE 13, FIGURE 22, MAP 23. Type: Burundi, Bururi, Gihofi (Mosso), 1350 m, 4 Mar. 1981, *Reekmans 9775* (BR, holotype; K, MO, PRE, UPS, isotypes).

EPONYMY

macrospathus, "with large spathes," alluding to the conspicuous, large overlapping floral bracts.

LATIN DIAGNOSIS

Plantae 45–110 cm altae, cormo (25–)30–40 mm in diametro tunicis fibrosis, spica (10–)12–26 florum congestis, bracteis 2.5–4 mm longis imbricatis scariosis, floribus malvinis, tepalo superiore 27–30 mm longo, filamentis c. 10 mm exsertis, antheris 10–11 mm longis.

FIGURE 22. *Gladiolus macrospathus*, × 0.5 (*Billiet 4181, Reekmans 1692*).

DESCRIPTION

Plants 45–110 cm high. CORM (25–)30–40 mm in diameter, tunics coriaceous or of moderately coarse vertical fibers. CATAPHYLLS usually dark brown or the uppermost green above ground. LEAVES four to seven, the lower three to five more or less basal, narrowly lanceolate, about two-thirds as long as the stem, rarely exceeding the base of the spike, (7–)10–18 mm wide, the margins and midribs hardly thickened, upper leaves cauline and shorter, imbricate and sheathing the stem up to the base of the spike. STEM stiffly erect, unbranched, 2–3 mm in diameter at the base of the spike.

SPIKE (10–)12- to 26-flowered, often congested, fairly straight, the axis concealed; BRACTS usually dry above or entirely by anthesis, pale straw-colored, firm, imbricate, 2–2.5 internodes long, 2.5–3.5(–4) mm long, the inner about two-thirds as long as the outer. FLOWERS lilac to mauve, the tepals more deeply colored distally, the lower lateral tepals sometimes with a dark purple diamond-shaped mark in the distal third, this fading on drying, the tube pale; PERIANTH TUBE 12–15 mm long, curving outward and widening above; TEPALS unequal, the dorsal 27–30 mm long, 15–18 mm wide, arched over the stamens, narrowed below, the flower thus windowed in profile, the lower three tepals narrowed below into claws, the limbs flexed downward and channeled, joined to the upper laterals for 6–8 mm and to one another for 1–2 mm, 16–20 mm long, 5–6 mm at the widest, in profile usually shortly exceeding the dorsal. FILAMENTS c. 15 mm long, exserted c. 10 mm from the tube; ANTHERS 10–11 mm long, purple or brown, pollen yellow. OVARY ellipsoid, 4–5 mm long, style arched over the filaments, dividing opposite the middle of the anthers, branches 3.5–4.5 mm long, reaching the upper third to apex of the anthers. CAPSULES ellipsoid, 12–15 mm long, hard and woody. SEEDS elliptic, the wing generally less developed on the sides, 5.5–6 × c. 3.5 mm.

FLOWERING TIME. February to mid April.

DISTRIBUTION & HABITAT

Gladiolus macrospathus is endemic to the small central African country of Burundi, where it occurs in light woodland or on forest verges. Judging from the collecting record, it is fairly common and conspicuous locally. It has not been found in any of the neighboring countries, Rwanda, Zaire, Uganda, or Tanzania.

DIAGNOSIS & RELATIONSHIPS

Generally a fairly robust species, with stems up to 1 m and more, *Gladiolus macrospathus* has lilac to mauve flowers flushed darker toward the tips of the tepals and conspicuously large, imbricate, straw-colored bracts that are about two internodes long. These distinctive bracts place the species in the small *G. gregarius* alliance and close to *G. gregarius*, the central species of the group. Flower color and the large size of the tepals and stamens separate *G. macrospathus* from the larger-flowered

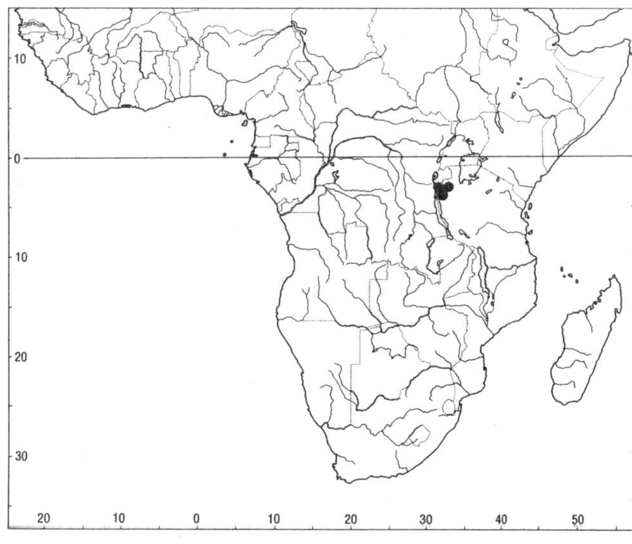

MAP 23. *Gladiolus macrospathus.*

forms of *G. gregarius,* the perianth of which usually has a dark-colored tube and conspicuous blackish markings on the lower lateral tepals. Unlike *G. gregarius, G. macrospathus* has relatively long style branches up to 4.5 mm long that reach the upper third of the anthers, whereas the style branches of *G. gregarius* are 2–3 mm long and seldom reach beyond the middle of the anthers.

SELECTED SPECIMENS

Burundi. Ruyigi: Kinyinya, 1300 m, 27 Mar. 1971, *Lewalle 5387* (G, K, MO); Mwishanga, 8 Apr. 1972, *Reekmans 1692* (BR, MO); km 5 on Rutana–Bukemba road, 24 Feb. 1977, *Reekmans 5705* (BR, K, MO, WAG). Gitega: Gitega, Mt. Songa, 1800 m, Oct. 1969, *Lewalle 1557* (BR, K, MO).

23. *Gladiolus microspicatus* Duvigneaud & van Bockstal ex Còrdova

MAP 24. Còrdova, Bull. Jard. Bot. Nat. Belgique 60: 326 (1990). Goldblatt, Fl. Zambesiaca 12(4): 80 (1993). Type: Zaire, Shaba, Tenke–Kwatebala, rocky summit of cupriferous hill, Jan. 1960, *Duvigneaud "5053GI,"* i.e., *5033GI* (BRLU, lectotype designated here; BR, BRLU, isotypes). Còrdova (1990) cited *Duvigneaud 5053GI* as the holotype but there are three sheets numbered *5033GI* at BRLU, none specifically annotated the holotype; from these I have chosen the specimen with the best preserved flowers as lectotype.

SYNONYMY

Gladiolus klattianus subsp. *angustifolius* van Bockstal, Bull. Soc. Roy. Bot. Belgique 96: 126 (1963). Type: Zaire, Shaba, Kasompi East, rocks on summit of metaliferous hill, 8 Feb. 1960, *Duvigneaud 5468* (BRLU, lectotype and isolectotypes, designated here).

EPONYMY

microspicatus, "with a small spike," alluding to the generally few flowers per inflorescence.

DESCRIPTION

Plants 24–35 cm high. CORM 10–20 mm in diameter, TUNICS of fine to coarse yellow brown to gray fibers, usually accumulating in a dense mass, often extending upward and with the dry cataphylls forming a thick neck. CATAPHYLLS more or less dry and brittle, often dark brown, the upper sometimes green above the ground. LEAVES (sometimes three or) four or five, the lower two to four more or less basal, linear (rarely lanceolate), reaching to at least the base of the spike or sometimes the lower shortly exceeding it, 1.5–3(–5–7) mm wide, the margins not thickened, decreasing in size above, the uppermost usually cauline. STEM erect, unbranched, c. 2 mm in diameter at the base of the spike.

SPIKE 4- to 10-flowered, inflexed at the base, straight, sheathed by the bracts; BRACTS green or becoming dry and brown apically, firm and erect, clasping the stem, usually imbricate, about two internodes long, occasionally (1–)1.5 internodes long, 18–25 mm long, the inner about two-thirds as long as the outer. FLOWERS either light to dark purple fading to cream in the throat and tube, the lower tepals each with a yellow median streak in the upper third, or uniformly dark purple; PERIANTH TUBE 10–12 mm long, curving outward and widening above; TEPALS unequal, the dorsal c. 18 mm long, 10–12 mm wide, arched over the stamens, the lower three united with the upper laterals for c. 4.5 mm and to each other for c. 2 mm, narrowed below into claws, horizontal or the limbs flexed downward distally, in profile usually shortly exceeding the upper. FILAMENTS 10–12 mm long, exserted 5–6 mm from the tube; ANTHERS c. 6 mm long, usually purple, pollen yellow. OVARY ellipsoid, 4–5 mm long; STYLE arched over the stamens, dividing opposite the lower third of the anthers, the branches 1–1.5 mm long. CAPSULES narrowly elliptic, 10–12 mm long, comparatively hard and woody; SEEDS angular, wingless but with a prominent membranous ridge on one end, c. 1.2 × 2.5 mm. CHROMOSOME NUMBER $2n = 22$.

FLOWERING TIME. January to mid March.

DISTRIBUTION & HABITAT

Gladiolus microspicatus is centered in southern Zaire, where it is most common in rocky grassland on soils that are enriched with heavy metals. Records indicate that it usually grows in open and sunny situations. This contrasts with the closely related *G. gregarius,* which favors light woodland, frequently on deeper soils. There are scattered records of *G. microspicatus* from elsewhere in Shaba Province and in Burundi and northeastern Zambia, where plants grow in well-drained, although seasonally moist, grassland or rock outcrops, sometimes in poor sandy soils.

DIAGNOSIS & RELATIONSHIPS

In overall appearance, *Gladiolus microspicatus* resembles a diminutive *G. gregarius* and almost certainly it is to this species that it is related. The two share a similar moderate-sized purple and cream flower with a pale throat, stiff, erect, imbricate bracts that clasp the stem, and short ellipsoid capsules that remain enclosed in the bracts until the latter decay. The major differences are the presence of a neck of fibers around the stem base, narrow leaves, usually 1–2 mm wide (sometimes up to 4 mm), and wingless seeds.

First described as variety *angustifolius* of *Gladiolus klattianus* (a synonym of *G. gregarius*), the plant was redescribed at species rank as *G. microspicatus* (Còrdova, 1990), without reference to the earlier varietal name, although including among the exsiccatae cited the type of variety *angustifolius.* As circumscribed here, *G. microspicatus* includes not only plants from heavy-metal-enriched soils of southern Shaba but some from elsewhere in Zaire as well as from adjacent northern Zambia and Burundi. Away from heavy-metal-enriched soils, plants are recorded as growing on sandy or stony sites in moist grassland or steppe. No capsulate specimens are known from such sites and seeds cannot be compared with those from such heavy-metal-enriched soils. However, the corms from non-heavy-metal-enriched sites lack the thick layer of fibers that form a neck around the stem base otherwise characteristic of *G. microspicatus.*

Until seeds of *Gladiolus microspicatus* from soils that do not have high concentrations of heavy metal have been examined, the possibility that these populations are independently derived from *G. gregarius* remains a tenable hypothesis. There seems no doubt that *G. microspicatus* is not simply an ecotypic variant of *G. gregarius.* Greenhouse-grown plants from Kasompi West, Shaba Prov-

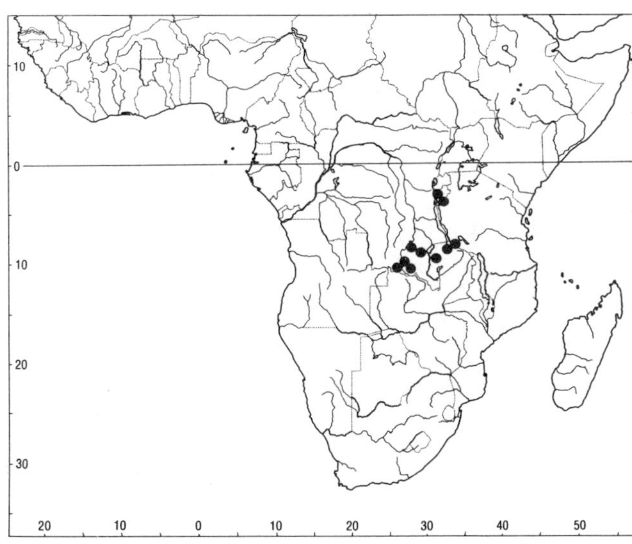

MAP 24. *Gladiolus microspicatus.*

ince, Zaire (*Duvigneaud & Timperman s.n.*), maintained their low stature, narrow leaves, and short bracts when cultivated in Belgium. Plants from Chabara, also in Shaba Province (*Bamps 846*), here included in *G. microspicatus,* have unusually broad leaves 5–7 mm wide but not reaching the base of the spike. In leaf width they seem to approach *G. gregarius,* but leaf height and other characteristics accord closely with *G. microspicatus.*

SELECTED SPECIMENS

Burundi: Kayanza: Rwegura, sun-baked rocks, 2000 m, 9 Jan. 1969, *Lewalle 3177* (MO). Bururi: Rutovu, 2100 m, *Lewalle 483* (BR); Rwamunyinya Hill, rocky grassland, 2000 m, 6 Mar. 1980, *Reekmans 8694* (K, MO, PRE). Gitega: road to Ruyigi, 1600 m, 10 Mar. 1968, *Lewalle 2981* (BR, K, MO).

Zaire. Shaba: Chabara, copper hill, 1430 m, 10 Jan. 1981, *Malaisse 11450* (BR, MO); Kamatanda, lightly copper-enriched steppe, summit of ore deposit, 9 Feb. 1960, *Duvigneaud 5483* (BRLU); Kasompi West, poisoned dambo south of metal-bearing hill, 8 Feb. 1960, *Duvigneaud 5464* (BRLU); Likasi Mine, on a copper deposit, 1400 m, 17 Mar. 1926, *Robyns 1697* (BR, K); Kolwezi, Dikuluwe, copper-rich rocky slope, 21 Jan. 1960, *Duvigneaud 5153* (BRLU); Kundelungu Plateau, 10 km north of Lualaba, Munwa River, c. 1600 m, 10 Jan. 1971, *Lisowski, Malaisse, & Symoens 12976* (BR, POZG, LSHI); Upemba National Park, Lufira trail at km 8, 26 Jan. 1960, *Bamps 846* (BR, plants with only three leaves, seeds globose).

Tanzania. Rukwa: Sumbawanga, Kalambo Falls, gritty sand, 11 Feb. 1965, *Richards 19645* (K, LISC, UPS).

Zambia. Luapula: Mbereshi–Kawambwa road, white sand, 1050 m, *Richards 12404* (BR, K). Northern: Rocky hill above Ndundu, sandy soil, 16 Feb. 1957, *Richards 8200* (K).

24. *Gladiolus harmsianus* Vaupel

PLATES 14, 15, FIGURE 23, MAP 25. Vaupel, Bot. Jahrb. Syst. 48: 540 (1913). Type: Angola, Huambo, Quiaca, 12°44′ S, 15°5′ E, 1360 m, Feb. 1908, *Wellmann 1586* (B, holotype; K, photo).

EPONYMY

harmsianus, named in honor of the German botanist, Hermann Harms, who described several species of tropical African *Gladiolus.*

DESCRIPTION

Plants (50–)80–120(–150) cm high. CORM 2.5–3 mm in diameter, tunics of medium to coarse fibers. CATAPHYLLS dark brown or becoming green above the ground. LEAVES seven or eight, the lower four or five more or less basal, lanceolate, half to two-thirds as long as the stem, 8–14 mm wide, the margins not thickened, upper leaves cauline and smaller than the basal. STEM stiffly erect, simple or with one branch, sheathed by leaf bases to the base of the spike, c. 2.5 mm at the base of the spike.

SPIKE 15- to 20-flowered, straight or barely flexed at the base; BRACTS green, firm, imbricate, 1.6–2(–3) internodes long, 25–42(–48) mm long, the inner about two-thirds as long as the outer. FLOWERS white, the upper tepal sometimes lightly flushed with pink, the lower laterals each sometimes with a pink median mark, the anthers dark purple; PERIANTH TUBE 30–35 mm long, curving outward and widening above; TEPALS unequal, the dorsal 25–35 mm long, c. 16 mm wide, hooded over the stamens, the lower three somewhat smaller than the upper. FILAMENTS 12–14 mm long, included in the tube; ANTHERS 8–12 mm long, dark purple. OVARY ellipsoid, c. 4 mm long; STYLE arched over the stamens, dividing opposite the upper half of the stamens, branches c. 4.5 mm long. CAPSULES narrowly el-

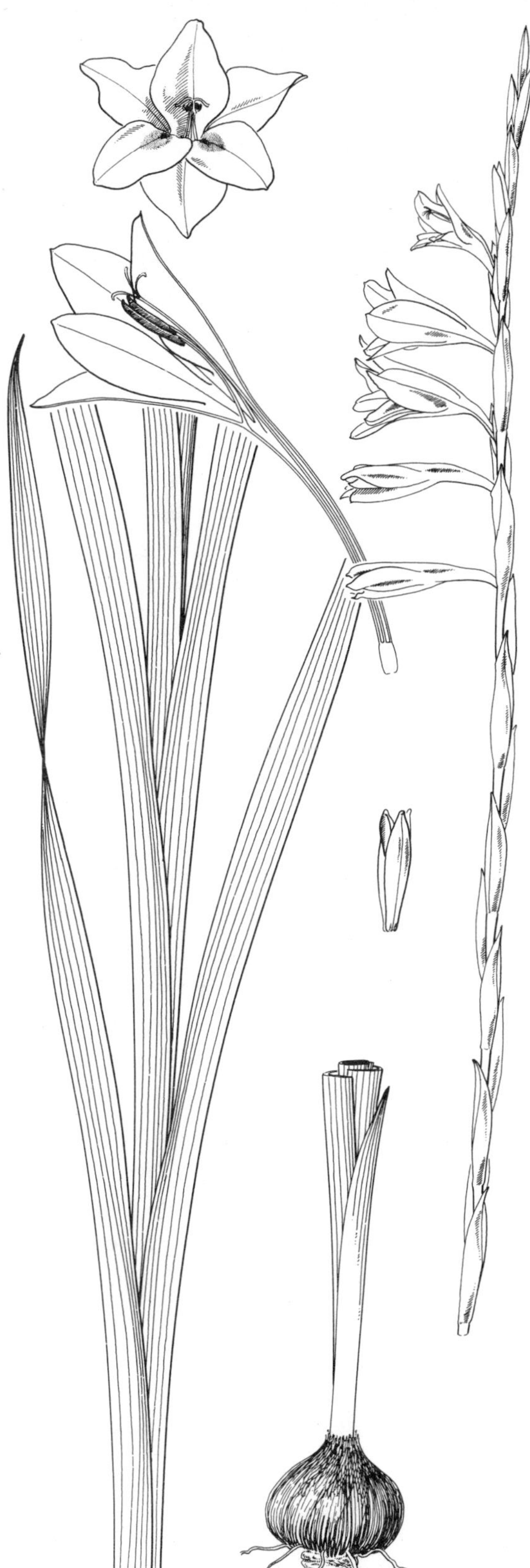

FIGURE 23. *Gladiolus harmsianus.* Corm, leaves, spike, and capsule, × 0.5; single flower, full size (*Bamps et al. 4443*).

lipsoid, 18–24 mm long, comparatively hard and woody; SEEDS oval, the wing sometimes poorly developed on one side, 4–5 × 3–4 mm.

FLOWERING TIME. February to April.

DISTRIBUTION & HABITAT

Restricted to western Angola, *Gladiolus harmsianus* occurs either at the edge of the main escarpment or on hills to the west of the escarpment at elevations between 1350 and 1550 m. The recorded localities are all from the center of the western Angolan highlands, either in Cuanza Sul, Benguela, or Huambo Provinces.

DIAGNOSIS & RELATIONSHIPS

The strong straight spike with large overlapping bracts some 2.5–3 internodes long and a fan of several basal leaves make clear the affinities of *Gladiolus harmsianus* with the vegetatively similar *G. gregarius.* The two species differ largely in their flowers, those of *G. harmsianus* having a long slender perianth tube 30–35 mm long and a white flower with the upper and lower tepals weakly differentiated. In addition, the anthers are dark pur-

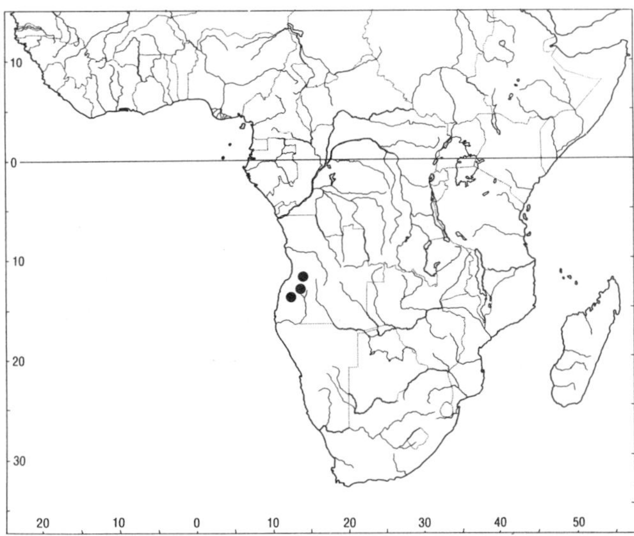

MAP 25. *Gladiolus harmsianus.*

ple and the style branches c. 4.5 mm long, substantially longer than those of *G. gregarius,* in which they are normally less than 2 mm long. *Gladiolus gregarius* and its close allies have a perianth tube 12–15 mm long and shorter than the upper tepals, usually at least the dorsal tepal purple and the lower three much narrower than the dorsal and with prominent nectar guides. It seems reasonable to suppose that *G. harmsianus* is a segregate of the widespread *G. gregarius* that is adapted to sphinx-moth pollination. The long tube, white perianth, and only weakly zygomorphic flowers are the characteristic signs of this pollination syndrome in Iridaceae (Vogel, 1954; Goldblatt, 1991). There seems to be strong selection for this type of pollination in western Angola, where another local endemic, *G. stenosiphon,* has apparently evolved from ancestors close to the southern Zairian species, *G. verdickii,* itself perhaps moth pollinated, although it has a shorter and broader perianth tube and considerably larger tepals.

ADDITIONAL SPECIMENS

Angola. Cuanza Sul: Quibala, Banza–Catumbi, 24 Mar. 1960, *Barbosa & Correia 8855* (BM, PRE, SRGH). Benguela: Quinjenje (14°56′ S, 11°50′ E), 1530 m, 12 Apr. 1973, *Bamps, Martins, & Matos 4443* (BR).

SECTION *HETEROCOLON* O. KUNTZE

O. Kuntze, Revisio Generum Plantarum 3(3): 66 (1898). Type: *Gladiolus pretoriensis* O. Kuntze.

Description

Plants medium, or small, usually unbranched. LEAVES linear or terete, the upper smaller and sometimes entirely sheathing, contemporary with the flowers. SPIKES usually inclined; BRACTS short, (6–)8–15 mm long, generally exceeding the perianth tubes. FLOWERS small, 12–27(–33) mm long, the lower tepals usually with contrasting nectar guides (actinomorphic and without markings in one species); PERIANTH TUBE (c. 4–)7–15 mm long; TEPALS equal or unequal, the dorsal usually slightly longer to half again as long as the tubes, the lower tepals much narrower and usually shorter than the dorsal, united basally and clawed below (subequal and not clawed in *Gladiolus actinomorphanthus*). CAPSULES more or less globose; SEEDS globose or with vestigial wings.

Comprising eight species as circumscribed here, section *Heterocolon* is restricted to southern tropical Africa and Transvaal Province, South Africa. It is most diverse in Shaba Province, Zaire, where species are especially frequent on soils that have high concentrations of heavy metals. The section was described by O. Kuntze (1898) to accommodate *Gladiolus pretoriensis,* which he regarded as unusual in both its narrow subterete leaves and wingless seeds. The inclusion here of several tropical African species is novel. Most of these species also have terete leaves and wingless seeds, but *G. gracillimus* and *G. pusillus* have nearly plane leaves and small, winged seeds. Their placement in the section rests on their close similarity to *G. robiliartianus* and *G. tshombeanus,* and if they are correctly treated, their leaf and seed characters represent reversals to the ancestral condition.

25. *Gladiolus ledoctei* Duvigneaud & van Bockstal

FIGURE 24, MAP 26. Duvigneaud & van Bockstal, Bull. Soc. Roy. Bot. Belgique 96: 219 (1963). Geerinck, Bull. Jard. Bot. Nat. Belgique 42: 280 (1972). Type: Zaire, Shaba, Tenke, “colline de

Goma," Feb. 1960, *Ledocte s.n.* (BRLU, holotype; BRLU, isotype).

SYNONYMY

Gladiolus fungurumeensis Duvigneaud & van Bockstal, Bull. Soc. Roy. Bot. Belgique 96: 219 (1963). Type: Zaire, Shaba, Fungurume, rocky ridge on summit of Hill I, soil rich in copper, 4 May 1957 (2 Mar. 1957 in the protologue), *Duvigneaud 2247* (BRLU, holotype).

EPONYMY

ledoctei, named to honor M. Ledocte, a geologist for a mining company in Upper Katanga, who assisted Paul Duvigneaud in his botanical studies in the region.

DESCRIPTION

Plants slender, 35–45 cm high, often growing in clumps. CORM 14–20 mm in diameter, tunics brownish, coriaceous or decaying into of medium to fine vertical fibers, extending upward in a neck. CATAPHYLLS obscured by the fibrous neck, membranous or dry, sometimes purplish, the inner longest and reaching shortly above the ground. LEAVES five or six, sometimes seven, the lower two basal and longest, with blades reaching at least to the base of the spike and sometimes exceeding it by up to 10 cm, the blades 1–2(–3) mm wide, either linear with heavily thickened and raised margins and midribs and narrow laminar grooves, or terete and four-sulcate, the remaining leaves cauline, imbricate, shorter and partly to entirely sheathing, the blades often thread-like, brown-streaked or dry entirely above, imbricate. STEM simple or with a single, often short branch, this always borne in the axil of the penultimate sheathing leaf, erect or flexed outward above the uppermost sheath, usually concealed for all but the upper 2–3 cm.

SPIKE (3–)5- to 8-flowered, inclined or suberect; BRACTS green, sometimes flushed purple,

FIGURE 24. *Gladiolus ledoctei.* Corm, leaves, spike, and capsules, × 1.3; single flower, full size (habit and spike, *Malaisse & Robbrecht 1908*; single flower, *Schaijes 1908*).

or purple at least on the veins, or evidently dry and brownish, 10–15 mm long, the outer sometimes with the apex filiform and attenuate, the inner as long or slightly shorter than the outer. FLOWERS usually pale lilac, pink, or cream to yellow, the lower tepals each with either a dark blue to purple or dark yellow diamond-shaped mark outlined in a darker color in the upper half; PERIANTH TUBE 8–12 mm long, curved outward and emerging between the bracts; TEPALS unequal, the dorsal largest and hooded over the stamens, c. 14 mm long, c. 10 mm wide, the upper laterals directed forward and curving outward distally, c. 10 mm long, the lower three held close together and more or less horizontal or tilted downward, united with the upper laterals for 3–4 mm, c. 1.5 mm long, narrowed below into claws, the limbs c. 3 mm wide. FILAMENTS 11–14 mm long, unilateral and arcuate, exserted 6–7 mm from the tube; ANTHERS parallel, 4.5–5.5 mm long, yellow. OVARY 1.5–2 mm long; STYLE arching over the filaments, dividing near the anther apices, the branches 1.5–2 mm long, ultimately reaching or exceeding the anther apices. CAPSULES more or less globose, 5–6 mm long, showing the outline of the seeds; SEEDS angular, c. 1 mm in diameter.

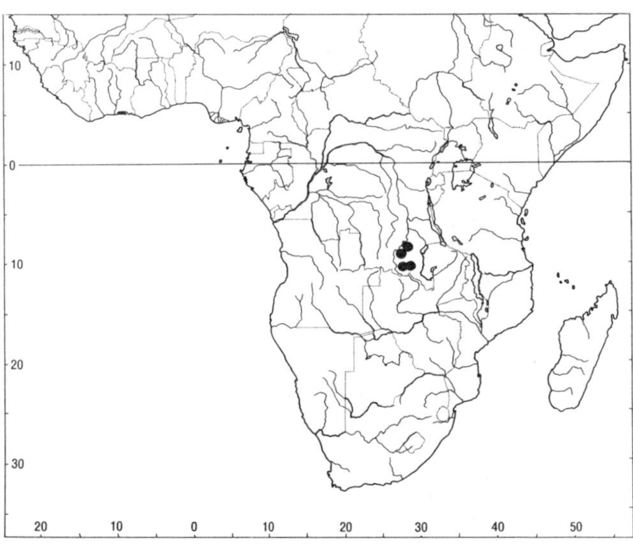

MAP 26. *Gladiolus ledoctei.*

FLOWERING TIME. Mainly February to April, sometimes in January or as late as June.

DISTRIBUTION & HABITAT

With a fairly wide distribution in Shaba Province, Zaire, *Gladiolus ledoctei* extends from the Upemba Plateau in the north across the Biano Plateau into the southern part of the province, where it seems to be restricted to rocky sites in areas with heavy-metal-enriched substrates. It seems to favor fairly dry sites in open savanna and stony grassland with stunted shrubs. The flowering time is unusually extended in this species. February to April seems to be the norm for populations on heavy-metal-enriched soils, but even there plants have been collected as early as January and as late as May. Plants from normal soils on the Manika Plateau have been collected in flower from March to June. Despite this unexpectedly extended flowering period, there seems little or no difference in the morphology of plants that bloom in the early, middle, or late part of the season.

DIAGNOSIS & RELATIONSHIPS

As circumscribed here, *Gladiolus ledoctei* includes a range of forms that have in common two or three linear to terete basal leaves and two to five cauline and partly to entirely sheathing leaves that conceal all but the upper 2–3 cm of the stem, even sometimes reaching to the middle of the spike. In addition, the plants have brownish and membranous to fibrous tunics that accumulate in a neck around the base, small but relatively long-tubed flowers, and globose capsules with angular seeds. Duvigneaud and van Bockstal (in Duvigneaud & Denaeyer-de Smet, 1963) recognized two species with largely the same habit, *G. ledoctei* from "colline de Goma," a slightly more slender, unbranched plant with one or two sheathing leaves and a flexed stem, and a more robust *G. fungurumeensis* from Fungurume, with thicker leaves, three or four sheathing leaves, the uppermost often reaching the base of the erect spike, and sometimes with a short lateral branch. There seems little reason to continue to separate the two. *Gladiolus ledoctei* is probably no more than a local variant, known only from the type locality, of a fairly widespread species that extends beyond heavy-metal-enriched sites into interior Shaba Province. In this expanded view of *G. ledoctei* I include plants from the Upemba Plateau with a habit virtually identical with the type of *G. fungurumeensis* but having linear leaves with thickened and raised margins and midribs. The plants of heavy-metal soils are seen as variants of this form, with the leaves narrower and the midribs more heavily thickened so that they are terete and four-sulcate. There is also some variation in flower color. The type forms of both *G. ledoctei* and *G. fungurumeensis* are pale bluish with purple markings, but there are populations of both linear- and terete-leaved plants that have yellow flowers with deeper yellow markings on the lower tepals.

Outside Shaba, thus excluding the endemics of heavy-metal soils, *Gladiolus ledoctei* appears to be most closely related to the Transvaal and eastern Botswanan *G. pretoriensis,* which is similar in habit and has both basal laminate leaves with either linear or terete blades, sometimes on the same plants, and partly to entirely sheathing cauline leaves, the latter less closely set than in *G. ledoctei. Gladiolus pretoriensis* also has partly to entirely dry floral bracts, angular seeds, and brownish corm tunics that accumulate in a neck at the base. There seems to be a trend for reduction in leaf number among the Shaban species related to this group. Thus *G. robiliartianus* has just three foliage leaves and no sheathing leaves, and *G. tshombeanus* has only two leaves, both largely sheathing.

SELECTED SPECIMENS

Zaire. Shaba: Kamfundwa, c. 7 km northwest of Kambole, cupriferous rock outcrop, 1410 m, 8 May 1983, *Schaijes 1908* (BR); Shinkolobwe, copper-enriched grassland, 5 June 1987, *Malaisse 11809* (BR); Kamoya No. 1, road from Kambove to Kakanda at km 2, 1450 m, 5 Apr. 1990, *Brooks et al. 109* (K, MO); Fungurume, cupriferous hill, 1200 m, 15 Feb. 1978, *Malaisse & Gregoire 6* (BR); Kwatebala, 31 May 1980, *Malaisse 10902* (BR); Kakanda, savanna, 22 Mar. 1974, *Malaisse 7697* (BR); Fungurume, cupriferous hill between Likasi and Kolwezi, 16 June 1975, *Breyne 2389* (BR); Fungurume, steppe, 1170 m, *Malaisse 7740* (BR); Fungurume, Hill IV, 1 Feb. 1977, *Malaisse 9210* (BR); Mupapala, copper-enriched steppe-savanna, 12 Apr. 1980, *Malaisse 10569* (BR); Fungurume, Hill I, south slope, "fleurs rose," 4 May 1957, *Duvigneaud 3013G* (BRLU); Petit Fungurume, without date, *Duvigneaud 3347* (BRLU); Upemba National Park, Munte Valley, 900 m, 23 June 1959, *de Wilde 752* (BR); Kalulumba Escarpment, 987 m, 12 Mar. 1949, *de Witte 58744* (BR); Upemba National Park, Ganza, between the Lukoka and Loie Rivers, 6 June 1949, *de Witte 6749* (BR).

26. *Gladiolus pungens* Duvigneaud & van Bockstal ex Còrdova

MAP 27. Cordòva, Bull. Jard. Bot. Nat. Belgique 60: 328 (1990). Type: Zaire, Shaba, schist escarpment north of Mitwaba, north slope of the Kalumengongo Valley, c. 1400 m, 21 Apr. 1957, *Symoens 3563* (*Duvigneaud & Timperman 3563* in the protologue) (BRLU, holotype).

EPONYMY

pungens, "sharp, piercing," referring to the stiff and fairly sharp-pointed leaf tips.

DESCRIPTION

Plants 60–70 cm high. CORM unknown. CATAPHYLLS dry and dark brown, accumulating in a neck around the base. LEAVES eight or nine, the lower three or four more or less basal and with long blades, some of these, at least, reaching to about the top of the spike, terete, 1–1.8 mm in diameter, the midribs and margins heavily thickened and with four narrow longitudinal grooves, the upper leaves cauline and largely to entirely sheathing, tapering to filiform apices, imbricate and sheathing the stem almost to the base of the spike. STEM unbranched, erect, c. 1 mm in diameter at the base of the spike.

SPIKE with about five flowers, evidently erect, lightly flexuose; BRACTS more or less dry and light brown at anthesis, c. 10 mm long, the inner slightly smaller than the outer. FLOWERS zygomorphic, white to pink, throat and narrow part of the tube purple, the lower tepals evidently lacking contrasting markings; PERIANTH TUBE obliquely funnel-shaped, c. 13 mm long; TEPALS unequal, the dorsal largest, c. 15 mm long, 9 mm wide, arching over the stamens, the lower three smallest, united for c. 2 mm, narrow below and abruptly expanded in the upper half, c. 12 mm long, in profile about a long as the upper. FILAMENTS unilateral and arcuate, c. 12 mm long, exserted 7 mm from the tube; ANTHERS c. 5 mm long, evidently yellow. OVARY obovoid, c. 3 mm long; STYLE arched over the stamens, dividing near the anther apices, the branches c. 2 mm long. CAPSULES unknown; SEEDS (according to Còrdova, 1990) flattened and winged, c. 3 mm in diameter but not present in material cited.

FLOWERING TIME. March and April.

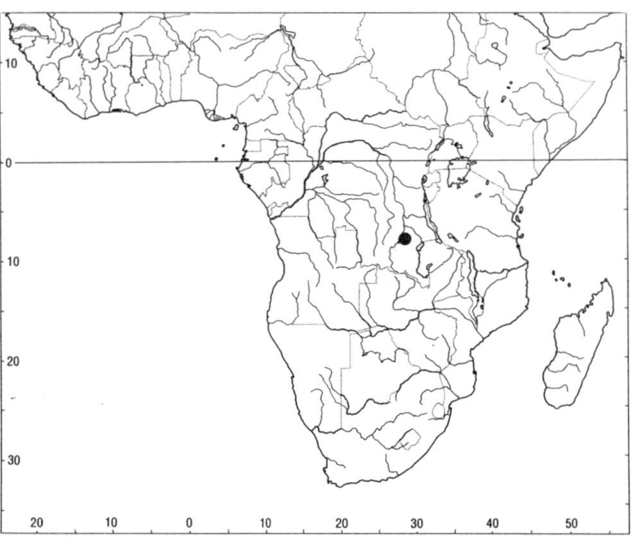

MAP 27. *Gladiolus pungens.*

DISTRIBUTION & HABITAT

As it is known from only one collection it is difficult to assess *Gladiolus pungens.* Its recorded range is along the northern slopes of the Kalumengogo Valley north of Mitwaba in northern Shaba Province, Zaire. The habitat is described as a rocky schist plateau.

DIAGNOSIS & RELATIONSHIPS

The several terete basal leaves and four or five cauline, partly sheathing leaves with attenuate filiform apices and small whitish to pink flowers that appear to lack contrasting markings on the lower tepals distinguish *Gladiolus pungens* from two or three other Zairian species that also have terete leaves, among which perhaps the most closely related is *G. ledoctei.* According to Còrdova (1990), *G. pungens* has flattened and winged seeds that contrast with the angular seeds of *G. ledoctei* and other species with similar terete leaves. Since the type and only specimen cited by Còrdova is in flower and lacks capsules it is difficult to place much credence in her assertion. If the seeds prove to be angular like those of *G. ledoctei,* it will be necessary to reassess whether it should continue to be regarded as distinct from that species.

SPECIMENS

Known only from the type, cited above.

27. *Gladiolus actinomorphanthus* Duvigneaud & van Bockstal

FIGURE 25, MAP 28. Duvigneaud & van Bockstal, Bull. Soc. Roy. Bot. Belgique 96: 123 (1963). Geerinck, Bull. Jard. Bot. Nat. Belgique 42: 271 (1972). Type: Zaire, Shaba, Kolwezi, Dikuluwe, meadow with *Vellozia* on very toxic soil south of cupriferous hill, Jan. 1960, *Duvigneaud 5141* (BRLU, holotype; BRLU, isotype).

EPONYMY

actinomorphanthus, "with an actinomorphic flower," referring to equal and similarly oriented tepals of the virtually regular (actinomorphic) flower.

DESCRIPTION

Plants slender, 30–50 cm high. CORM 12–18 mm in diameter, tunics of medium to fine, mostly vertical fibers, extending upward in a neck. CATAPHYLLS apparently two, the inner longer and reaching shortly above the ground. LEAVES five to seven, the lower two basal and longest, reaching the base of the spike, the blades terete, 1–1.5 mm in diameter, with four narrow longitudinal grooves, the other leaves cauline, partly to almost entirely sheathing, progressively shorter above, the free portions slender and bifacial or terete, brown-streaked or dry entirely above, imbricate. STEM simple or occasionally with one branch, erect, flexed outward at the base of the spike, only the upper 2–3 cm below the spike visible, c. 1 mm in diameter at the base of the spike.

SPIKE (5–)7- to 15-flowered, inclined or suberect; BRACTS green or flushed purple, (5–)7–9 mm long, the inner as long or slightly shorter than the outer. FLOWERS pale yellow or cream, unmarked, actinomorphic except for the apically outcurving tube; PERIANTH TUBE 4–5 mm long, narrowly and obliquely funnel-shaped, curved outward and emerging between the bracts; TEPALS equal, all directed forward and curving outward in the distal third, 6–7 mm long, c. 3 wide. FILAMENTS c. 7 mm long, symmetrically arranged, exserted c. 1 mm from the tube; ANTHERS centrally arranged around the style, 3–3.5 mm long. OVARY 1.5–2 mm long; STYLE central, dividing opposite the middle of the anthers, the branches c. 1 mm long, reaching to about the anther apices. CAPSULES broadly obovoid, c.

FIGURE 25. *Gladiolus actinomorphanthus.* Corm, leaves, and flowering spike, full size; single flowers and capsule, × 2; seeds, × 7; transverse section of leaf, much enlarged (*Schaijes 2491, 2729*).

8–10 mm long, showing the outline of the seeds; SEEDS compressed, with vestigial wings on the angles, or wing absent, c. 1.5 mm in diameter. CHROMOSOME NUMBER $2n = 28$.

FLOWERING TIME. December to late February in the middle of the wet season, also in May and June, 2 months into the dry season.

DISTRIBUTION & HABITAT

Endemic to southwestern Shaba Province in Zaire, *Gladiolus actinomorphanthus* is largely restricted to heavily mineralized soils, most often those with high concentrations of copper. In such sites plants are usually found among rocks in low grass. The plants are not absolutely restricted to such soils. They have also been collected near Kolwezi in woodland on rocky, nonmineralized soils (*Schaijes 5151–5164*). The southwestern part of Zaire was explored botanically comparatively late, and this species was apparently discovered only in 1957 by Paul Duvigneaud in the course of his studies of the flora of areas with high concentrations of heavy metals (Duvigneaud & Denaeyer-de Smet, 1963).

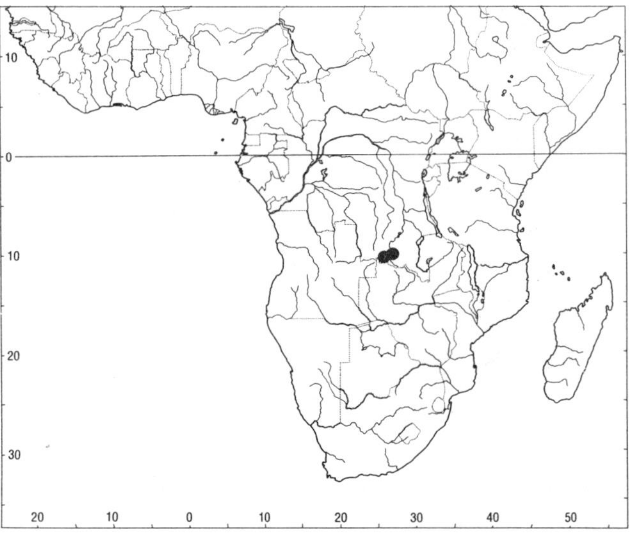

MAP 28. *Gladiolus actinomorphanthus.*

DIAGNOSIS & RELATIONSHIPS

No tropical African species of *Gladiolus* is more distinctive than *G. actinomorphanthus.* Its small, pale yellow, actinomorphic, urn-shaped flowers and symmetrically disposed stamens are unique in the genus. The coarsely fibrous corm tunics, a neck of fibers around the stem base, slender terete leaves with four narrow longitudinal grooves, and wingless seeds are virtually identical with the southern African *G. pretoriensis* and with three other Shaban endemics of mineral-enriched soils. It is clear that despite its superficially simple flowers the species is not primitive, as the actinomorphic flower suggests, but a specialized relative of these derived species, adapted to unusual soils or edaphically dry sites.

Late-flowering plants, blooming in June, *Schaijes 2491B* and *Schaijes 5090,* otherwise closely resembling *Gladiolus actinomorphanthus,* are puzzling. Although populations of the same species flowering several months apart is very unusual in *Gladiolus,* there seems little to separate the late-flowering plants morphologically from those flowering from late December to February except for somewhat drier sheathing leaves, 6–12 flowers per spike, and an occasionally branched individual. The flowers of what may be called the two races appear identical. Provisionally, the late-flowering plants must be regarded as a constituting an ecological race, not surprising in an area where rapid evolution seems to have occurred in *Gladiolus.* Late-flowering plants are marked below with an asterisk (*).

SELECTED SPECIMENS

Zaire. Shaba: 12.2 km west of Kolwezi, 1375 m, 8 Dec. 1985, *Schaijes 2729A* (BR); Kolwezi–Wakipinji road at km 97.3, rock outcrop, 950 m, 20 Feb. 1987, *Billiet & Jadin 4205* (BR); Dikuluwe, west of Kolwezi, cupriferous hill, 1400 m, 17 Feb. 1982, *Malaisse & Robbrecht 2381* (BR); Dikuluwe, among rocks on metaliferous hill, 8 May 1957 (fruit), *Duvigneaud 3088* (BRLU); Kolwezi, Mupine, poisoned dambo at the foot of metaliferous hill, 21 Jan. 1960, *Duvigneaud 5171* (BRLU); Mupine, summit of hill with mineralized rocks of copper and cobalt, 21 Jan. 1960, *Duvigneaud 5174* (BRLU); Chabara, near the foot of metaliferous hill, 27 Jan. 1960, *Duvigneaud 5277* (BRLU); 6.3 km west of Nzilo, c. 1200 m, 11 May 1985, *Schaijes 2491** (BR, K); 1 June 1991, *Schaijes 5090** (MO).

28. *Gladiolus robiliartianus* Duvigneaud

FIGURE 26, MAP 29. Duvigneaud, Bull. Soc. Roy. Bot. Belgique 91: 132 (1959). Geerinck, Bull. Jard. Bot. Nat. Belgique 42: 287 (1972). Type: Zaire, Shaba, Swambo, Jan. 1957, *Derricks 6* (BRLU, holotype).

SYNONYMY

Gladiolus duvigneaudii van Bockstal, Bull. Roy. Soc. Belgique 96: 126 (1963). Type: Zaire, Shaba, Chabara Hill, uranium-bearing rocks, Jan. 1960, *Duvigneaud 5137* (BRLU, holotype; BRLU, isotypes).

EPONYMY

robiliartianus, named in honor of H. Robiliart, an official of the mining company in what was at the time Upper Katanga, who assisted Paul Duvigneaud in his research on the heavy-metal-tolerant plants of the region.

DESCRIPTION

Plants slender, 25–40 cm high, growing in compact clumps. CORM 9–12 mm in diameter, tunics of brownish coriaceous to fibrous layers, the fibers moderately coarse to fine and mostly vertical, extending upward in a neck. CATAPHYLLS evidently two (possibly a third disintegrated), dry and brown, the inner longer and reaching shortly above the ground. LEAVES three, the lower two more or less basal and longest, exceeding the stem by 10–20 cm, tightly sheathing the stem for half to

FIGURE 26. *Gladiolus robiliartianus,* full size (*Bamps & Malaisse 8228, Malaisse & Robbrect 2188*).

three-quarters of its length, terete, 0.5–1 mm in diameter, with four narrow grooves running the length of the blade, third leaf similar but shorter, inserted in the upper third of the stem. STEM unbranched, erect below, abruptly flexed outward above the sheath of the uppermost leaf, only the upper 2–3 cm below the spike visible, c. 1.5 mm in diameter at the base of the spike.

SPIKE (3–)5- to 8-flowered, generally inclined or suberect; BRACTS green or becoming dry above, sometimes the upper half rust-colored, 9–12 mm long, the inner as long or slightly shorter than the outer. FLOWERS usually pale blue to lilac or pink with the lower tepals paler to nearly white distally and each with a dark blue to purple diamond-shaped mark in the distal half, or dark purple and apparently unmarked; PERIANTH TUBE 7–8 mm long, curved outward and emerging between the bracts; TEPALS unequal, the dorsal largest and hooded over the stamens, 12–14 mm long, c. 10 mm wide, the upper laterals directed forward, c. 9 mm long, the lower three held close together and more or less horizontal or tilted downward, united with the upper laterals for c. 5 mm, c. 9 mm long, narrowed below into claws, c. 3 wide. FILAMENTS c. 8 mm long, unilateral and arcuate, exserted 6.5 mm from the tube; ANTHERS parallel, c. 4.5 mm long. OVARY 1.5–2 mm long, style arching over the filaments, dividing opposite the base of the anthers, the branches 1–1.5 mm long, reaching to the middle of the anthers. CAPSULES globose, 5–6 mm long; SEEDS angular to more or less globose, c. 1.6 mm at the longest axis.

FLOWERING TIME. Late December and January, sometimes as late as February, thus early to mid wet season.

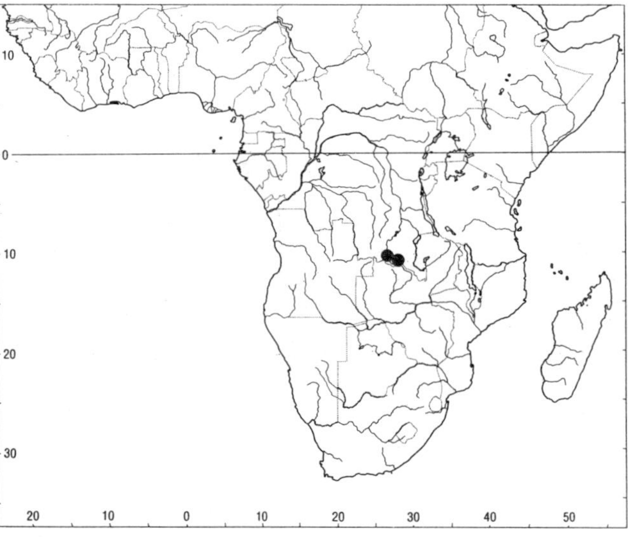

MAP 29. *Gladiolus robiliartianus.*

DISTRIBUTION & HABITAT

Gladiolus robiliartianus is restricted to southern Shaba Province, Zaire, where it occurs on heavy-metal-enriched soils in the Kolwezi District. Plants grow in rocky grassland and light woodland in well-drained situations. Their tolerance for high concentrations of copper allows them to survive in areas contaminated by mining activities.

DIAGNOSIS & RELATIONSHIPS

Among the small-flowered *Gladiolus* species of southern Shaba, *G. robiliartianus* stands out in consistently having just three foliage leaves, all with well-developed terete and more or less filiform blades, at least the lower two substantially exceeding the spike. The third leaf is sometimes rather reduced but is never entirely sheathing, and the lack of entirely sheathing leaves as well as a strongly flexed stem and inclined spike immediately distinguish it from *G. ledoctei,* which has a nearly erect spike and at least one and usually three or four sheathing leaves in addition to at least two and as many as four laminate leaves.

Van Bockstal (in Duvigneaud & Denaeyer-de Smet, 1963) separated *Gladiolus duvigneaudii* from *G. robiliartianus* largely on the basis of the former's pink rather than bluish flowers with somewhat different markings and slightly longer leaves. Both consistently have three laminate leaves, and the differences in leaf length between the types of the two species seem to me minimal. Flowers vary in color in populations discovered since, from bluish to purple, thus extending the range of variation in a character that is frequently variable within species. The distinctions seem trivial and I concur with Geerinck (1972) in regarding the two as conspecific.

SELECTED SPECIMENS

Zaire. Shaba: Chabara Hill, Kolwezi District, meadow on cupriferous rocks, 1450 m, 21 Jan. 1986, *Bamps & Malaisse 8228* (BR, K, MO); Chabara, cupriferous hill, 10 Jan. 1981, *Malaisse 11434* (BR), 23 Jan. 1981, *Malaisse 11531* (BR); 45 km east-northeast of Kolwezi, 25°50′ S, 10°32′ E, cupriferous rocks, 15 Feb. 1982, *Malaisse & Robbrecht 2188* (BR); Chabara, malachite-rich soil on the flank of cupriferous hill, 27 Jan. 1960, *Duvigneaud 5280* (BRLU); Swambo, *Vellozia* steppe on cupriferous hill, 5 Dec. 1959, *Duvigneaud 4405G* (BR, BRLU); between Menda and Mindingi, Atashyo dambo, forest at the edge of the dry belt, *Duvigneaud 4452* (BRLU); Mindingi, rocks mineralized with copper and cobalt, 6 Feb. 1960, *Duvigneaud 5440* (BRLU).

29. *Gladiolus tshombeanus* Duvigneaud & van Bockstal

FIGURE 27, MAP 30. Duvigneaud & van Bockstal, Bull. Soc. Roy. Bot. Belgique 96: 126 (1963). Type: Zaire, Shaba, between Baya and Welgelegen, Dec. 1959, *Duvigneaud 4371* (BRLU, holotype).

SYNONYMY

Gladiolus peschianus Duvigneaud & van Bockstal, Bull. Soc. Roy. Bot. Belgique 96: 126 (1963). Type: Zaire, Shaba, Luiswishi, steppe on the slope of copper-bearing hill, 4 Dec. 1959, *Duvigneaud 4389* (BRLU, holotype; BRLU, isotypes).

Gladiolus tshombeanus subsp. *parviflorus* Duvigneaud & van Bockstal, Bull. Soc. Roy. Bot. Belgique 96: 126 (1963). Type: Zaire, Shaba, Lubumbashi (Elisabethville), Étoile Mine, meadow on malachite rubble, 2 Dec. 1959, *Duvigneaud 4330* (BRLU, holotype; BRLU, isotypes).

EPONYMY

tshombeanus, named in honor of Moïse Tshombe, prime minister of Katanga in the mid 1960s when the province seceded from Zaire.

DESCRIPTION

Plants slender, 15–30(–40) cm high, evidently growing in clumps. CORM 9–12 mm in diameter, tunics of firm coriaceous layers, the outer decaying into coarse fibers from the base, reddish brown. CATAPHYLLS usually brownish, the inner longest and reaching shortly above the ground, usually dry at anthesis, accumulating with the corm tunics in a neck around the base. LEAVES two, the lower one basal, tightly sheathing the stem for half to three-quarters of its length, the upper leaf inserted in the upper half of the stem, sheathing for more than half its length, usually reaching to about the middle of the spike, usually exceeding the lower leaf, the blades terete, 2–6 cm long, four-grooved. STEM erect, unbranched, sheathed for most of its length, sharply flexed outward above the sheath of the upper leaf.

SPIKE 4- to 7-flowered, inclined 45°; BRACTS green or more often dry above and membranous, sometimes completely dry toward the end of flowering; 8–10(–14) mm long, the inner slightly shorter than the outer. FLOWERS usually pale blue to lilac, the lower lateral tepals each with a dark purple mark at the base of the limb, sometimes dark purple with the lower lateral tepals pale in the lower half and with purple marginal spots; PERIANTH TUBE (6–)8–10 mm long, curved outward and emerging between the bracts; TEPALS unequal, the dorsal largest and hooded over the stamens, 12–14 mm long, (6–)7–9 mm wide, the upper laterals directed forward, the lower three held close together and united with the upper lat-

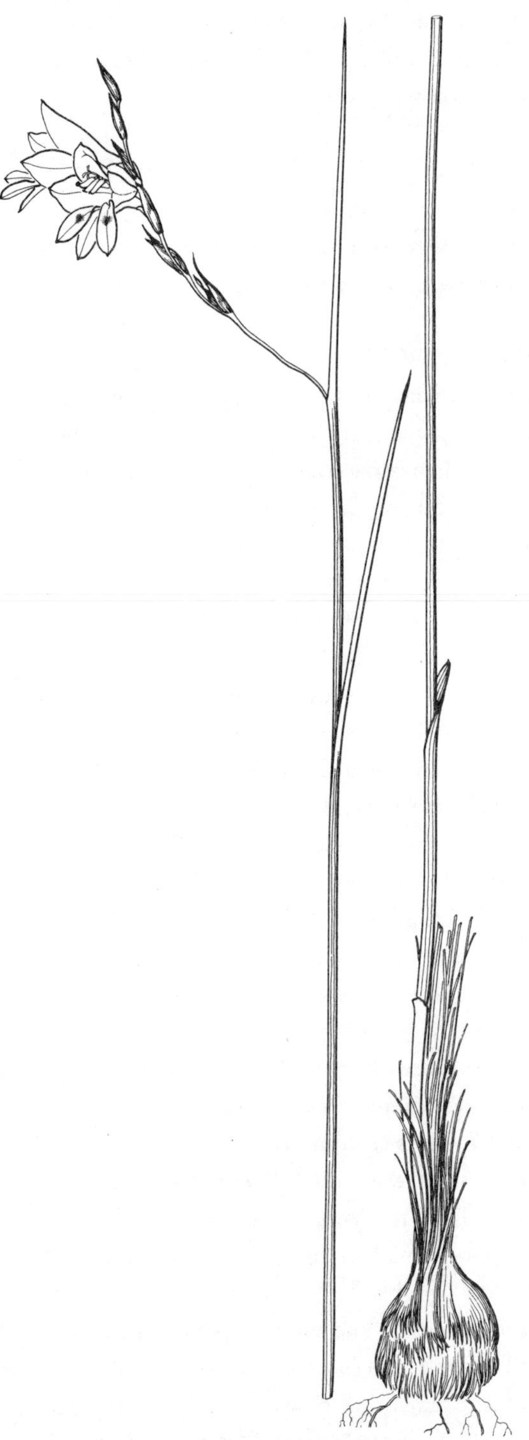

FIGURE 27. *Gladiolus tshombeanus,* full size (*Duvigneaud 4s806*, *Poelman 105*).

erals for 3–5 mm and to themselves for 1–2 mm, more or less horizontal or tilted downward, (5–) 7–9 mm long, narrowed below into claws, the limbs abruptly widened to c. 2 mm. FILAMENTS (7–)9–10 mm long, unilateral and arcuate, exserted 3–5 mm from the tube; ANTHERS parallel, 5–6 mm long, yellow. OVARY c. 1.5 mm long; STYLE arching over the filaments, dividing opposite the middle of the anthers, branches 1.5–2 mm long. CAPSULES nearly globose, c. 5 mm long; SEEDS angular, c. 1.2 mm in diameter, the testa fairly loose and forming ridges at the angles.

FLOWERING TIME. November and December.

DISTRIBUTION & HABITAT

Gladiolus tshombeanus is restricted to southern Shaba Province, Zaire, in the vicinity of Lubumbashi, where it is recorded largely from heavy-metal-enriched soils that occur throughout this region. Collection information indicates that it grows in open rocky sites or occasionally in poisoned (and apparently seasonal) dambos. Duvigneaud and Denaeyer-de Smet (1963) have recorded in detail the distribution of the three main variants of the species: the dark-purple-flowered form occurs to the north of Lubumbashi at Luiswishi, the smaller and light-purple-flowered plant sometimes called subspecies *parviflorus* (not recognized here but such plants are marked with an asterisk (*) in the list of Selected Specimens, below) exclusively at the Étoile Mine near Lubumbashi, and the typical light-purple-flowered form at Lubumbashi and at several localities to the west. Additional collecting since their work has established the presence of the dark purple-flowered form also at Étoile (*Malaisse 12822*). Another specimen (*Duvigneaud 3014G*), if correctly labeled, documents dark-purple-flowered plants at Fungurume.

DIAGNOSIS & RELATIONSHIPS

Gladiolus tshombeanus is easily confused with *G. gracillimus,* a widespread southern tropical African species of seasonally wet, often inundated habitats, and the two have been regarded as a single species (Geerinck, 1972). Their similarities include the distinctive slender habit, flowering stem consistently bearing two short-bladed leaves, and relatively small flowers. In *G. tshombeanus* the lower of the two leaves is basal, sheathing for most of its length and terminating in a short, terete, four-grooved blade; the upper leaf is inserted in the upper third of the stem and also has a short, terete, four-grooved blade. The general pattern is almost identical in *G. gracillimus,* but the leaf blades are linear with strongly raised margins and midrib and the lower leaf often exceeds the upper, the reverse of the condition in *G. tshombeanus.* The two species differ more obviously in their corm tunics, which in *G. tshombeanus* are reddish brown, coriaceous, and decay into a coarsely fibrous mass that accumulates with time into thick neck around the base of the stem. In *G. gracillimus* the corm tunics are finely fibrous and blackish, never brown, and do not accumulate to a significant degree. The spikes of the two also differ, *G. tshombeanus* generally has more flowers, (4–)6–10, than *G. gracillimus,* which has spikes of 2–4(–6) flowers. Perhaps most significantly, the seeds of the two species are different. *Gladiolus tshombeanus* has angular seeds in small globose capsules, whereas *G. gracillimus* has flattened, winged seeds typical of the genus.

The brown, densely accumulating corm tunics that form a neck around the base, the terete leaf blades, and the angular seeds are also characteristic of several Shaban species of section *Heterocolon,* and *Gladiolus tshombeanus* is evidently a vegetatively specialized member of the section. It is probably most closely allied to *G. robiliartianus,* which differs significantly only in having three leaves, the lower two more or less basal and all with terete, four-grooved blades.

Floral variation in *Gladiolus tshombeanus* is noteworthy. The type form has flowers 23–25 mm long, colored pale bluish with purple nectar guides and a purple flush to the dorsal tepals. Plants from Luiswishi and the Étoile Mine have flowers of comparable size with dark purple flowers and slightly differently shaped nectar guides (Duvigneaud & Denaeyer-de Smet, 1963: 219) outlined in white, and a smaller-flowered plant from Étoile Mine has a perianth with the same color and markings as they type form. The discrete pattern of local variation suggests the differentiation of distinct local races, but differentiation has not proceeded far enough to merit taxonomic recognition. Comparable variation occurs in several other species of *Gladiolus,* notably the related *G. atropurpureus* and *G. gracillimus,* in both of which there are races with light bluish and dark purple flowers.

HISTORY

Although apparently first collected in 1937 by R. P. Salesiens, *Gladiolus tshombeanus* was only identified as a distinct species in 1963 by Duvigneaud and van Bockstal, who documented its occurrence mostly on heavy-metal-enriched soils. At the same time Duvigneaud recognized *G. peschianus,* which

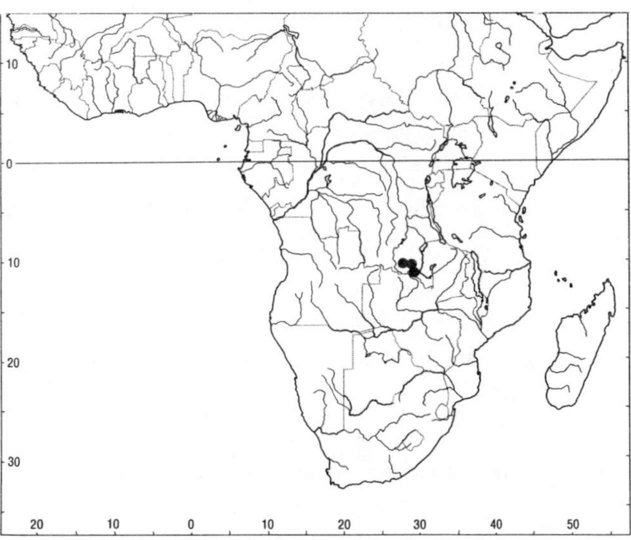

MAP 30. *Gladiolus tshombeanus.*

I regard as a local color form. Geerinck (1972) considered both conspecific with *G. gracillimus,* a treatment that for the reasons outlined above I prefer not to follow.

SELECTED SPECIMENS

Zaire. Shaba: Between Baya and Welgelegen, Dec. 1959, *Duvigneaud 4371* (BRLU); Luiswishi, meadow on copper-enriched soil, 9 Nov. 1959, *Duvigneaud 4030G* (BRLU); Luiswishi, *Cryptosepalum* savanna on slope of copper-rich hill, 4 Dec. 1959, *Duvigneaud 4390G1* (BRLU); 25 km north of Lubumbashi near farm St. Hubert, 15 Nov. 1969, *Lisowski 81515* (POZG); Kasonta Nord, along footpaths in grassland on copper-rich soil, 10 Nov. 1959, *Duvigneaud 4055G* (BRLU); Kasonta Nord, grassland on copper-enriched soil, 2 Dec. 1959, *Duvigneaud 4405G1* (BRLU)); Lupoto, poisoned dambo, 10 Nov. 1959, *Duvigneaud 4062G1* (BRLU); Lupoto hill at old copper workings, 27 Nov. 1957, *Schmitz 60332* (BR); Tilwezembe, rocky summit, on metal-bearing substrate, 5 Dec. 1959, *Duvigneaud 4588G* (BRLU), 15 Dec. 1959, *Duvigneaud 4585G* (BRLU); Karavia, copper mine, poisoned dambo, 10 Nov. 1959, *Duvigneaud 4067G* (BRLU); Étoile, 1260 m, copper-enriched steppe-savanna, 25 Nov. 1983, *Malaisse 12822* (BR); to the south of Étoile Mine, in a large poisoned dambo greening after fire, 7 Nov. 1959, *Duvigneaud 4015** (BRLU); Lubumbashi (Elisabethville), Étoile Mine, Dec. 1938, *Russel 49** (K), Dec. 1959, *Duvigneaud 4336** (BRLU); Étoile, among schist rocks, 23 Nov. 1974, *Malaisse 7966** (BR). Keyberg, fairly dry rocky soil, 28 Nov. 1947, *Schmitz 1086* (BR); Keyberg, open woodland in valley bottom, 6 Dec. 1956, *Detilleux 216* (BR); near the temporary university buildings, Lubumbashi, 14 Nov. 1961, *Poelman 105* (BR, K); Lubumbashi, 1937, *Salesiens 1002** (BR); Fungurume, disturbed savanna on the lower slopes of the hill, 4 May 1957, *Duvigneaud 3014G* (BRLU).

Smaller-flowered plants, sometimes recognized as subspecies *parviflorus,* are marked above with an asterisk (*).

30. *Gladiolus gracillimus* Baker

PLATE 16, FIGURE 28, MAP 31. Baker, Kew Bull. 1895: 74 (1895); Fl. Trop. Africa 7: 363 (1898). Geerinck, Bull. Jard. Bot. Nat. Belgique 42: 275 (1972). Goldblatt, Fl. Zambesiaca 12(4): 83 (1993). Type: Zambia, Fwambo (Lake Tanganyika), in 1893, *Carson 118* (K, holotype).

SYNONYMY

Gladiolus aphanophyllus Baker, Fl. Trop. Africa 7: 363 (1898). Type: Tanzania, Tanganyika Plateau, Urungu, *Carson s.n.* (K, holotype, mounted with two spikes of *G. unguiculatus*).

EPONYMY

gracillimus, "most graceful," alluding to the graceful, slender form of the plants.

DESCRIPTION

Plants small, 15–30(–40) cm high. CORM 9–12 mm in diameter, tunics of fine netted fibers, occasionally extending upward in a weakly developed neck. CATAPHYLLS membranous, the upper longest, reaching shortly above the ground and then green or flushed purple. LEAVES two, the lower basal, tightly sheathing the stem for half to three-quarters of its length, the blade usually reaching to about the top of the spike, the upper leaf inserted shortly below the spike, shorter or sometimes exceeding the apex of the lower leaf, blades linear, 3–5 cm long, 0.5–1 mm wide, the margins and midrib somewhat to moderately thickened and with two fairly narrow grooves on each surface. STEM erect, unbranched, sheathed for most of its length, flexed sharply outward above the sheath of the upper leaf.

SPIKE 2- to 4(–7)-flowered, inclined 45°, flexuose; BRACTS green, sometimes flushed purple, 7–10 mm long, the inner slightly shorter than the outer. FLOWERS usually pale blue to lilac, occasionally white, the lower three tepals each with a dark blue to purple diamond-shaped marking in

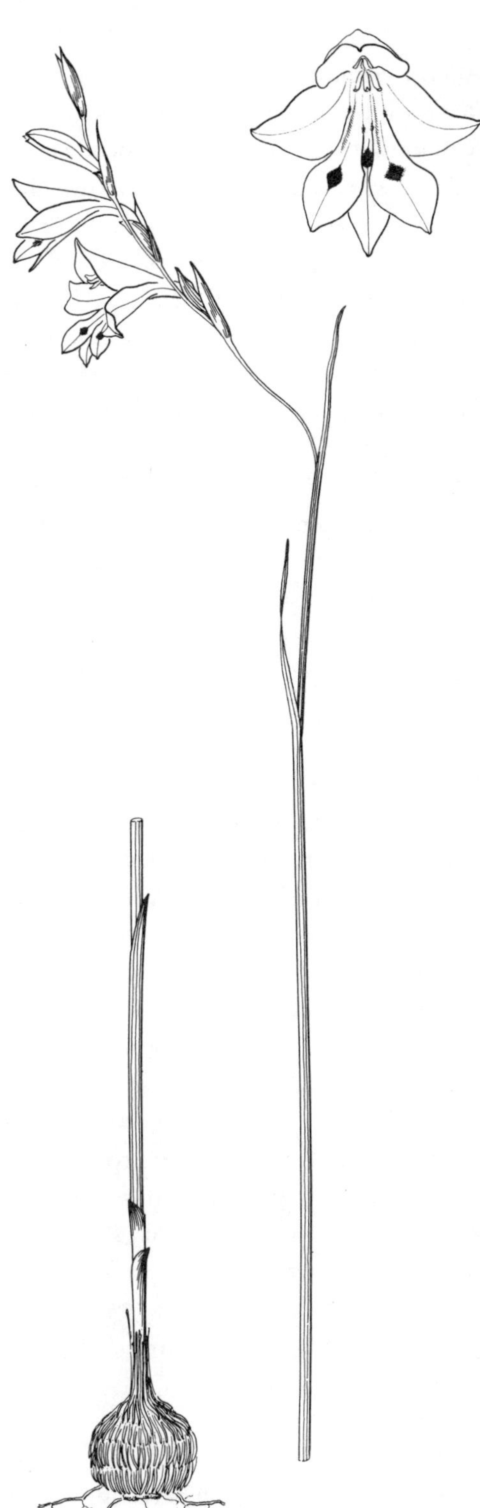

FIGURE 28. *Gladiolus gracillimus.* Corm and flowering spike, × 0.5, single flower, full size (*Goldblatt et al. 8138*).

the upper half, or uniformly purple, or pale yellow to greenish with deep yellow markings on the lower three tepals; PERIANTH TUBE 7–10 mm long, curved outward and emerging between the bracts; TEPALS unequal, the dorsal largest and hooded over the stamens, (8–)10–15 mm long, 8–11 mm wide, the upper laterals directed forward, the lower three united for 3–4 mm, held close together, more or less horizontal or tilted downward, 7–12 mm long, 4–5 mm at the widest, in profile much exceeding the upper, abruptly narrowed below into claws. FILAMENTS 6–8 mm long, unilateral and arcuate, usually exserted 3–4 mm from the tube; ANTHERS parallel, 4.5–6 mm long, usually yellow. OVARY c. 2 mm long; STYLE arched over filaments, reaching to about the apex of the anthers, branches 1.5–2 mm long, extending beyond the anthers. CAPSULES obovate-ellipsoid, 7–8 mm long; SEEDS oblong-elliptic, (3.5–)5 × 2.5 mm, broadly (to narrowly) winged.

FLOWERING TIME. November and December, occasionally in early January, at the end of the dry season or early in the wet season.

DISTRIBUTION & HABITAT

Fairly widely distributed across central Africa, *Gladiolus gracillimus* extends across the northern Zambian provinces of Luapula and Northern Province into western Tanzania and northern Malawi. In Tanzania it is restricted to the Ufipa Plateau and in Malawi it ranges no farther south than Mzimba. It occurs almost exclusively in wet habitats either permanently or seasonally waterlogged, where it blooms at the end of the dry season when the surrounding vegetation is burnt or grazed to the ground, and its flowers, borne on relatively short stems, make a good display. By the time the capsules mature, the plants are often in saturated soil or even in standing water, and under such conditions the seeds are probably readily dispersed by water rather than by wind despite the broad membranous wing.

DIAGNOSIS & RELATIONSHIPS

The unusual habit of *Gladiolus gracillimus,* of low, slender stature and with only two leaves, the lower sheathing the stem for most of its length and with a short linear blade that usually exceeds the short upper leaf, is shared by two other species, *G. pusillus* from northern Zambia and Shaba Province, Zaire, and *G. tshombeanus* from heavy-metal-enriched sites in southern Shaba. *Gladiolus pusillus* has very small yellow to white flowers with yellow nectar guides and is easily distinguished from *G. gracillimus* despite their similar fine netted fibrous corm tunics. *Gladiolus tshombeanus* is more readily confused with it, and only the brown coarsely fibrous or partly coriaceous corm tunics that extend upward in a neck, terete, four-grooved leaf blades, and angular rather than winged seeds separate the two. The flowers are fairly typical in shape for the smaller-flowered species of *Gladiolus,* although they exhibit a fair range of variation in size. The upper tepal is always largest and hooded over the stamens, and the lower three tepals are narrow, united for a short distance, and distinctly clawed below.

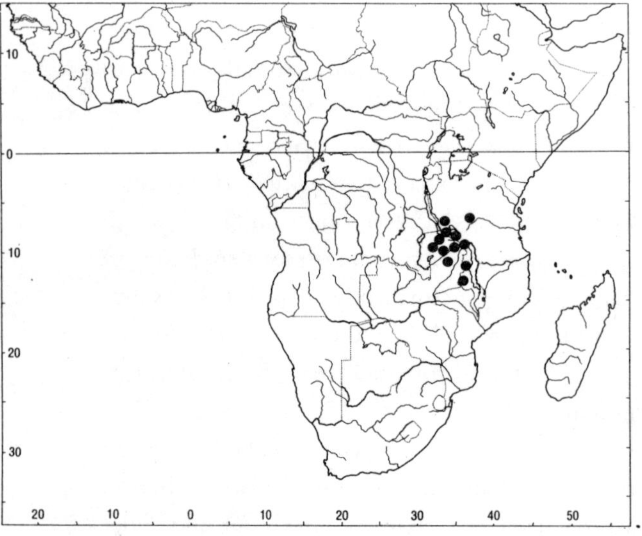

MAP 31. *Gladiolus gracillimus.*

Almost certainly, the three species are closely related for it is difficult to see this very specialized growth form having arisen independently in separate lineages. *Gladiolus tshombeanus* strongly resembles several other *Gladiolus* species in Shaba in its corm tunics and narrow, usually terete leaves, and there seems little doubt that its relationships (discussed more fully under that species) lie with this group of *Gladiolus* species, all of which additionally have angular rather than winged seeds. Thus *G. gracillimus* is interpreted here as having arisen from ancestors adapted to heavy-metal-enriched soils. Presumably, this ancestral adaptation allowed it to move into the semiaquatic habitat unusual for corm-bearing species. In this scenario, the presence of winged seeds in *G. gracillimus* (and *G. pusillus*) must be interpreted as a reversal of the loss of this feature.

SELECTED SPECIMENS

Malawi. Northern: Karonga, Mafinga foothills, wet dambo, 19 Dec. 1964, *Robinson 6316* (K, SRGH); 16 km east of Chitipa, grassy meadow, 27 Dec. 1970, *Pawek 4182* (K, MAL, SRGH); Mzimba, 10 km east of Mzambazi, wet bottom, 30 Dec. 1975, *Pawek 10662B* (K, MO).

Tanzania. Iringa: Ruaha National Park at Magangwe Ranger Post, 1330 m, 14 Dec. 1972, *Bjornstad 2090* (EA, UPS). Rukwa: Sumbawanga, Malonje plateau, old Sumbawanga road, marshy grassland, *Richards 15891* (K); 25 km south of Sumbawanga, wet dambo, 3 Jan. 1962, *Robinson 4888* (K, SRGH); near Tatanda, dry dambo along river, 11 Nov. 1986, *Goldblatt, Brummitt, & Lovett 8138* (DSM, K, MO, PRE). Mbeya: Mbozi, 1700 m, 19 Nov. 1932, *Davies 704* (K).

Zambia. Luapula: Kasama, Kilolo village, Mpanda River, grassy swamp, 13 Dec. 1964, *Richards 19364* (K, MO); 8 km north of Kasama, seasonal dambo, 7 Jan. 1967, *Anton-Smith 181,827* (K, SRGH). Northern: Mbala District, Kambole road, open bush, 1500 m, 1 Jan. 1955, *Richards 3819* (K); Mbala, Nmcole, 900 m, stony hillside, 18 Dec. 1956, *Richards 7308* (MO); near Nakatali farm, burnt grassland, swampy in rains, 14 Nov. 1965, *Richards 20655* (BR, K, LISC).

31. *Gladiolus pusillus* Goldblatt

MAP 32. Goldblatt, Fl. Zambesiaca 12(4): 85 (1993). Type: Zambia, Luapula, Kawambwa District, M'tunatusha River, damp white sand, 1290 m, 28 Nov. 1961, *Richards 15422* (K, holotype; BR, SRGH, isotypes).

EPONYMY

pusillus, "small, diminutive," alluding to the very small flowers, no more than 16 mm long.

DESCRIPTION

Plants slender, 15–30(–40) cm high. CORM 7–12 mm in diameter, tunics of fine netted fibers sometimes accumulating in a weakly developed neck around the base. CATAPHYLLS pale and membranous, the upper longest, reaching shortly above the ground and then green or sometimes flushed with purple. LEAVES two, the lower one basal, tightly sheathing the stem for most of its length, often reaching to about the top of the spike, the upper leaf inserted shortly below the spike, the apex usually shorter than but occasionally exceeding the lower leaf, blades linear, 2–5 cm long, c. 0.5 mm wide, the margins and midrib somewhat to heavily thickened and with two fairly narrow grooves on each surface. STEM erect, unbranched, sheathed for most of its length, flexed sharply outward above the sheath of the upper leaf.

SPIKE (2–)4- to 9-flowered, inclined about 30°, lightly flexuose; BRACTS green, sometimes flushed purple, 6–9 mm long, generally slightly more than one internode long, thus usually overlapping the base of the next bract, the inner slightly shorter than the outer. FLOWERS pale yellow or sometimes whitish or pink, the keels sometimes red to brown, the lower three tepals each with a deep yellow mark in the upper half often outlined greenish or brown; PERIANTH TUBE 3.5–4.5 mm long, curved outward and emerging between the bracts; TEPALS unequal, the dorsal largest and hooded over the stamens, 8–10 mm long, c. 6 mm wide, the upper laterals directed forward, the lower three joined to the upper laterals for 3 mm and to themselves for c. 2 mm, closely aligned, more or less horizontal, 5–7 mm long, 1.5 mm at the widest, in profile much exceeding the upper, abruptly narrowed below into claws. FILAMENTS c. 5 mm long, unilateral and arcuate, exserted c. 2.5 mm from the tube; ANTHERS parallel, 4.5–5 mm long, shortly apiculate, reaching or barely exceeding the apices of the upper lateral tepals, yellow. OVARY c. 2 mm long; STYLE arching over the filaments, dividing opposite the middle of the anthers, the branches c. 1 mm long, not reaching the anther apices. CAPSULES broadly obovoid, 5–6 mm long; SEEDS more or less oblong, wing best developed distally and hardly at all laterally, 4 × 2 mm.

FLOWERING TIME. November and December, occasionally in January, at the end of the dry season or early in the wet season.

DISTRIBUTION & HABITAT

Gladiolus pusillus has a fairly narrow range in tropical Africa, being restricted to southern Zaire and adjacent Zambia where it occurs in Copperbelt, Luapula, and parts of Northern Province. Like its close relative, *G. gracillimus,* it occurs almost exclusively in wet habitats either permanently or seasonally waterlogged, where it blooms at the end of the dry season when the surrounding vegetation is burnt or grazed to the ground.

DIAGNOSIS & RELATIONSHIPS

Although known since at least 1933, when it was collected by Paul Quarré in what was then Katanga Province of Zaire, *Gladiolus pusillus* was included in *G. gracillimus* until 1993, when it was recognized as a separate species (Goldblatt, 1993). It is notable, however, that the Belgian botanist, Paul Duvigneaud, who had considerable field knowledge of *Gladiolus,* considered it a separate species and annotated some herbarium specimens with the unpublished name *G. opheliae* Duvigneaud & van Bockstal. That *G. pusillus* was not separated from *G. gracillimus* sooner was clearly due to their virtually identical, and highly unusual, vegetative morphology—a slender habit, fine fibrous corm tunics, only two leaves, the lower sheathing most of the stem, and a strongly flexed spike. The tiny yellow to white flowers of *G. pusillus,* with the upper tepal 8–10 mm long, contrast strikingly with the larger pale blue-lilac to occasionally dark purple flowers of *G. gracillimus,* with the upper tepal 11–15 mm long. The two can be distinguished by other differences in flower size, including a perianth tube 3.5–4 mm long and anthers reaching or exceeding the upper lateral tepals in *G. pusillus,* contrasting with a tube 7–9 mm long and anthers not reaching the upper lateral tepal apices in *G. gracillimus.* Both *G. pusillus* and *G. gracillimus* have winged seeds, unlike the vegetatively similar Zairian endemic, *G. tshombeanus,* which favors well-drained soils often having high concentrations of heavy metals.

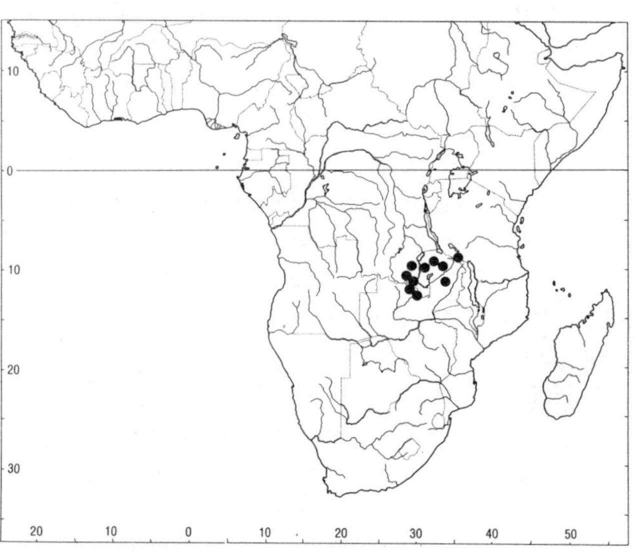

MAP 32. *Gladiolus pusillus.*

SELECTED SPECIMENS

Zaire. Shaba: Kundelungu Plateau, near source of the Lofoi, 1600 m, 7 Jan. 1971, *Lisowski 81986* (POZG); Luapula Valley near Kiniama, 11 Dec. 1970, *Lisowski 81651* (POZG); pastures near Karavia, Nov. 1933, *Quarré 3634* (BR); 6 km north of Poste Katshupa, 1680 m, 11 Nov. 1981, *Malaisse 12047* (BR); vicinity of Lubumbushi, dambo at edge of the Matwebo, 25 Oct. 1970, *Lisowski 733C* (BR); Keyberg, 7 km south-southwest of Lubumbashi, 23 Nov. 1948, *Schmitz 2123* (BR)); 20 km northwest of Lubumbashi, Kilibati dambo, 10 Dec. 1957, *Schmitz 5809* (BR).

Zambia. Copperbelt: Chingola, moist peaty dambo, 28 Nov. 1959, *Fanshawe 5267* (K, NDO), Chingola, dambo on peaty granitic sand, 27 Nov. 1959, *Duvigneaud 4236* (BRLU); Nkana, damp marsh, Jan. 1937, *Brenan 47* (K). Luapula: Mbereshi–Kawambwa road, 1050 m, 18 Jan. 1960, *Richards 12398* (K). Northern: 8 km east of Kasama, damp ground, 4 Dec. 1961, *Robinson 4716* (K, PRE, SRGH); Mporokoso District, dambo 120 km northwest of Kasama, 1400 m, 17 Oct. 1967, *Simon & Williamson 1084* (K, SRGH); Shiwa Ngandu, wet dambo, 21 Dec. 1964, *Robinson 6327* (K, SRGH);

Tanzania. Rukwa: Uvinza–Mpanda road, 23 Nov. 1962, *Verdcourt 3441A* (K).

SECTION *HEBEA* (PERSOON) BENTHAM & J. D. HOOKER

Bentham & J. D. Hooker, Genera Plantarum 3(2): 710 (1883). *Gladiolus* group *Hebea* Persoon, Synopsis Plantarum 1: 44 (1805). Type: *G. alatus* Linnaeus (lectotype designated here).

Synonymy

Subgenus *Hebea* (Persoon) Baker, J. Linn. Soc., Bot. 16: 177 (1878). Although the basionym was not directly cited, it is reasonable to assume that this was intended as a combination.

Description

Plants medium or small, usually unbranched. LEAVES lanceolate, linear, rarely terete, the upper smaller and sometimes entirely sheathing, the sheaths open to the base, contemporary with the flowers. SPIKES usually inclined; BRACTS short in tropical African representatives, 7–16 mm long, generally about as long or slightly longer than the perianth tubes. FLOWERS small (larger in some southern African species), 24–32 mm long, the lower tepals usually with contrasting nectar guides; PERIANTH TUBES 9–12 mm long in tropical African species; TEPALS usually all more or less clawed (at least unusually narrow in the lower half), the dorsal tepals about half again as long as the tubes (much longer in some southern African species), lower tepals narrower and usually shorter than the dorsal, clawed below. ANTHERS obtuse, without apical appendages. CAPSULES oblong-ellipsoid; SEEDS winged.

There are only two species of section *Hebea* in tropical Africa, one in southern Angola, and the other in Zimbabwe, this shared with southern Africa. The section is very diverse in the southern and western Cape, South Africa, and has radiated extensively in semiarid habitats in that area of winter rainfall and summer drought.

32. *Gladiolus permeabilis* subsp. *edulis* (Burchell ex Ker Gawler) Obermeyer

FIGURE 29, MAP 33. Obermeyer, J. S. African Bot., Suppl. 10: 135 (1972). Van Wyk & Malan, Field Guide Wild Flowers Witwatersrand Pretoria 150, pl. 352 (1988). Goldblatt, Fl. Zambesiaca 12(4): 75 (1993). *Gladiolus edulis* Burchell ex Ker Gawler, Edwards's Bot. Reg. 2: pl. 169 (1817); Baker, Flora Capensis 6: 161 (1896); Fl. Trop. Africa 7: 373 (1898). Sölch, Prodromus Fl. Südwestafrika 155: 4 (1969). Type: South Africa, Cape, Pellat Plains, *Burchell 2240* (K, holotype; B, isotype).

SYNONYMY

Gladiolus remotiflorus Baker, Bull. Herb. Boissier, Sér. 2, 1: 867 (1901). Type: South Africa, Transvaal, between "Poster" and Trigaardsfontein, *Rehmann 6618* (Z, holotype, not seen).

EPONYMY

permeabilis, "able to pass through," referring to the gap between the dorsal and upper lateral tepals so that in profile one can see through the flower; *edulis,* "edible," reflecting the explorer William Burchell's observation that the corms were eaten by the native people who he met in the northern Cape, South Africa.

DESCRIPTION

Plants 30–60 cm high. CORM 10–12 mm in diameter, tunics coarsely fibrous. CATAPHYLLS pale and more or less membranous, the upper green or purplish above the ground. LEAVES three to six, at least the lower two or three inserted at or below ground level, these shorter to about as long as the stem, occasionally longer, linear, 1–2.5(–3) mm wide, the midrib and margins lightly, rarely moderately, thickened and hyaline, the upper two or three leaves cauline, shorter than the basal and sometimes entirely sheathing. STEM simple or branched, 1–2 mm in diameter at the base of the spike.

SPIKE 4- to 8-flowered; BRACTS apparently green to gray below, membranous and becoming dry and pale above, 7–12 mm long, the inner usually shorter than the outer, the outer acute to somewhat attenuate, the inner forked apically. FLOWERS whitish to dull cream, gray or mauve, flushed with gray-blue on the dorsal tepal, the keels dark purple to maroon, the lower lateral tepals yellow in the upper half, sometimes intensely sweet-scented; PERIANTH TUBE widening and curving outward near the apex, 8–12 mm long; TEPALS unequal, long-attenuate and narrowed into claws below, the dorsal largest, hooded over the stamens, 16–20 mm long, often less when dry, the lower three joined for c. 2 mm with the upper lateral tepals and to themselves for 3–5 mm, 13–15 mm long, the limbs abruptly expanded, more or less horizontal or flexed downward distally, the cusps often twisted. FILAMENTS arcuate, 13–14 mm long, inserted in the middle of the tube, exserted for 9–10 mm; ANTHERS 5–6 mm long, purple. OVARY narrowly obovoid, c. 2.5 mm long; STYLE arching over the filaments, dividing between the middle and apex of the anthers, the branches 1.5–2 mm long, much expanded above. CAPSULES nearly globose to obovoid, 6–10 mm long; SEEDS oval, broadly winged, 4–5 × 2.5–3.5 mm. CHROMOSOME NUMBER $2n = 30$.

FLOWERING TIME. February to April.

DISTRIBUTION & HABITAT

Gladiolus permeabilis subsp. *edulis* is widespread in dry areas of southern Africa, including the Orange Free State, Transvaal, and southern and northern Cape of South Africa, and Namibia (Lewis et al., 1972). It also extends locally into southern tropical Africa, where it occurs in eastern Botswana and fairly widely in both in the drier western and inte-

FIGURE 29. *Gladiolus permeabilis* subsp. *edulis*. Corm, leaves, and flowering spike, full size; single flower, × 1.3 (corm, leaves, and spike, *Goldblatt & Manning 8806;* single flower, *Hanekom 1801*).

rior half of Zimbabwe. There, it occurs at isolated sites in the north, near Rusape and on the Great Dyke. Plants generally favor deep, sandy soils, mostly Kalahari sands, but also grow in open, rocky sites, usually among grasses in light woodland. At least some populations have flowers that are so strongly fragrant that the presence of the species can often be detected before plants are seen.

A population from the Great Dyke in northern central Zimbabwe (*Philcox & Müller 9079*), found on a rocky ridge in this area of serpentine- and other heavy-metal-enriched soils, is unusually robust and may be found on further investigation to be a separate species. Although it has the typical habit and flower coloring of subspecies *edulis,* the tepals lack the characteristic long tapering apices, and the rather broad leaves have unusual, heavily thickened margins.

DIAGNOSIS & RELATIONSHIPS

Gladiolus permeabilis subsp. *edulis* is generally easy to recognize by the combination of linear basal leaves with blades seldom exceeding 2 mm in width, and a spike of small rather dull-colored flowers, the tepals of which are all very narrow below and are darkly marked purple to maroon on the keels. Viewed in profile, there is a wide gap or window between the bases of the dorsal and upper lateral tepals, except in subspecies *wilsonii* (Baker) G. Lewis. This is a useful diagnostic characteristic, although it is shared by a few other species, notably the western Angolan *G. fenestratus,* which has terete leaves, and the widespread *G. unguiculatus.* The absence of the window in subspecies *wilsonii,* as well its broad tepals and hooded dorsal tepal, makes the inclusion of this southeastern African plant in *G. permeabilis* questionable. As is typical of section *Hebea,* the lower tepals of *G. permeabilis* are united for a short distance and extend outward, so that in side view they appear to exceed the dorsal tepal although they are actually shorter in length when the united portion is excluded from the measurement. In tropical Africa and most of summer-rainfall southern Africa, including Namibia, the apices of the tepals of *G. permeabilis* are drawn into long tapering cusps, the characteristic that serves to distinguish subspecies *edulis* from subspecies *permeabilis,* which is restricted to the winter-rainfall area of the southern Cape and has acute tepals.

Gladiolus permeabilis, described by Daniel de la Roche (1768: 27 & pl. 2), was based on plants of uncertain origin, but floral morphology makes it clear that they were collected in the winter-rainfall southwestern part of southern Africa. As noted by Baker (1896), *G. permeabilis* and *G. edulis,* described by J. B. Ker in 1817, differ from one another mainly (if not exclusively) in the long-cuspidate tepals of the latter.

The value of treating *Gladiolus edulis* as separate from *G. permeabilis* is doubtful, and the subspecific status accorded the former by Lewis et al. (1972) is followed here, despite the variation in cusp length that makes assignment of some collections to subspecies somewhat arbitrary. As discussed above, I disagree with Lewis et al.'s treatment of *G. perme-*

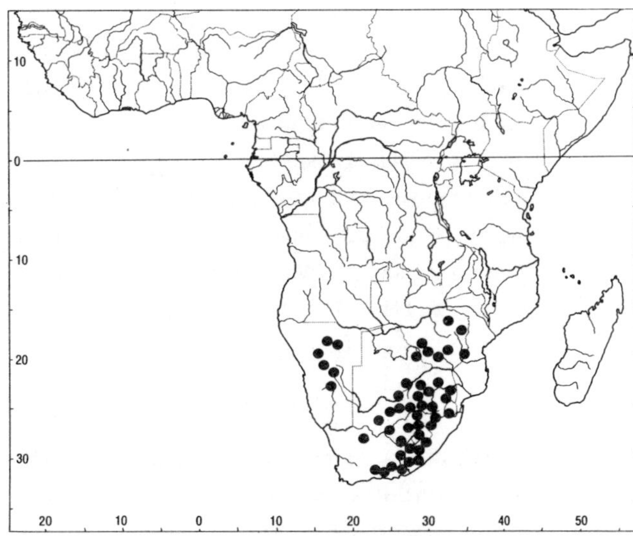

MAP 33. *Gladiolus permeabilis* subsp. *edulis.*

abilis as far as their inclusion of subspecies *wilsonii* is concerned, and the latter is probably a distinct species although no doubt closely related to *G. permeabilis.* As currently circumscribed, *G. permeabilis* is extremely variable and the southern African populations need to be reexamined critically before changes are made to Lewis et al.'s (1972) treatment.

HISTORY

Gladiolus edulis was first recorded by the English botanist and explorer William Burchell during his travels into the interior of southern Africa. Plants that he collected in the northern Cape near Litakun were later grown in England, and the species was described in 1817 by Ker after a painting had been completed. It was first regarded as a subspecies of *G. permeabilis* after 1972, when critical studies by Lewis et al. (1972) showed how closely it resembled the latter. *Gladiolus permeabilis* subsp. *edulis* appears to have been recorded in Zimbabwe only in 1909 and much later in Botswana, where it seems to be rare.

SELECTED SPECIMENS

Botswana. East: University Campus, Gaborone, 1000 m, 23 Apr. 1975, *Mott 899* (SRGH), *Mott 274* (UCBG). Ngwaketsi: southeast of Ramotwswa, 24°52′ S, 25°50′ E, *Hansen 3126* (C, PRE). Central: near Tsessebe Station, mixed woodland on sand, 6 Jan. 1974, *Ngoni 247* (SRGH).

Zimbabwe. Manicaland: woodland between Chipinga and Mt. Selinda, 900 m, 16 Feb. 1962, *Chase 7624* (SRGH). Mashonaland East: Rusape District, 21 Feb. 1940, *Hopkins 7533* (SRGH). Mashonaland West: Lomagundi, Great Dyke, rocks on summit ridge, 8 Apr. 1981, *Philcox & Müller 9079* (K, SRGH). Masvingo: Victoria, 1909, *Monro 841* (BM, SRGH). Matabeleland North: Nyamandhlovu, Pasture Research Station, red sandy loam, Feb. 1954, *Plowes 1680* (K, SRGH); Shangani, 10 Apr. 1943, *Feiertag s.n.* (SRGH 45433). Matabeleland South: Belingwe District, 30 km west of West Nicholson, 17 Mar. 1964, *Wild 6405* (K, SRGH); Matobo, sparse grassland, Besna Kobila Farm, Feb. 1961, *Miller 7718* (SRGH).

33. *Gladiolus fenestratus* Goldblatt, new species

FIGURE 30, MAP 34. Type: Angola, Huila, Chela Mountains, 16 km northwest of Humpata, road to Turnevala, near Cascada Spring, 20 Apr. 1968, *Kers 3196* (PRE, holotype; S, isotype).

EPONYMY

fenestratus, "windowed," referring to the space between the dorsal and upper lateral tepals so that in profile there is a window or gap through which one can see.

LATIN DIAGNOSIS

Plantae 35–50 cm altae, foliis (3–)4–5 ± teretibus 4-sulcatis, spica 6–10 florum inclinata, bracteis 12–16 mm longis, floribus roseis luteis notatis, tubo perianthii 9–10 mm longo, tepalis inaequalibus superiore grandiore 15–17 mm longis, filamentis c. 12 mm longis c. 8 mm exsertis, antheris 4.5–6.5 mm longis.

DESCRIPTION

Plants 35–50 cm high. CORM c. 2 cm in diameter, tunics of coarse, dark gray reticulate fibers, sometimes forming a weakly developed neck around the base. CATAPHYLLS membranous, often dark brown, especially above the ground. LEAVES sometimes three, normally four or five, the lower two or three basal and with long blades about as long as or slightly exceeding the spikes, the blades oval to terete in transverse section, 2–3 mm wide, the margins and midrib strongly thickened and raised and with two narrow grooves on each surface, the grooves up to 1 mm wide, the upper leaves shorter and partly to entirely sheathing. STEM simple or rarely with one branch, inclined above the sheath of uppermost leaf, 1.5 mm in diameter at the base of the spike.

SPIKE 6- to 10-flowered, inclined; BRACTS green entirely or becoming purplish above, 12–16 mm long, the inner slightly shorter than the outer. FLOWERS pink, the lower lateral tepals each deep yellow in the lower half of the limb, edges in dark red; PERIANTH TUBE 9–10 mm long, curving outward between the bracts, widening near the mouth; TEPALS unequal, lanceolate, unusually narrow at the base, the dorsal hooded over the stamens, 15–17 mm long, c. 10 mm wide, the upper laterals smaller, directed forward and ultimately curving outward, the three lower tepals united with the upper laterals for (2–)3 mm, and to themselves for 1(–2) mm, narrowed below into claws, the limbs abruptly expanded, 12–13 mm long, 3.5 mm at the widest, horizontal below, the limbs inclined toward the ground, in profile slightly exceeding the dorsal. FILAMENTS c. 12 mm long, exserted for c. 8 mm; ANTHERS 4.5–6.5 mm long, yellow. OVARY 2.5 mm long; STYLE usually dividing just below the base of the anthers, sometimes

FIGURE 30. *Gladiolus fenestratus.* Corm, leaves, flowering spike, and capsule, × 0.67; cross-section of leaf, much enlarged (*Kers 3196*).

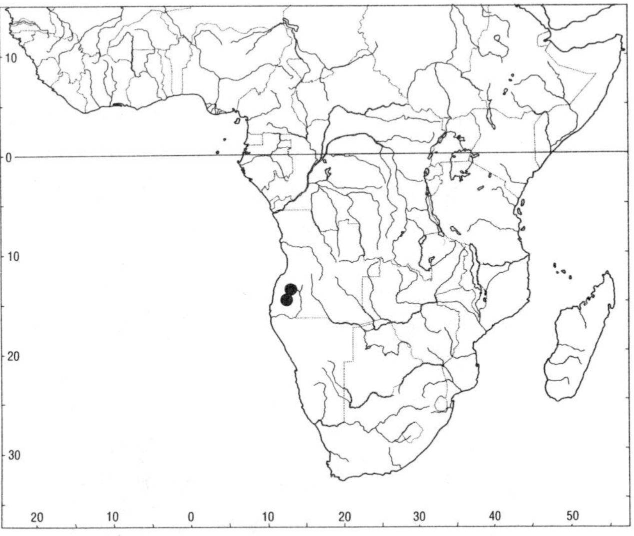

MAP 34. *Gladiolus fenestratus.*

near the middle, the branches c. 1.5 mm long, seldom reaching the middle of the anthers. CAPSULES and SEEDS unknown.

FLOWERING TIME. April and May.

DISTRIBUTION & HABITAT

Gladiolus fenestratus is endemic to southwestern Angola, where it was first collected by Eugène de Kindt in 1897. Only three more collections are known, making this one of the rarest species of the genus, although it is apparently common locally. As far as is known, it grows on well-drained rocky sites on the Huila Plateau and Chela Mountains at elevations of c. 2000 m.

DIAGNOSIS & RELATIONSHIPS

Gladiolus fenestratus is readily distinguished by its terete leaf blades, inclined spike, and moderate-sized flowers, the tepals of which are strongly narrowed in the lower part so that in profile the flowers are windowed. The flowers are pale pink with yellow nectar guides. In the shape of the tepals it resembles the southern African and southern tropical African *G. permeabilis,* a species not recorded in Angola. The more widespread form of the latter, subspecies *edulis,* which occurs in areas of summer rainfall, has linear leaves with more or less plane blades, slightly thicker in the midline, and dull-colored cream and blue-gray flowers with the tepals tapering into long attenuate apices. Except in these features, *G. fenestratus* resembles *G. permeabilis* subsp. *edulis* fairly closely and it is to *G. permeabilis* that *G. fenestratus* is probably most closely related. The resemblance of the terete, four-grooved leaves to those of *G. pretoriensis* is probably convergent. *Gladiolus pretoriensis* and species closely related to it (section *Heterocolon*) have at least partly dry bracts, brownish corm tunics, and irregularly angular seeds. The seeds of *G. fenestratus* are unknown. Should they prove to be angular, a reappraisal of the relationships of *G. fenestratus* will be necessary.

SELECTED SPECIMENS

Angola. Huila: open places on Humpata side of Chela Ridge, 2000 m, 4 May 1909, *Pearson 22806* (K); rocky ground, Chiringizo (Tyiringizo), 1740 m, Apr. 1897, *de Kindt 786* (COI, P); Lubango (Sá da Bandeira), Tchiviringuiro, rocky soil, 3 May 1965, *Santos 1099* (PRE).

SUBGENUS *OPHIOLYZA* KLATT

Klatt, Abh. Naturforsch. Ges. Halle 15: 340 (Ergänzungen 6) (1882), not *Ophiolyza* Salisbury (1866) (= *Gladiolus* section *Hebea*). Type: *Gladiolus natalensis* Reinwardt ex J. D. Hooker (= *Gladiolus dalenii* van Geel) (lectotype designated here).

Description

Plants usually large, with stems thick and straight, usually unbranched, rarely one- or two-branched. LEAVES lanceolate to linear, rarely nearly terete, the upper smaller and sometimes entirely sheathing. SPIKES erect, rarely inclined; BRACTS fairly long, (12–)20–80 mm long, about as long or slightly shorter than the perianth tubes (much shorter in section *Acidanthera*). FLOWERS large to medium, (25–)35–200 mm long, nectar guides, when present, usually longitudinal and on the lower half of each of the three lower tepals; PERIANTH TUBE (7–)20–150 mm long, usually about as long or longer than the bracts (in section *Acidanthera* about twice as long); TEPALS subequal or the lower smaller than the upper but these not distinctly clawed and rarely united below, the dorsal tepals about as long or shorter than the tubes. ANTHERS rounded apically or with short to long apiculate appendages. CAPSULES usually large, (10–)15–35 mm long; SEEDS (6–)8–11 × (4–)5–7 mm, the wing always well developed.

Subgenus *Ophiolyza* is widespread in Africa south of the Sahara, where there are more than 70 species, three shared with the Arabian Peninsula, and one of them also with Madagascar. The subgenus is especially well represented in tropical Africa and is diverse there. I recognize six sections in subgenus *Ophiolyza,* five of which are represented in tropical Africa, and one, section *Cardinales,* is restricted to southern Africa. Two of the five tropical African sections occur nowhere else. Section *Tenuibracteus* is restricted to eastern tropical Africa, and section *Decoratus* ranges from northern Mozambique to Ethiopia and Ivory Coast. Of the remaining, section *Ophiolyza* occurs throughout tropical Africa and extends to western Arabia, Madagascar, and eastern southern Africa; section *Acidanthera,* present almost throughout tropical Africa, extends locally into southern Arabia; and section *Blandus* is shared with southern Africa.

Although diverse, both florally and vegetatively, subgenus *Ophiolyza* includes a range of mostly large-flowered species with relatively long floral bracts and perianth tubes. Most of the species placed in the subgenus also have large capsules and seeds, and erect stems and spikes. In contrast, smaller-flowered subgenus *Gladiolus* has relatively smaller capsules and seeds, short-tubed flowers, and the upper part of the stems or the spikes are inclined. The tepals are also usually united below for some distance in subgenus *Gladiolus* and are normally narrowed below into claws. A few species assigned to subgenus *Ophiolyza* have flowers smaller than expected, and their placement in the subgenus is a consequence of their presumed relationships to one or another more typical member of the subgenus. Thus *G. salmoneicolor*, with flowers c. 25 mm long and a tube c. 7 mm long, seems misplaced according to these criteria, but the species has a strong resemblance to *G. serenjensis*, in turn apparently related to *G. oligophlebius,* a typical member of section *Decoratus* of subgenus *Ophiolyza.* Likewise, the diminutive Ethiopian *G. lithicola,* with flowers 33–36 mm long and a tube c. 18 mm long, accords better in leaf and other floral features with subgenus *Ophiolyza.* It has apiculate anthers and leaf venation that match section *Decoratus* of the subgenus.

SECTION *BLANDUS* (BAKER) GOLDBLATT
New Combination and Status

Basionym: *Gladiolus* group *Blandi* Baker, J. Linn. Soc. Bot. 16: 176 (1878). Type: *G. blandus* Aiton (= *G. carneus* D. Delaroche, a southern African species), lectotype designated here.

Synonymy

Subgenus *Hyptissa* Klatt, Abh. Naturforsch. Ges. Halle 15: 342 (Ergänzungen 8) (1882), not *Hyptissa* Salisbury (1866), the identity of the latter uncertain. Type: *Gladiolus blandus* Aiton (= *G. carneus* D. Delaroche), lectotype designated here.

Description

Plants erect, branched or unbranched. LEAVES lanceolate to linear, the upper smaller and partly to entirely sheathing. SPIKES erect or occasionally inclined; BRACTS generally slightly shorter than the tubes, the inner only slightly shorter than the outer. FLOWERS large to medium, 32–125 mm long in tropical African species; PERIANTH TUBE (10–)15–50 mm long, generally longer than the bracts, much shorter in a few species; TEPALS subequal or the dorsal larger, the dorsal tepals about as long or shorter than the tubes. ANTHERS rounded at apex.

A pan-African section with nine tropical species, mostly of eastern Africa, section *Blandus* extends from Mozambique to Ethiopia, but one species occurs in western Angola. There are several more species in South Africa, concentrated in the southwestern Cape. Most of the species have white to pink flowers (red in *Gladiolus rupicola*). Among the tropical African species of section *Blandus, G. bellus* and *G. usambarensis* have notably long perianth tubes, about twice as long as the bracts. These species superficially resemble members of section *Acidanthera* in their predominantly white perianths and long tubes, but in section *Acidanthera* the anthers are always long-apiculate and the inner bracts much shorter than the outer.

34. *Gladiolus zambesiacus* Baker

PLATES 17, 18, FIGURE 31, MAP 35. Baker, Handbook Irideae 212 (1892); Fl. Trop. Africa 7: 364 (1898). Goldblatt, Fl. Zambesiaca 12(4): 71 (1993). Type: Malawi, Shire Highlands, near Blantyre, 1887, *Last s.n.* (K, lectotype, so designated by Geerinck on the sheet, the more complete material); Malawi (or Mozambique), "mountains east of Lake Nyasa," probably 1883, *Johnson s.n.* (K, syntype, identity uncertain, the plants depauperate and outside the recorded range).

SYNONYMY

Gladiolus buchananii Baker, Handbook Irideae 212 (1892); Fl. Trop. Africa 7: 362 (1898). Type: Malawi, summit of Direndi (Ndirandi), 1500 m, without date, *Buchanan s.n.* (K, holotype; E, isotype).

EPONYMY

zambesiacus, "from the Zambezi country," the general name for southeastern tropical Africa in the 19th century.

DESCRIPTION

Plants 45–75 cm high. CORM c. 15 mm in diameter, tunics of papery to finely fibrous layers. CATAPHYLLS pale and membranous, the upper brown or becoming greenish or purple above the ground, usually at least sparsely pubescent, often densely so. LEAVES six or seven, the lower four more or less basal and largest, reaching to the base or middle of the spike, the blades narrowly lanceolate to nearly linear, (2–)3.5–6 mm wide, the sheaths especially of the lowermost often with a

sparse or rarely dense short pubescence, the blades occasionally sparsely puberulent, the midrib and margins lightly thickened, upper leaves cauline and decreasing in size above, the uppermost leaves usually entirely sheathing. STEM simple or rarely with one branch, c. 2.5 mm in diameter at the base of the spike.

SPIKE 6- to 12-flowered, flexed below the first flower and inclined toward the ground; BRACTS green, lanceolate, attenuate, firm-membranous, green or flushed purple, becoming dry only after flowering, (20–)25–35(–45) mm long, the inner slightly shorter than the outer. FLOWERS pink to mauve, fading to whitish in the throat, the lower three tepals each cream in the lower half; PERIANTH TUBE obliquely funnel-shaped, curving outward near the apex, c. 10 mm long; TEPALS unequal, the upper three largest, lanceolate, 22–26 mm long, often less when dry, the dorsal inclined over the stamens, c. 12 mm wide, the lower three inclined toward the ground, the lower laterals smallest, c. 18 × 8–10 mm. FILAMENTS 10–12 mm long, exserted 4–5 mm from the tube; ANTHERS 7–8(–10) mm long, apices obtuse, yellow. OVARY ellipsoid, c. 5 mm long; STYLE arching over the filaments, usually dividing at or beyond the anther apices, the branches c. 4 mm long, extending beyond the anthers, expanded above. CAPSULES obovoid-ellipsoid, 15–20 mm long; SEEDS more or less oval, 5–6 × 3–3.5 mm. CHROMOSOME NUMBER $2n = 30$.

FLOWERING TIME. Mostly in March and April but sometimes as late as May.

DISTRIBUTION & HABITAT

Gladiolus zambesiacus is well known from the higher elevations of the Shire Highlands of southern Malawi. It has been reported from most of the higher mountains of the area, including Ndirandi, Zomba, Malosa, Chiradzulu, and the Mulanje Massif. A collection from a montane site in nearby western Mozambique, *Torre & Correia 14721*,

FIGURE 31. *Gladiolus zambesiacus.* Corm, leaves, and flowering spike, × 0.5; single flower, full size; cross-section of leaf, × 2 (*Goldblatt 9092, 9099*).

probably also belongs in *G. zambesiacus.* Those plants have unusually narrow and entirely glabrous leaves, c. 3 mm wide, with thickened margins and midribs. Typically favoring exposed habitats, *G. zambesiacus* is usually found growing in thin soil in rocky crevices that bake completely dry during the winter months but are seasonally moist although well drained during the wet summer. The flowers of *G. zambesiacus* are regularly visited by large solitary bees of the family Anthophoridae, and the species must be assumed to be primarily adapted for bee pollination.

DIAGNOSIS & RELATIONSHIPS

Moderate-sized to fairly large pale pink flowers with a cream throat and cream lower tepals tipped with pink, yellow anthers, and several well-developed leaves, nearly always minutely pubescent on the basal leaf sheaths and lower blades, characterize *Gladiolus zambesiacus.* The inclined spike, fairly short perianth tube in relation the length of the dorsal tepal, and the narrow but comparatively long lower tepals, all plesiomorphic, make the immediate affinities of this species difficult to assess, except for an obviously close similarity with the western Angolan *G. chelamontanus.* The two species can be distinguished largely by differences in their foliage. The leaves of *G. zambesiacus* are plane, broad, and relatively soft-textured, whereas in *G. chelamontanus* the leaves are linear and have strongly thickened margins and midribs. The floral morphology suggests that *G. zambesiacus* belongs in subgenus *Ophiolyza,* in which its fairly short perianth tube points to a position in section *Blandus.*

Gladiolus zambesiacus is easily confused with the superficially similar *G. erectiflorus.* The latter, which occurs widely across southern tropical Africa but does not extend as far south as southern Malawi, also has fairly large flowers, often of a pinkish color. The tepals of *G. erectiflorus* spread more widely than those of *G. zambesiacus,* are subequal (the three lower tepals of *G. zambesiacus* are much smaller than the upper), and are conspicuously veined with pink to red. In addition, the anthers of *G. erectiflorus* are dark red to purple, unlike the yellow anthers of *G. zambesiacus,* and the upper leaves are not reduced to short sheaths but have long linear blades.

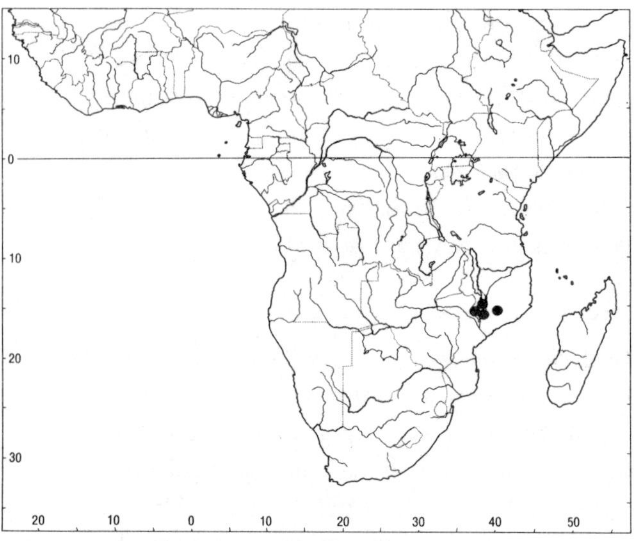

MAP 35. *Gladiolus zambesiacus.*

HISTORY

Described by J. G. Baker in 1892, *Gladiolus zambesiacus* was based on two specimens from separate collections, both of unusually slender plants. The flowers of the more complete plant, gathered from near Blantyre, Malawi, correspond fairly well with modern collections of the species from highland areas of southern Malawi, and the leaf sheaths still show the microscopic pubescence so characteristic of *G. zambesiacus.* The second specimen, from "mountains east of Lake Nyasa" (possibly in Mozambique), has leaves lacking pubescence on the sheaths, and its identity is uncertain owing to its generally poor state of preservation. A third collection of *G. zambesiacus,* made at "Direndi," almost certainly Mt. Ndirandi near Blantyre, formed the basis for *G. buchananii,* also described by Baker in

1892. There is no doubt that this species is identical with *G. zambesiacus,* although the type specimen now lacks mature flowers.

SELECTED SPECIMENS

Malawi. Southern: Zomba Plateau, 5 June 1946, *Brass 16248* and *16246* (NY); Zomba Plateau, near turn to Chingwe's Hole, wet flush, 16 Mar. 1970, *Brummitt 9167* (K, MAL); Zomba Mountain, upper slopes, 14 May 1963, *Wild 6260* (K, SRGH); Malosa, rocky slopes above Domasi Valley just below plateau, 15 Apr. 1991, *Goldblatt 9099* (MO); Ntonya Hill, Zomba district, rock slabs, *la Croix 682* (K); Ndirandi summit in rocky grassland, *la Croix 479* (K); Mt. Chiradzulu, above Lisao Forest, 1430 m, 13 Mar. 1977, *Brummitt et al. 14857* (K, MAL); Mt. Mlanje, Litchenya, 1460 m, 1 Jan. 1984, *Johnson-Stewart 265* (E).

Mozambique. Zambézia: Gurué Mountains, road to Namúli Peak, 1300 m, 21 Feb. 1966, *Torre & Correia 14721* (LISC).

35. *Gladiolus chelamontanus* Goldblatt, new species

FIGURE 32, MAP 36. Type: Angola, Huila, Tundevala, between the falls and lookout, 6 May 1962, *Barbosa & Moreno 10217* (COI, holotype; LISC, isotypes).

EPONYMY

chelamontanus, "from the Chela Mountains," the southwestern Angolan mountain range where the species occurs.

LATIN DIAGNOSIS

Plantae 45–70 cm altae, foliis 6–7 linearibus marginibus costisque incrassatis, cataphyllis et foliis superioribus scabridis, spica 6–10 florum erecta, bracteis 25–45 mm longis, floribus roseis luteis notatis, tubo perianthii c. 10 mm longo, tepalis inae-

FIGURE 32. *Gladiolus chelamontanus.* Corm, leaves, and flowering spike, full size; cross-section of leaf, × 3 (*Kers 3299*).

qualibus superioribus grandioribus, 25–30 mm longis, filamentis c. 15 mm longis 10 mm exsertis, antheris c. 12 mm longis.

DESCRIPTION

Plants 45–75 cm high. CORM 15–20 mm in diameter, tunics of firm-membranous layers decaying into vertical fibrous strips, at first below, later above as well. CATAPHYLLS pale and membranous, the upper longest and green or becoming purple above the ground, usually at least densely scabrid in parts. LEAVES six or seven, the lower four more or less basal and largest, usually exceeding the spike by 5–15 cm, the blades linear, (1.5–)3–5 mm wide, the midrib and margins heavily thickened and raised over the leaf surface, the upper leaves cauline and decreasing in size above, the blades progressively shorter and less strongly thickened, or vestigial, imbricate and sheathing the stem almost to the base of the spike, usually microscopically scabrid in the intercostal areas. *Stem* evidently erect, unbranched, c. 2 mm in diameter at the base of the spike.

SPIKE 6- to 10-flowered, evidently erect; BRACTS narrowly lanceolate, firm-membranous, green or flushed purple, becoming dry after flowering, 25–35(–45) mm long, the inner slightly shorter than the outer. FLOWERS pink to mauve, the lower tepals marked with yellow below; PERIANTH TUBE obliquely funnel-shaped, curving outward near the apex, c. 10 mm long; TEPALS unequal, the upper three largest, lanceolate, 25–30 mm long, often less when dry, the dorsal inclined over the stamens, c. 12 mm wide, the lower three inclined toward the ground, the lower laterals smallest, 15–18 mm long. FILAMENTS c. 15 mm long, exserted 10 mm from the tube; ANTHERS c. 12 mm long, apices obtuse, without appendages, yellow. OVARY ellipsoid, c. 6 mm long; STYLE arching over the filaments, usually dividing opposite the middle of the anthers, the branches c. 4 mm long, not reaching the anther apices, expanded above. CAPSULES and SEEDS unknown.

FLOWERING TIME. Mid April to June.

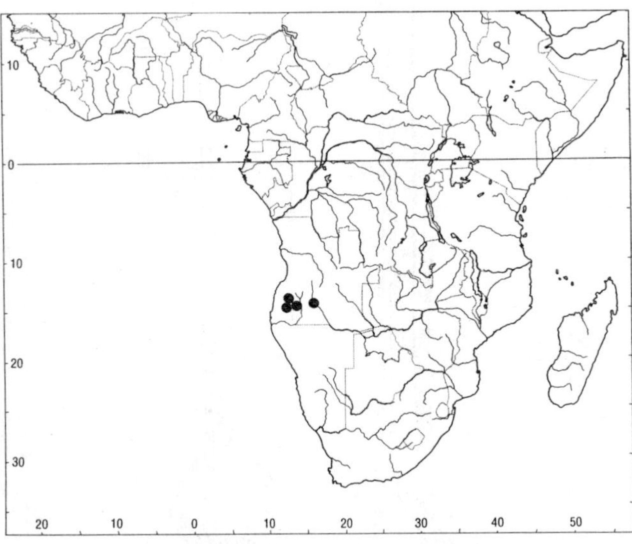

MAP 36. *Gladiolus chelamontanus.*

DISTRIBUTION & HABITAT

Gladiolus chelamontanus is confined to the southwestern highlands of Huila Province, Angola. Like the apparently related eastern African *G. zambesiacus,* it favors rocky outcrops and grows in thin soils and in rock crevices. Records indicate that it has a very localized distribution and is never abundant, but the species is far from being adequately known.

DIAGNOSIS & RELATIONSHIPS

In its general appearance, including a moderate stature, medium-sized pink to purple flowers, and attenuate bracts, *Gladiolus chelamontanus* resembles most closely the southeastern tropical African *G. zambesiacus.* Its primary distinguishing features are the linear leaves with thickened and raised margins and midrib, and the scabrid texture of the sheaths of the upper leaves. In both these features, as well as in having an erect spike, it differs from *G. zambesiacus.* The sheathing and imbricate cauline leaves and erect rather than inclined spike recall *G. erectiflorus,* but the flowers lack the unusual purple anthers and dark tepal venation of that species, instead having yellow anthers and a uniformly pink perianth with yellow marks on the lower tepals. *Gladiolus chelamontanus* is also unusual in the comparatively large anthers, c. 12 mm long, that exceed the style branches, in contrast to both *G. erectiflorus* and *G. zambesiacus.*

ADDITIONAL SPECIMENS

Angola. Huila: 15 km southwest of Lubango (Sá da Bandeira) toward Tundevala, near Cascada, 20 Apr. 1968, *Kers 3299* (S); Humpata, 1800 m, June 1936, *Gossweiler s.n.* (COI); Humbia, Chela, 1300 m, 17 May 1937, *Gossweiler 10949* (COI, K); between Quipungo and Lubango, 26 May 1937, *Carrisso & Sousa 194* (BM); Quilemba, c. 1950 m, 4 June 1937, *Exell & Mendonça 2520* (BM); Rio Cutato, Cubango, Ganguelas, 2 May 1906, *Gossweiler s.n.* (COI, K).

36. *Gladiolus balensis* Goldblatt, new species

PLATE 19, FIGURE 33, MAP 37. Type: Ethiopia, Bale, 5–6 km from Ghinir on the road to Robi via Sof Omar, basalt outcrops, 1750–1850 m, 29 May 1983, *Gilbert, Ensermu, & Vollesen 7931* (K, holotype; C, ETH, UPS, isotypes).

EPONYMY

balensis, "from Bale," a southeastern province of Ethiopia.

LATIN DIAGNOSIS

Plantae 50–60 cm altae, cormo 18–22 mm diametro, foliis 3 linearo–lanceolatis, caule simplici, spica 5–9 florum, bracteis 20–27(–40) mm longis, floribus albis carneis notatis, tubo perianthii 18–20 mm longo, tepalis inaequalibus, dorsali 22 × 15 mm, inferioribus c. 24 mm longis, filamentis c. 14 mm longis, antheris 8–10 mm longis.

DESCRIPTION

Plants 50–60 cm high. CORM 18–22 mm in diameter, the tunics membranous or breaking into vertical fibers. CATAPHYLLS pale and membranous below ground, green above the ground. LEAVES three, only the lowermost basal, this longest, the upper two leaves inserted above ground level and shorter than the basal, the blades narrowly lanceolate to linear, initially short and none reaching the base of the spike, at the end of flowering often longer and the basal leaf sometimes as long as the spike, the margins and midribs lightly thickened. STEM erect, unbranched, 2–3 mm in diameter at the base of the spike.

SPIKE 5- to 9-flowered, straight and erect; BRACTS initially green, sometimes becoming membranous and dry above, 20–27(–40) mm long, the inner only slightly shorter than the outer. FLOWERS white, usually pink in the midline of the upper lateral tepals and flushed pink on fading, the lower lateral tepals each with a yellow median streak outlined distally in light pink to purple; PERIANTH TUBE 18–20 mm long, expanded in the upper 5 mm; TEPALS unequal, the dorsal longest and arched over the stamens, broadly lanceolate, c. 22 mm long, 15 mm wide, upper laterals slightly smaller, the lower three tepals c. 24 mm long, the lower laterals 6–7 mm wide, the lower median c. 10 mm wide. FILAMENTS c. 14 mm long, exserted 4–5 mm from the tube; ANTHERS 8–10 mm long, yellow. OVARY c. 4 mm long; STYLE dividing just beyond the anther apices, the branches extending well past the anthers, 3–4 mm long. CAPSULES and SEEDS unknown.

FLOWERING TIME. May and early June.

DISTRIBUTION & HABITAT

Known from only two collections, *Gladiolus balensis* appears to be a local endemic of the mountains of northern Bale Province in southern Ethiopia. It grows in rock outcrops on forested mountain slopes. The first record of the species was made in May 1980, and the second in 1983, also in May.

DIAGNOSIS & RELATIONSHIPS

Relatively tall, *Gladiolus balensis* has flowers that most closely resemble those of *G. boranensis,* another southern Ethiopian species, but whereas the latter has pink to red flowers, those of *G. balensis* are white with yellow nectar guides on the lower tepals. The foliage of the two species is quite different. *Gladiolus balensis* has only three leaves, the lowermost of which is longest, sometimes exceeding the spike, but the other leaves are quite short, whereas *G. boranensis* has six to seven leaves, at least three of which are about as long as the spike or longer. The general form of the flowers corresponds to that for subgenus *Ophiolyza.* The short perianth tube, only about half as long as the upper tepals, is unusual for the subgenus but not uncommon in section *Blandus.* Additional collections are needed before the species is fully understood.

ADDITIONAL SPECIMENS

Ethiopia. Bale: 20 km on the Sof Omar–Goro

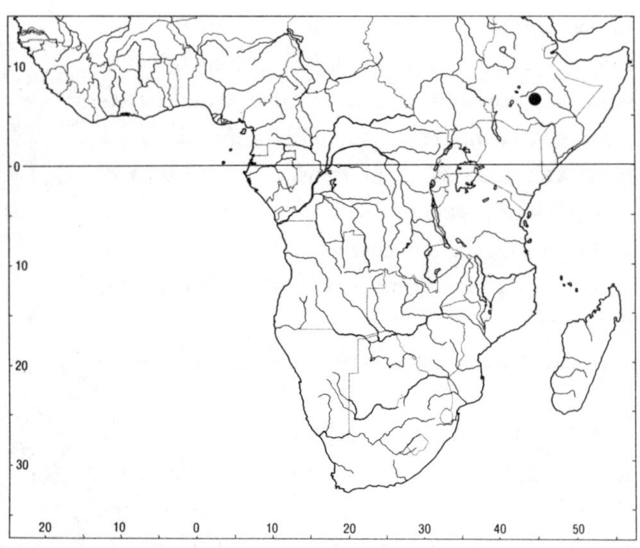

MAP 37. *Gladiolus balensis.*

FIGURE 33. *Gladiolus balensis.* Corm, leaves, and flowering spike, full size; single flower, × 1.3 (*Gilbert et al. 7931*).

road, limestone escarpment, c. 1800 m, 18 May 1980, *Thulin, Hunde, & Tadesse 3830* (K, UPS)

37. *Gladiolus negeliensis* Goldblatt, new species

PLATE 20, FIGURE 34, MAP 38. Type: Ethiopia, Sidamo, 14 km south of Negeli on the road to Melka Guba, 1500 m, 18 May 1982, *Friis, Tadesse, & Vollesen 3060* (ETH, holotype; C, K, UPS, isotypes).

EPONYMY

negeliensis, "from Negeli," a town in Sidamo Province, southern Ethiopia.

LATIN DIAGNOSIS

Plantae 15–30 cm altae, cormo 9–14 mm diametro, folia 3 usitate caulem brevioribus marginibus costisque leviter incrassatis, laminis 2.5–4 mm diametro, spica 2–5 florum, bracteis 20–27(–35) mm longis, floribus albis vel carneis, tubo perianthii 30-40 mm longo, tepalis subaequalibus 25–28 mm longis, filamentis 8–10 mm longis, antheris 7–8 mm longis.

DESCRIPTION

Plants 15–30 cm high. CORM 10–14 mm in diameter, the tunics membranous, becoming lacerated at base and apex with age, ultimately irregularly fibrous. CATAPHYLLS membranous, the uppermost green above the ground. LEAVES usually three, or fewer, all basal or the upper inserted on the lower part of the stem, the blades narrowly lanceolate to linear, reaching to about the middle of the stem or, in short plants, reaching the apex of the spike, 2.5–4 mm wide, the margins and midribs lightly thickened. STEM unbranched, generally flexed outward at the base of the spike or above the sheath of the upper leaf, 1.5–2 mm in diameter at the base of the spike.

FIGURE 34. *Gladiolus negeliensis,* full size (*Thulin, Hunde, & Tadesse 3630*).

SPIKE 2- to 5-flowered; BRACTS green, 20–27 (–35) mm long, the inner about two-thirds as long as the outer. FLOWERS white to pale pink, the tepals each with a median pink streak and the lower three with greenish to yellow markings, the throat often streaked with pink; PERIANTH TUBE 30–40 mm long, obliquely funnel-shaped, the narrow part 25–30 mm long and reaching or exceeding the apices of the bracts; TEPALS lanceolate, the dorsal 25–28 × 10–12 mm, the lower three tepals 25–27 × 8–10 mm. FILAMENTS 8–10 mm long, exserted 4–5 mm from the tube; ANTHERS 7–8 mm long, yellow. OVARY oblong, 4–5 mm long; STYLE dividing near the anther apices, or sometimes beyond them, the branches 4–5 mm long, spreading beyond the anthers. CAPSULES and SEEDS unknown.

FLOWERING TIME. April and May.

DISTRIBUTION & HABITAT

First collected in 1974 by the English botanist, M. G. Gilbert, *Gladiolus negeliensis* appears to be a fairly local endemic of southern Ethiopia. The three known collections are all from the vicinity of Negeli in central Sidamo Province. The species grows in open flat grassland, sometimes waterlogged in the rainy season, at elevations of 1500 to 1700 m.

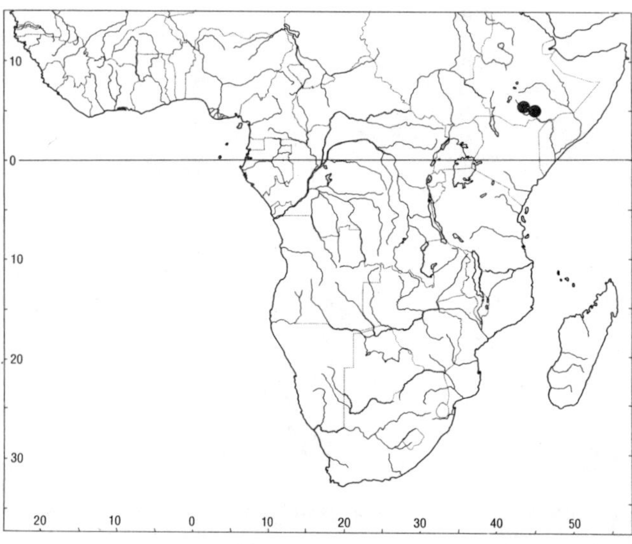

MAP 38. *Gladiolus negeliensis.*

DIAGNOSIS & RELATIONSHIPS

Although the relationships of *Gladiolus negeliensis* are uncertain, the species can readily be distinguished from other eastern and northeastern African species of *Gladiolus* by its slender form and moderate-sized white to pale pink flowers, which are produced at the same time as the leaves. Plants consistently produce only three leaves, all with relatively well-developed blades. The leaves are unusually short, not normally reaching the base of the spike at flowering time, although they may increase in length later in the season. The flowers are also unusual in having a slender but comparatively long tube, as long to slightly longer than the tepals. For want of a better understanding of its relationships, the species is placed here next to a second white-flowered Ethiopian species, *G. balensis,* in section *Blandus* of subgenus *Ophiolyza.* Both *G. negeliensis* and *G. balensis* remain among the most phylogenetically puzzling of the Ethiopian species of the genus.

ADDITIONAL SPECIMENS

Ethiopia. Sidamo: 14 km southeast of Negeli, just beyond junction of road to Wachile, 1590 m, 6 May 1974, *Gilbert 3334* (C, EA, ETH, K); 15 km from Negeli on the road to Filtu, 1700 m, 12 May, 1980, *Thulin, Hunde, & Tadesse 3630* (ETH, K, MO, UPS).

38. *Gladiolus mensensis* (Schweinfurth) Goldblatt

MAP 39. Goldblatt in Stork & Lebrun, Enumeration Plantes Fleurs Afrique Tropicale 3: 91 (1995).

SYNONYMY

Tritonia mensensis Schweinfurth, Bull. Herb. Boiss. 2: Appendix 2: 86 (1894). Type: Eritrea, Gheleb, 2200 m, 20 Apr. 1891, *Schweinfurth 1188* (G, lectotype designated here; B, C, G, K, isolectotypes); Eritrea, Gheleb, 2200 m, *Schweinfurth 1552* (not located at herbaria consulted for this study, perhaps lost).

EPONYMY

mensensis, "from Mensa," a region of northern Eritrea.

DESCRIPTION

Plants 25–50(–80) cm high. CORM 12–14 mm in diameter, the tunics of pale, fairly fine reticulate fibers. CATAPHYLLS membranous. LEAVES four or five, the lower two or three more or less basal, seldom reaching beyond the middle of the spike, the blades linear, 2–4 mm wide, the upper two or three leaves much shorter than the basal and largely to entirely sheathing. STEM evidently erect, unbranched, c. 2 mm in diameter at the base of the spike.

SPIKE 4- to 7-flowered, evidently erect; BRACTS 14–20(–25) mm long, evidently green below but dry and membranous apically, the outer usually slightly exceeding the inner. FLOWERS pink or white, nectar guides unknown; PERIANTH TUBE c. 15 mm long, obliquely funnel-shaped; TEPALS evidently more or less equal, possibly the dorsal slightly larger, 18–20 mm long, c. 8 mm wide, their orientation uncertain. FILAMENTS 9–12 mm long, exserted 4–5 mm from the tube; ANTHERS 6–7 mm long. OVARY oblong, c. 3 mm long; STYLE arched over the stamens, dividing just beyond the anther apices, the branches c. 2.5 mm long. CAPSULES obovoid-ellipsoid, c. 13 mm long; SEEDS c. 5 × 4 mm, the wing well developed.

FLOWERING TIME. April.

DISTRIBUTION & HABITAT

Poorly known, *Gladiolus mensensis* appears to be restricted to the mountains north of Asmara around Gheleb in northern central Eritrea. According to Schweinfurth, who made the type collection in 1891 and who later described the species, *G. mensensis* grows in open sites in rocky grassland.

DIAGNOSIS & RELATIONSHIPS

Gladiolus mensensis is one of the smaller northeastern African species of *Gladiolus.* The plants are normally 25–50 cm tall and have white to pale pink flowers 33–35 mm long. Although originally described as *Tritonia mensensis,* there is no doubt that the species belongs in *Gladiolus.* Ripe capsules on some plants in the type collection have broadly winged seeds typical of *Gladiolus* and unknown elsewhere in the Iridaceae. The fairly small flowers have a short perianth tube, c. 15 mm long, normally exceeding the bracts, and the filaments are exserted from the tube. So generalized are its characteristics that it is difficult to assess the relationships of the species. Provisionally, I place *G. mensensis* in section *Blandus* of subgenus *Ophiolyza,* closest to other smaller-flowered northeastern African species of the genus, but I am not at all confident that this reflects its true relationships.

ADDITIONAL SPECIMENS

Eritrea: Amasen, Mt. Lesa, 6 Apr. 1902, *Pappi 4638* (MO).

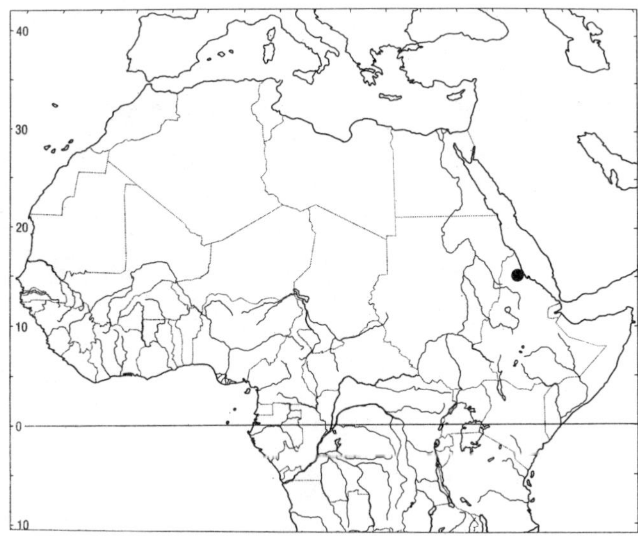

MAP 39. *Gladiolus mensensis.*

39. *Gladiolus boranensis* Goldblatt, new species

Map 40. Type: Ethiopia, Sidamo, Mega, 2100 m, 22 Oct. 1962, *Mooney 9815* (ETH, holotype; WAG, isotype).

EPONYMY

boranensis, "from Borana," an old name for the southern region of Ethiopia.

LATIN DIAGNOSIS

Plantae 38–55 cm altae, cormo 18–20 mm diametro, foliis 6–7, inferioribus 3–4 linearibus caulem breviter excedentibus marginibus costisque incrassatis, superioribus (1–)2–3 elaminatis, caule eramoso, spica 5–10 florum, bracteis 25–30(–40) mm longis infra viridibus supra siccis, floribus roseis vel rubris, tubo perianthii c. 15 mm longo, tepalis subaequalibus 24–32 mm longis, filamentis 12–15 mm longis, antheris 8–10 mm longis.

DESCRIPTION

Plants 38–55 cm high. Corm 18–20 mm in diameter, the tunics membranous, becoming irregularly broken with age. Cataphylls membranous to firm, pale but usually turning purplish above the ground. Leaves six or seven, the lower three or four basal and longest, usually exceeding the spike by 5–15 cm, the blades linear, 2–4(–6) mm wide, the midrib and margins moderately thickened, the upper (1–)2–3 leaves short and largely to entirely sheathing, usually without blades. Stem erect, unbranched, 2–3 mm in diameter at the base of the spike.

Spike 5- to 10-flowered, straight and erect; bracts green below, dry and brownish above, 25–30(–40) mm long, the inner about two-thirds as long as the outer. Flowers pale to deep pink, pale in the throat and toward the bases of the lower tepals; perianth tube c. 15 mm long, obliquely funnel-shaped; tepals apparently nearly equal or the dorsal slightly larger, 24–32 × c. 15 mm, the lower 3 tepals 24–30 × c. 13 mm. Filaments 12–15 mm long, exserted c. 5 mm from the tube; anthers 8–10 mm long. Ovary oblong, c. 5 mm long; style arching over the stamens, dividing near the anther apices, the branches c. 4 mm long. Capsules and seeds unknown.

Flowering Time. September and October.

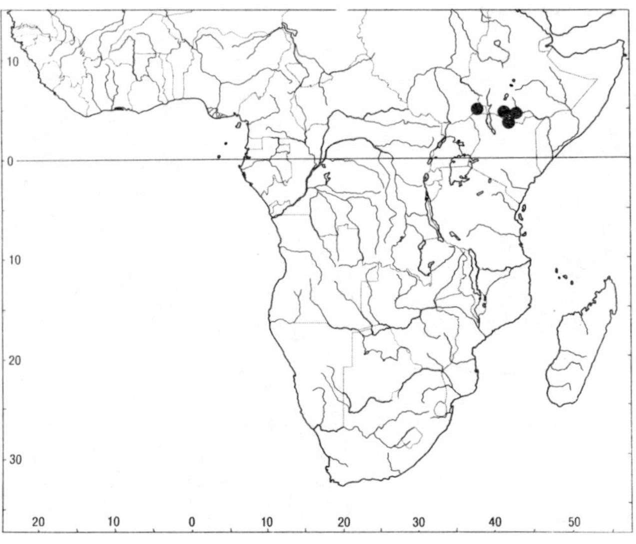

Map 40. *Gladiolus boranensis.*

DISTRIBUTION & HABITAT

Gladiolus boranensis appears to be a rare endemic of the mountains of southeastern Sudan and southern Ethiopia. There is just one record from Sudan, from the Dongotona Mountains, and only two from southern Ethiopia. A fourth collection is from Furroli, on the Ethiopia–Kenya border, but according to the collection information, actually from Kenya. Plants grow at elevations of 1800 to 2100 m in Ethiopia, in juniper forest or *Commiphora* scrub. The Sudan collection is from 2400 m on rocky slopes.

DIAGNOSIS & RELATIONSHIPS

Gladiolus boranensis is far from adequately known, and it is by no means certain that all the specimens cited here actually belong to this species. The species is defined by its moderate-sized pink to reddish flowers with a fairly short perianth tube, c. 15 mm long, normally slightly shorter than the upper tepals, and tall stems with six to seven narrowly lanceolate to linear leaves. The lower three or four leaves are about as long or slightly longer than the spikes, and the upper leaves are short and entirely sheathing. The floral bracts are fairly long and exceed the perianth tube. Although first collected in southern Sudan in 1941, and in the early 1950s in Ethiopia and Kenya, *G. boranensis* has remained either unrecognized or confused with *G. roseolus,* a central and northern Ethiopian species, probably not immediately related to it. *Gladiolus boranensis* is provisionally referred to section *Blandus,* and *G. roseolus* to section *Ophiolyza,* both subgenus *Ophiolyza.*

ADDITIONAL SPECIMENS

Sudan. Sharq el Istiwaiya: Imogadung, Dongotona Mountains, fell field of peak, c. 2400 m, 28 Oct. 1941, *Myers 14188* (K).

Ethiopia. Sidamo: Yavello, juniper forest, 1800 m, 17 Sept. 1953, *Bally 9226* (K, G).

Kenya. Eastern: Furroli, lava plateau, 1900 m, 20 Sept. 1952, *Gillett 13956* (K).

40. *Gladiolus rupicola* Vaupel

PLATE 21, FIGURE 35, MAP 41. Vaupel, Bot. Jahrb. Syst. 48: 541 (1913). Type: Tanzania, western Usambara Mountains, Wurunigebiet, 1600 m, *Engler 1094* (B, holotype; K, photo).

EPONYMY

rupicola, "growing on rocks," alluding to the habitat, rocky outcroppings in otherwise forested country.

DESCRIPTION

Plants 50–75 cm high. CORM 18–25 mm in diameter, tunics brown, membranous, becoming torn irregularly from the base, possibly always with small cormlets around the base. CATAPHYLLS membranous and brown. LEAVES several, about as long as the stems, lanceolate to linear, 8–15 mm wide, the lower three to five basal and longest, the upper cauline and progressively shorter. STEM unbranched, sometimes inclined or drooping, c. 3 mm in diameter at the base of the spike.

SPIKE 3- to 5-flowered; BRACTS green, lanceolate-attenuate, 30–40(–50) mm long, the inner shorter than the outer. FLOWERS red to deep pink, the lower three tepals white to cream at the base and each with a narrow cream to yellow median streak; PERIANTH TUBE 25–30 mm long, narrowly funnel-shaped, the lower part slender for 20 mm, abruptly expanded above; TEPALS unequal, the dorsal largest, more or less arched over the stamens, (25–)40–55 mm long, 22–30 mm wide, upper lateral tepals 25–40 mm long, c. 15 mm wide, the lower three tepals more or less horizontal, 22–35 mm long. FILAMENTS 18–45 mm long, exserted from the tube for 20–30 mm; ANTHERS (6–)12–14 mm long, yellow, obtuse and without apiculate appendages. OVARY 6–7 mm long, ellipsoid, STYLE arched over the stamens, dividing at or beyond the anther apices, branches 4–6 mm. CAPSULES ovate to ellipsoid, 15–24 mm long; SEEDS elliptic, 8–10 × c. 5 mm.

FLOWERING TIME. Mainly May to September, but also November to February.

DISTRIBUTION & HABITAT

With a fairly wide distribution across East Africa, *Gladiolus rupicola* extends from southwestern Tanzania across that country to the Taita Hills in southeastern Kenya. Its is exclusively a montane species, generally occurring above 2000 m, although it may grow at considerably lower elevations in the Taita Hills. Plants may be found in all the higher mountains of Tanzania and appear to be most common in the Uluguru and Usambara Mountains, where the most robust plants occur. Both the Kenyan and southwestern Tanzanian populations have smaller flowers and at first appear to be a different species. Floral form and proportion are, however, identical with the larger-flowered plants, and available information suggests that they are best regarded merely as smaller-flowered *G. rupicola*.

DIAGNOSIS & RELATIONSHIPS

Gladiolus rupicola may be distinguished by the combination of bright red flowers, well-exserted anthers without apiculate appendages, and a dorsal tepal noticeably longer than the upper laterals. The leaves and stems may be erect or inclined, depending on whether the plants grow on cliffs or in open grassland. The several lanceolate leaves are unremarkable

The species is easily confused with members of section *Decoratus,* especially *G. decoratus* and *G. oligophlebius,* which have similarly colored, fairly large flowers. *Gladiolus rupicola,* however, has obtuse anthers, lacking the apiculate appendages characteristic of section *Decoratus,* and it also has firmer-textured leaves. Thus despite its general similarity to species of section *Decoratus, G. rupicola* does not belong to that alliance. The immediate relationships of *G. rupicola* within section *Blandus* are not clear.

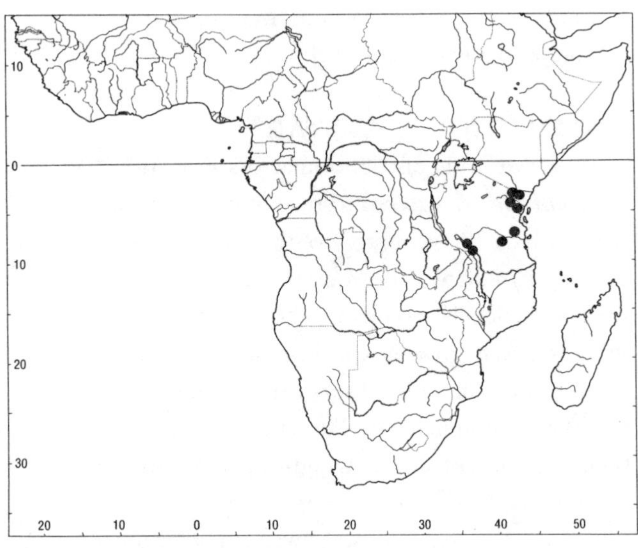

MAP 41. *Gladiolus rupicola.*

FIGURE 35. *Gladiolus rupicola,* full size (*Thomas 3706*).

SELECTED SPECIMENS

Kenya. Taita: Voi District, Dulun, Mt. Kisigau, c. 1200 m, July 1937, *Dale 3802* (K); Taita Hills, summit of Vuria, 2200 m, 17 Sept. 1979, *Lavranos & Newton 17640* (MO).

Tanzania. Tanga: eastern Usambaras, Nov. 1916, *Peter 18426* (B); Lushoto, Mazumbai rainforest, just west of tea estate, 1900 m, 18 Nov. 1974, *Balslev 300* (B, MO, NY); Mlingi Central Peak, eastern Usambara Mountains, 7 Oct. 1936, *Greenway 4695* (EA, PRE); Lushoto, Shagayu Peak, Shagai Forest, 2230 m, 15 May 1953, *Drummond & Hemsley 2533* (B, EA, K, S, SRGH). Morogoro: Uluguru Mountains, northwest aspect, 1150 m, 18 Sept. 1932, *Schlieben 2706* (BM, BR, G, P, S, Z); Uluguru Mountains, eastern slopes, 1500 m, 25 Feb. 1933, *Schlieben 3576* (B); Uluguru Mountains, Lupanga Peak, 1800 m, 22 May 1933, *Burtt 4708* (K); Lukwangule Plateau, western slopes, 2040–2400 m, 6 Dec. 1969, *Harris & Pocs 4134* (EA); Mwanihana Forest Reserve above Sanje, ridge top grassland above forest, c. 1400 m, 7/50 36/55, *Thomas 3706* (K, MO); Ukaguru Mountains, east slopes of Mamiwa Mountain, 2000 m, crevices of vertical rocks, 31 May 1978, *Thulin & Mhoro 2785* (UPS); Mnyera Mountain, Ukagurus, hanging from steep rocks, 1900 m, 1 June 1978 (flower and fruit), *Thulin & Mhoro 2834* (UPS). Mbeya: Station Kyimbila, 1913, *Stolz 2187* (C, G, LD, S, WAG, Z); upper waterfall, Kimani River, Mbeya, 21 June 1980, *Nicholson s.n.* (K). Iringa: Njombe–Rungwe, bank of Lumakaria River, in moss, 2000 m, 3 Oct. 1929, *Wigg 5* (K).

41. *Gladiolus usambarensis* W. Marais ex Goldblatt, new species

PLATE 22, FIGURE 36, MAP 42. Type: Tanzania, Lushoto District, western Usambara Mountains, World's View, 1.5 km west of Golongolo, 1900 m, rock ledges at top of cliff, 4 July 1953, *Drummond & Hemsley 2852* (K, holotype; B, BR, S, isotypes).

EPONYMY

usambarensis, from the Usambara Mountains, northeastern Tanzania, the source of the type collection.

LATIN DIAGNOSIS

Plantae 20–40 cm altae, cormo 15–20 diametro, foliis usitate 4–6 inferioribus 3–5 basalibus anguste lanceolatis ad linearibus 5–12 mm latis, caule eramoso, spica (2–)3–5 florum, bracteis viridis (20–) 30–50 mm longis, floribus usitate pallide roseis, tubo perianthii 35–50(–60) mm longo, tepalis subaequalibus 30–32 × 12–14 mm, filamentis c. 16 mm longis ex tubo c. 7 mm exsertis, antheris c. 8 mm longis.

DESCRIPTION

Plants 20–40 cm high, erect or inclined to trailing from cliffs. CORM 15–20 mm in diameter, the tunics membranous to coriaceous, reddish brown. CATAPHYLLS membranous, green above the ground, sometimes decayed and broken. LEAVES four to six, or more, the lower three to five basal, reaching at least to the base of the spikes, or slightly exceeding them, narrowly lanceolate to more or less linear, 5–12 mm wide, firm-textured, the midribs raised, the upper leaves cauline and progressively reduced in size. STEM erect or trailing, unbranched, 2–2.5 mm in diameter at the base of the spike.

SPIKE (2–)3- to 8-flowered, straight or lightly flexuose; BRACTS green, 20–30(–50) mm long, the inner about two-thirds as long as the outer. FLOWERS zygomorphic, usually white, sometimes pale pink, greenish on the reverse of the tube, the lower three tepals each with a narrow red stripe in the lower midline and yellow in the throat; PERIANTH TUBE 35–50(–60) mm long, straight, slender below, widening slightly in the upper part; TEPALS subequal, lanceolate, the lower laterals slightly smaller than the others, 30–32 mm long, 12–14 mm wide, extended forward and gradually curved outward in the upper half. FILAMENTS c. 16 mm long, exserted c. 7 mm from the tube; ANTHERS c. 8 mm long, yellow. OVARY ovoid, 4–5 mm long; STYLE straight and held above the stamens, dividing shortly beyond the anther apices, the branches c. 4 mm long. CAPSULES (11–)20–25 mm long, ellipsoid-obovoid; SEEDS ovate, c. 6 × 4 mm.

FLOWERING TIME. Almost any month but evidently more often in September to November.

DISTRIBUTION & HABITAT

Gladiolus usambarensis, as the name suggests, occurs in northeastern Tanzania, where it appears to be centered in the western Usambara Mountains, an area that receives high rainfall almost throughout the year. There are also populations on Mt. Kisigau in the Taita Hills of Kenya, a short distance to the east of the Usambaras, and to the west and south of the Usambaras. In western Tanzania there are scattered records from the Ngorongoro Mountains of more slender plants with narrow leaves. Flowering in these areas occurs in March and April. An isolated population (*Schlieben 4197*) to the south, from the Nguru Mountains in east-

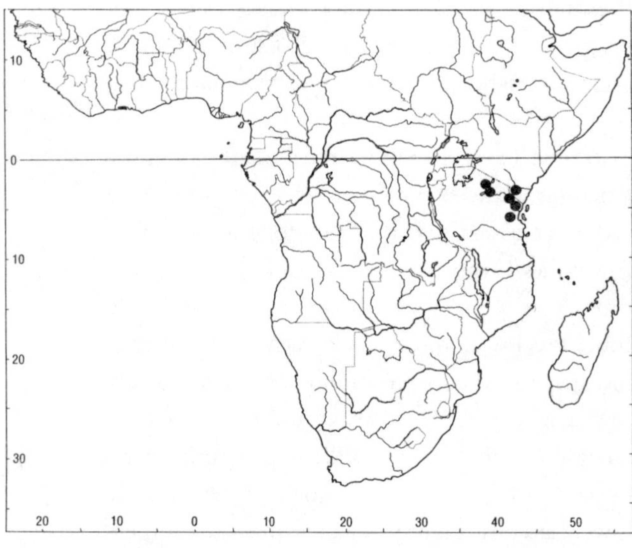

MAP 42. *Gladiolus usambarensis.*

FIGURE 36. *Gladiolus usambarensis.* Corm, leaves, and spike, × 0.67; single flower, full size (*Bally & Smith 14790, Woodruff & Smith s.n.*).

ern central Tanzania, was collected in late flower and fruit in July. Those plants stand out in having nearly globose capsules, 12–14 mm long, unlike the more typical ovoid-ellipsoid capsules found in populations to the north. Plants generally grow in montane habitats, generally above 2000 m, in rocky sites, sometimes on cliffs, where they occur in tussocks of grass. In the Taita Hills they occur at lower elevations of 1200 to 1400 m.

DIAGNOSIS & RELATIONSHIPS

A long-tubed white flower with nearly equal tepals, the lower three of which are striped with a narrow red band in the midline, readily distinguishes *Gladiolus usambarensis.* The perianth tube is normally 40–50 mm long, always somewhat exceeding the tepals, and sometimes substantially so. Plants have a delicate, almost fragile appearance, although the leaves are firm-textured. The affinities of the species are uncertain. It resembles most closely the southern African *G. microcarpus,* which grows on rocks, sometimes near streams and waterfalls in the Drakensberg. That plant species has bright pink flowers, attenuate tepals, and ribbed leaves, so there is no likelihood of confusion. The resemblance may be entirely coincidental. *Gladiolus usambarensis* may also be confused with the eastern African *G. pauciflorus,* which also has long-tubed flowers, but the perianth in that species is usually cream, the tepals are unequal, and the tube is curved and wider than in *G. microcarpus.*

Gladiolus usambarensis was collected first in the 1920s and was well-represented in herbaria by 1955 although it remained unnamed. This may have been due in part to confusion with long-tubed species like *G. candidus* and *G. murielae,* plants that were until 1973 referred to the genus *Acidanthera,* now *Gladiolus* section *Acidanthera.* Unlike the members of section *Acidanthera,* which have anthers with long attenuate anther apiculi, *G. usambarensis* has blunt-tipped anthers without apiculate appendages.

SELECTED SPECIMENS

Kenya. Taita: Kisigau Hill, 30 km south of Voi, 1200 m, 13 June 1967, *Archer 547* (BR, K); Mt. Kisigau, on rocks, 1200 m, July 1937, *Dale 3839* (K); Mt. Kisigau, 14 Nov. 1969, *Bally 13579* (EA, K).

Tanzania. Arusha: Ngorongoro District, Angata Kiti, 24 Mar. 1966, *Herlocker 405* (K); Ngata Kiti Valley, 13 Apr. 1968, *Turner s.n.* (EA 13989); Ngorongoro, crater floor in grassland, Apr. 1941, *Bally 2321* (K). Tanga: Lushoto District, western Usambara Mountains, top of Ndamauwilo, c. 2200 m, Oct. 1922, *Grant 24* (K); western Usambaras, east of Magamba, c. 2280 m, 11 Sept. 1945, *Greenway 7537* (K, N, PRE); Mkusi Peak, exposed rocks, 1800 m, 5 Sept. 1944, *Greenway 7035* (BR, K, PRE); Mkusi Valley, grass tussocks on rock faces, c. 1650 m, Sept. 1953, *Eggeling 6706* (BR, K, PRE). Bagamoyo: Nguru Mountains, Mssimbaberg, 23 July 1933 (flower and fruit), *Schlieben 4197* (BR, G, P, S, Z).

42. *Gladiolus bellus* C. H. Wright

PLATE 23, FIGURE 37, MAP 43. C. H. Wright, Kew Bull. 169 (1906). Brenan, Mem. New York Bot. Gard. 9: 85 (1954). Goldblatt, Fl. Zambesiaca 12(4): 95 (1993). Type: Malawi, Mt. Mulanje, Tuchila Plateau, 1800 m, May 1901, *Purves 4* (K, lectotype designated by Goldblatt, 1993: 95, the specimen with spike and leaves; K, isolectotype); Zomba Plateau, date unknown, *Whyte s.n.* (BM?, syntype, not seen); Mulanje (Mlanji), 1800 m, date unknown, *Mahon s.n.* (location unknown, syntype, not seen).

EPONYMY

bellus, "beautiful," referring to the large white flowers.

DESCRIPTION

Plants 60–90 cm high. CORM c. 15 mm in diameter, tunics of brittle papery layers, the outer be-

FIGURE 37. *Gladiolus bellus,* full size, whole plant much reduced (*Chapman & Chapman 7790*).

coming irregularly broken, rarely subfibrous, reddish brown. CATAPHYLLS pale and membranous below ground, green to purple above ground, the upper largest and up to 15 cm long. LEAVES four or five, the lower three or four basal or nearly so, the blades narrowly lanceolate to more or less linear, 7–12 mm wide, at least reaching the base of the spike, sometimes shortly exceeding it, firm textured with the midrib and margins lightly thickened, the upper one or two leaves cauline and much shorter than the basal. STEM simple, usually inclined or drooping, 3–4 mm in diameter below the first flower.

SPIKE 3- to 8-flowered; BRACTS green, becoming dry and brown above, 40–50 mm long, the inner slightly shorter than the outer. FLOWERS white, the lower tepals each with a dark red to violet spade shaped mark in the lower midline and usually with dark streaks in the throat; PERIANTH TUBE cylindric below, the upper part curved outward and flared, (42–)50–90 mm long, the wider upper part c. 12 mm long; TEPALS nearly equal, broadly obovate, 30–35 mm long, 20–22 mm wide. FILAMENTS 22–28 mm long, inserted at the top of the cylindric part of the tube, exserted for 5–8 mm from the tube; ANTHERS 9–12 mm long, apices with obscure appendages less than 0.2 mm long. OVARY c. 5 mm long; STYLE arching over the stamens, dividing just beyond the anther apices, branches c. 4.5 mm long. CAPSULES oblong-ellipsoid, 30–35 mm long; SEEDS oval to oblong, 9–11 × 5–7 mm, broadly winged. CHROMOSOME NUMBER $2n = 60$.

FLOWERING TIME. March to June, sometimes into July.

DISTRIBUTION & HABITAT

Gladiolus bellus is restricted to the Mulanje Massif in southern Malawi, a mountain complex that reaches to almost 3000 m. The range rises abruptly above the surrounding plain and is separated from the nearest highlands to the north and west by a distance of some 40 km. The species appears to be relatively common in rocky sites as well as in wet grassland above 1800 m, although it has been collected at lower elevations. Early collections by Alexander Whyte and by Harry Johnson from Mt. Zomba are probably incorrect, for the plant has not been found again in this well-collected area. *Gladiolus bellus* flowers toward the end of the rainy season from late March until June and even into July when temperatures on Mt. Mulanje reach to below freezing at night.

DIAGNOSIS & RELATIONSHIPS

Its long perianth tube, 4.2–8.8 cm long and about half as long again to twice as long as the bracts, and white perianth with large conspicuous markings on the lower tepals, distinguish *Gladiolus bellus* from nearly all other species of the genus. It seems morphologically closest to the northern Tanzanian and Kenyan *G. usambarensis,* which has white flowers with subequal tepals and a perianth tube nearly as long. It seems unlikely that these

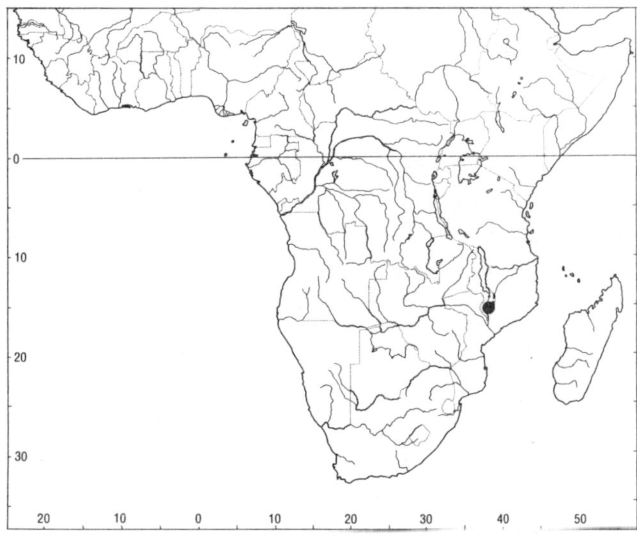

MAP 43. *Gladiolus bellus.*

two species are closely related to section *Acidanthera* despite their similar long-tubed white flowers, as they do not have the long-apiculate anthers that characterize all species of section *Acidanthera*. Moreover, the shape of the tepal markings in *G. bellus* are spathulate and located in the middle of the tepals, quite different from the basal lanceolate streaks in section *Acidanthera*. The floral similarities between *G. bellus* and section *Acidanthera* are probably convergent rather than an indication of a close phylogenetic relationship.

SELECTED SPECIMENS

Malawi. Southern: Mt. Mulanje, Apr. 1906, *Adamson 361* (E, K, P); Mt. Mlanje, Luchenya Saddle, among rocks, May 1963, *Wild 6181* (K, LISC, PRE); Mlanje, Chambe Basin, 2040 m, 14 June 1962, *Richards 11692* (K); path from Tuchila Hut to head of Ruo basin, 6 Apr. 1970, *Brummitt 9647* (K, MAL, PRE, SRGH); Likhabula Valley, 1300 m, 21 Mar. 1986, *Chapman & Chapman 7363* (E, MO); Luchenya Plateau, stream bank, 5 July 1946, *Brass 16669* (NY).

SECTION *TENUIBRACTEUS* GOLDBLATT
New Section

Type: *Gladiolus erectiflorus* Baker.

Latin Diagnosis

Plantae erectae eramosae, foliis anguste lanceolatis vel linearibus supra angustioribus et caulibus vaginantibus, caule spicaque erecto eramoso, bracteis attenuatis, floribus 20–100 mm longis, tubis perianthii 12–45 mm longis brevioribus vel longioribus bracteis, tepalis subaequalibus dorsalibus tubis excedentibus, antheris obtusis vel breviter apiculatis.

Description

Plants erect and unbranched, mostly large. LEAVES narrowly lanceolate to linear, the upper progressively more slender but with long blades concealing the stem almost to the base of the spikes. SPIKES erect; BRACTS fairly long, the apices attenuate, longer or shorter than the perianth tubes. FLOWERS large to small, 20–100 mm long, the lower tepals with cream to yellow basal nectar guides; PERIANTH TUBE 12–45 mm long, shorter or longer than the bracts, much longer in one species; TEPALS subequal, the dorsal usually exceeding the tube (much shorter in *G. stenosiphon*), the lower tepals about as long as the dorsal. ANTHERS obtuse or shortly apiculate.

An exclusively tropical African section of five species, section *Tenuibracteus* extends from Malawi and Angola in the south to Ethiopia in the north. The most characteristic features of the alliance are the narrow and apically attenuate floral bracts, sometimes dry above at anthesis, and the several long, overlapping leaves, even the uppermost of which have long blades. The flowers are generally pink to white or sometimes cream, often very conspicuously veined in dark pink to red.

43. *Gladiolus erectiflorus* Baker

PLATES 24, 25, FIGURE 38, MAP 44. Baker, Kew Bull. 1895: 293 (1895); Fl. Trop. Africa 7: 367 (1898); Geerinck, Bull. Jard. Bot. Nat. Belgique 42: 274 (1972), in part, excluding *Gladiolus linearifolius* and *G. verdickii*. Goldblatt, Fl. Zambesiaca 12(4): 85 (1993). Type: Zambia, south of Lake Tanganyika, Liendwe, 1894, *Carson 1/1894* (K, holotype).

SYNONYMY

Gladiolus nyikensis Baker, Kew Bull. 283 (1897); Fl. Trop. Africa 7: 367 (1898). Type: Malawi, "Nyika Plateau, 6000–7000 ft," 1896, *Whyte s.n.* (K, holotype; B, G, isotypes).

Gladiolus venulosus Baker, Kew Bull. 282 (1897); Fl. Trop. Africa 7: 366 (1898). Type: Malawi, near Chitipa (Fort Hill), "Nyasa-Tanganyika Plateau, 3500–4000 ft," July 1896, *Whyte s.n.* (K, holotype; B, G, isotypes).

EPONYMY

erectiflorus, "erect-flowered," referring to the stiffly erect spikes, unlike many other species of *Gladiolus,* which have spikes inclined slightly toward the ground.

DESCRIPTION

Plants (35–)60–90 cm high. CORM 12–18 mm in diameter, tunics of papery to matted fibrous layers, fragmenting irregularly or breaking into vertical strips below, producing slender runners from the base, these bearing one or a few cormlets each. CATAPHYLLS membranous, the upper greenish or brown above the ground. LEAVES five or six, the lower three more or less basal and largest, reaching to about the base of the spike, narrowly lanceolate to nearly linear, 2–5 mm wide, the midrib and margins lightly thickened, upper leaves cauline and decreasing in size above, imbricate and sheathing the stem almost to the base of the spike. STEM erect, unbranched, c. 2 mm in diameter below the spike.

SPIKE 4- to 8-flowered, erect, rarely flexed at the base and weakly inclined; BRACTS lanceolate-attenuate, green below, membranous above, becoming dry and light brown toward the apices, 20–30(–40) mm long, the inner slightly shorter than the outer. FLOWERS cream to pale pink or purple, conspicuously veined pink to red or purple, the color darkening on fading, rarely apparently uniformly pale pink to whitish, the lower lateral or all three lower tepals each with a broad lanceolate median cream to yellow mark in the lower center and sometimes a pink median streak above; PERIANTH TUBE narrowly funnel-shaped, widening and curving outward near the apex, 16–18 mm long; TEPALS more or less equal or the dorsal slightly larger, lanceolate, (25–)30–35 mm long, often less when dry, inclined over the stamens, the lower three tepals united with the upper laterals for c. 2 mm, straight, usually inclined weakly toward the ground. FILAMENTS arcuate, 12–14 mm long, ultimately exserted 6–7 mm from the tube; ANTHERS 7–8 mm long, apices with minute subacute appendages less than 0.2 mm long, dark purple, pollen whitish. OVARY ellipsoid, c. 3.5 mm long; STYLE arching over the filaments, dividing at about the apex of the anthers, branches c. 4 mm long, extending beyond the anthers, expanded in the upper half. CAPSULES obovoid-ellipsoid, somewhat three-lobed above, 10–12 mm long; SEEDS oval to irregularly rounded, the wing well developed or vestigial at the proximal end, c. 4 × 2–3 mm. CHROMOSOME NUMBER $2n = 30$.

FLOWERING TIME. Mostly in April and May, but sometimes as early as February and as late as July.

DISTRIBUTION & HABITAT

Relatively widespread in southern tropical Africa, *Gladiolus erectiflorus* extends from the Mwinilunga District of northwestern Zambia across the Copperbelt and southern Zaire into northern Malawi, western and southern Tanzania, and northern Mozambique. It favors rocky sites in hilly country and grows in miombo (*Brachystegia*-dominated) woodland, usually in areas of somewhat higher rainfall where the thin, organic-rich soils remain moist throughout the growing season. Plants growing in full sun are usually smaller and have fewer flowers than those in shaded sites.

DIAGNOSIS & RELATIONSHIPS

Gladiolus erectiflorus is striking in having relatively large flowers with the subequal tepals typically heavily veined in red to purple on a whitish to cream or pale pink background. It is difficult to mistake the species because of the very characteristic flower coloration, although in some populations the veining is poorly developed. In addition, it is notable in having the upper leaf blades overlapping and sheathing the stem to the base of the spike and in the narrow, rather attenuate bracts that are often dry above at flowering. The capsules, 10–12 mm long, are comparatively small for a fairly large-flowered species. The small capsules, overlapping upper leaves, and attenuate bracts are shared with *G. verdickii,* of southern Zaire and adjacent Zambia, and the western Angolan *G. stenosiphon*. These three species appear to form a closely related group, otherwise isolated in the genus. *Gladiolus verdickii* has larger flowers than those of *G. erectiflorus,* either cream with yellow markings or with purple venation rather like that found in *G. erectiflorus. Gladiolus stenosiphon* has pure white flowers that stand out in having a long perianth tube, 45–50 mm long.

In plants from the north of its range in northeastern Shaba Province, Zaire, and in some from northern Zambia, the typical heavy tepal venation is weakly developed or absent, but there seems no reason to doubt the identity of such plants. A few collections from the Mugila Plateau in eastern Shaba (*Lisowski 81,665, 81,658,* POZG) have linear

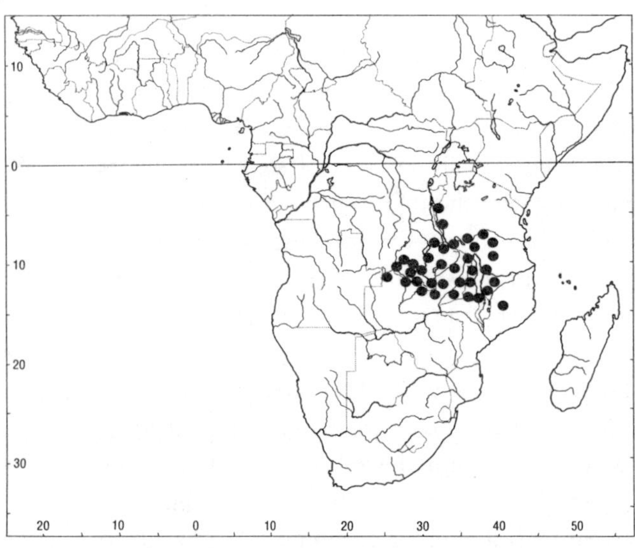

MAP 44. *Gladiolus erectiflorus.*

FIGURE 38. *Gladiolus erectiflorus.* Corm, leaves, and flowering spike, × 0.67; single flower, full size (*Goldblatt 9091, Pawek 9206*).

leaves with the margins and midribs heavily thickened. In leaf features these plants resemble *G. debeerstii,* but the flowers are substantially larger and accord rather with *G. erectiflorus.* Until more is known about these plants, they too are included in *G. erectiflorus.* A collection from the Mpika District of central Zambia (*Fanshawe 2015*) appears to represent an unusual variant of the species. The flowers are evidently uniformly white to very pale pink and have little or none of the dark veining so characteristic of the species.

HISTORY

Gladiolus erectiflorus appears to have been discovered by Alexander Carson in 1884 at Fwambo at the southern end of Lake Tanganyika, in what is now Zambia. This is surprisingly late for so common and widespread a species, but I have been unable to find an earlier record. It was described by J. G. Baker in 1895 and was based solely on Carson's gathering. A second collection, made by Charles Meller on the Nyika Plateau in Malawi, was described by Baker as *G. nyikensis* in 1897, and a third collection, from the Misuku Hills in northern Malawi, was described by Baker as *G. venulosus.* Neither *G. nyikensis* nor *G. venulosus* differs in any significant way from typical *G. erectiflorus,* and both must be regarded as synonyms. It was not until 1927 that *G. erectiflorus* was discovered in Shaba Province, Zaire, and it was first recorded in Tanzania only in 1950. It has now been amply collected and its wide range across central Africa is well established. Geerinck (1972) expanded the circumscription of *G. erectiflorus* to include the related but larger-flowered *G. verdickii.* As I have explained above, although closely related the two species differ substantially, and they are best treated as separate species.

SELECTED SPECIMENS

Zaire. Shaba: Road to Étoile, Farm Prince Leopold, 4 May 1927, *Quarré 355* (BR); Kipopo, c. 19 km northwest of Lubumbashi, 1230 m, 4 July 1982, *Schaijes 1446* (BR); Kafubu, 30 Apr. 1957, *Detilleux 936* (BR); Marungu Plateau, Mt. Lusale, 27 June 1957, *Duvigneaud 3739G* (BR); Marungu Plateau, source of the Lunanga, near Kasiki, 2200 m, 12 June 1969, *Lisowski, Malaisse, & Symoens 6422* (LSHI); Mugila Plateau, Mt. Kamabale, 1550 m, 15 May 1971, *Lisowski 81658* (POZG).

Zambia. Northwestern: Kabompo Gorge, Mwinilunga District, 19 Apr. 1965, *Robinson 6654* (K, UPS). Copperbelt: Ndola, 20 May 1953, *Fanshawe 19* (BR, K); Mufulira, 31 May 1934, *Eyles 8234* (K, SRGH); Kitwe, Parklands, 1250 m, 8 June 1961, *Linley 144* (K, LISC, MO, SRGH). Central: Mpika District, 7 Feb. 1955, *Fanshawe 2015a* (EA). Northern: Kambole Escarpment, dry rocks near Ngozi Falls, 6 June 1957, *Richards 10055* (BR, EA, K, SRGH); Mbala District, c. 1850 m, among rocks, 15 Mar. 1931, *Gamwell 17* (BM). Central: Serenje, 40 km north of Mpika, 15 June 1960, *Leach & Brunton 10047* (BM, SRGH). Eastern: Katete, St. Francis Hospital, 21 Apr. 1957, *Wright 190* (BR, K).

Tanzania. Iringa: Mufindi District, Madibira road just before escarpment, 1350 m, 5 Mar. 1987, *Lovett & Congdon 1930* (DSM, MO); Kigoma: Bugwe–Ubende Ntikimu, Apr. 1929, *Grant s.n.* (BM). Mbeya: Sumbawanga, Mpanda to Ufipa, between Katavi and Kisi, red soil, woodland, 25 Jan. 1950, *Bullock 2345* (B, K, S); Karimba Mt., Ubingu, 1700 m, 8 Apr. 1980, *Leedal 5890a* (EA). Rovuma: Unangwa Hill, c. 5 km east of Songea, 1170 m, 23 Mar. 1956, *Milne-Redhead & Taylor 9283* (EA).

Malawi. Northern: 30 km west of Mzuzu, Lunyangwa River Bridge, 18 Apr. 1974, *Pawek 8342* (BR, K, MAL, MO, SRGH, WAG); Karonga, Nyanga Hill, c. 760 m, 4 Apr. 1954, *Williamson 251* (BM). Central: 5 km north of the Dwanga River, 8 June 1938, *Pole Evans & Ehrens 620* (BR, E, K, P, PRE, S, SRGH); near Dedza, Chincherere Hill, 24 Apr. 1970, *Pawek 4701* (K, MAL, SRGH), 3 Apr. 1991, *Goldblatt 9091* (MO, MAL, PRE).

Mozambique. Niassa: Vila Cabral, open woodland, Dec. 1933, *Torre 320* (COI); Massangulo, Apr. 1933, *Gomes e Sousa 1396* (COI). Moçambique: slope of Mt. Ribaué, woodland on steep slopes, 19 July 1962, *Leach & Schelpe 11411* (K, PRE, SRGH); 9 km west of Namina, 2100 m, 23 May 1961, *Leach & Rutherford-Smith 10981* (K, MO, SRGH).

44. *Gladiolus verdickii* de Wildeman & Durand

PLATE 26, FIGURE 39, MAP 45. De Wildeman & Durand, Bull. Soc. Nat. Bot. Belgique 40: 29 (1901). Goldblatt, Fl. Zambesiaca 12(4): 86 (1993). Type: Zaire, Shaba, Lukafu, July 1900, *Verdick 612* (BR, holotype).

SYNONYMY

Gladiolus arnoldianus de Wildeman & Durand, Bull. Soc. Bot. Nat. Belgique 40: 27 (1901), included here with reservations. Type: Zaire, Shaba, Lukafu, Mutumgwe, May 1900, *Verdick 495* (BR, holotype).

EPONYMY

verdickii, named in honor of the Belgian plant collector, Edgard Verdick, who made many new discoveries in Zaire at the beginning of the 20th century, including this species.

DESCRIPTION

Plants 70–110 cm high. CORM 20–35 mm in diameter, tunics of papery to matted fibrous layers, fragmenting irregularly or breaking into vertical fibers below, producing short, flat, fasciated stolons from the base, these each bearing numerous cormlets. CATAPHYLLS membranous, the upper reaching well above the ground and then brown or becoming greenish. LEAVES five to seven, the lower three or four more or less basal and largest, reaching to about the base of the spike, the blades narrowly lanceolate to linear, (8–)10–15 mm wide, the midrib and margins hyaline and lightly thickened, upper leaves cauline and decreasing in size above. STEM unbranched or with one short branch, c. 3 mm in diameter at the base of the spike.

SPIKE 4- to 8-flowered, flexed outward at the base; BRACTS narrowly lanceolate, attenuate, apparently green below, membranous above and becoming dry and light brown toward the apices, (30–)40–45(–55) mm long, the inner slightly shorter than the outer and forked apically. FLOWERS white to yellow, sometimes conspicuously veined dark red to pink, the lower lateral tepals each with a cream to yellow mark in the lower center; PERIANTH TUBE narrowly funnel-shaped, widening and curving outward above, 25–35 mm long; TEPALS unequal, lanceolate-oblong, the dorsal largest, 45–50 mm long, often less when dry, inclined over the stamens, the upper laterals slightly shorter and spreading at right angles to the tube, the lower laterals smallest, 38–40 mm long. FILAMENTS arcuate, c. 30 mm long, exserted 14–18 mm from the tube; ANTHERS 7–8 mm long, apices with rudimentary appendages less than 0.2 mm long, dark purple. OVARY ellipsoid, 5.5–6.5 mm long; STYLE arching over stamens, dividing at about the apex of the anthers, the branches c. 3.5 mm long, spreading beyond the anthers, expanded in the upper third. CAPSULES obovoid-ellipsoid, 10–16(–20) mm long; SEEDS broadly elliptic, c. 5 mm × 4 mm.

FLOWERING TIME. Mostly in April and May, but sometimes as early as February and as late as July.

DISTRIBUTION & HABITAT

Restricted to the province of Shaba in Zaire and neighboring northwestern Zambia, *Gladiolus verdickii* occurs in woodland where it flowers toward the end of the wet season or more often at the beginning of the dry season, in May and June. Although it is represented by relatively few collections, it is fairly common in southern Shaba, but is known only from one collection from Zambia.

FIGURE 39. *Gladiolus verdickii.* Corm with fasciated stolons, leaves, flowering spike, and capsules, × 0.67; single flower and vertical section, full size (*Malaisse 7401, Schaijes 1912*).

DIAGNOSIS & RELATIONSHIPS

Gladiolus verdickii is distinguished by large flowers with tepals 45–50 mm long and an unusually long perianth tube, 25–35 mm long. The flowers range in color from whitish to cream with yellow markings on the lower tepals or have a light to heavily pink- to red-veined pattern on a pale background. The plants themselves are tall, sometimes exceeding 100 cm, and have firm-textured leaves usually 10–15 mm wide. The corms have a feature unique in *Gladiolus.* They bear several short, flat, fasciated stolons that bear numerous tiny terminal cormlets (Figure 39). In combination these characteristics make *G. verdickii* unmistakable. It is almost certainly closely related to the southern tropical African *G. erectiflorus.* That species has moderate-sized flowers with tepals 25–35 mm long and a tube 16–18 mm long, far smaller than in *G. verdickii.* Vegetatively, the two are similar in having comparatively closely set and narrow leaves covering the stem up to the base of the spike, and narrow, elongated floral bracts. Both species also produce cormlets from the base of the main corm, but in *G. erectiflorus* the cormlets are produced on slender runners of the type found occasionally in several genera of Iridaceae.

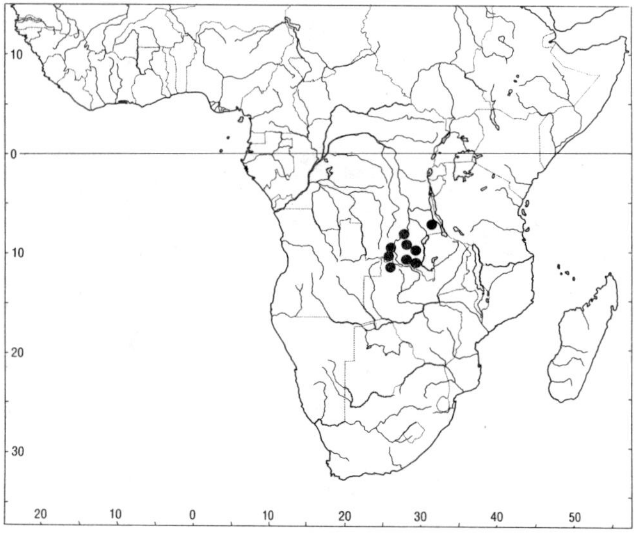

MAP 45. *Gladiolus verdickii.*

First collected in 1900 by Edgard Verdick and formally described shortly thereafter by Emile de Wildeman in 1901, *Gladiolus verdickii* has generally been confused with the related and superficially similar *G. erectiflorus.* The two were regarded as conspecific by Geerinck (1972) in a revision of *Gladiolus* in Zaire. No doubt closely related, the two species are separate. Together with the western Angolan *G. stenosiphon,* they constitute a close-knit and taxonomically isolated group. The latter species is readily distinguished within the alliance by its predominantly white flowers with a long perianth tube, 45–50 mm long, and comparatively short tepals, 25–28 mm long.

Gladiolus arnoldianus, known only from the type collection, is tentatively included in *G. verdickii,* but the tall plants have flowers with pure white tepals that have dark (purple?) apices and short stamens, the anther apices of which do not reach above the lower third of the tepals. The stamens of most populations of *G. verdickii* seem consistently longer and reach about the middle of the tepals. Until more is learned about *G. arnoldianus,* a species based on just two flowering spikes, it seems best to regard it as a variant of *G. verdickii.*

SELECTED SPECIMENS

Zaire. Shaba: Mugila Mountains, west slopes, 23 May 1908, *Kassner 2995* (BM, BR, E, HBG, K, P, Z); Mugila (Muhila) Plateau, Mt. Makala, 1670 m, 14 May 1971, *Lisowski 81648* (POZG); Lubumbashi (Elisabethville) to Likasi (Jadotville), in tall dry grass, May 1939, *Russel 55* (K); Keyberg, light woodland, Apr. 1951, *Schmitz 3554* (BR); Kasompo, 8 km west of Lubumbashi, woodland, 10 May 1955, *Schmitz 5143* (BR); near Kipopo, 25 km northwest of Lubumbashi, June 1971, *Thoen 4882* (BR); Kundelungu Mountains, July 1939, *Quarré 5590* (BR); Luiswishi, 31 Apr. 1972, *Malaisse 7401* (BR); Kyampongo, Upemba National Park, *de Wilde 731*

(BR); Masumbwe, Upemba National Park, 17 June 1949, *de Witte 6623* (BR); Tilwizembe, woodland, 8 May 1957, *Duvigneaud 3073* (BRLU); Kipopo, 9 May 1982, *Schaijes 1392* (BR); 7.8 km west-northwest of Kambove, between Kamfundwa and Shangolowe, 8 May 1983, *Schaijes 1912* (BR); 63.9 km northwest of Kolwezi, in 1985, *Schaijes 2502* (K).

Zambia. Northwestern: Mwinilunga District, Solwezi road, 79 km east of Mwinilunga and 10 km northwest of Lumwana Mission, termite mound in woodland, 14 May 1972, *Kornas 1785* (K).

45. *Gladiolus stenosiphon* Goldblatt, new species

MAP 46. Type: Angola, Huila, Lubango to Vila Arriaga, km 24, 13 Apr. 1973, *Bamps, Martins, & Matos 4484* (BR, holotype; K, WAG, isotypes).

EPONYMY

stenosiphon, "narrow tube," so named for the distinctive long, slender perianth tube.

LATIN DIAGNOSIS

Plantae 80–110 cm altae, cormis 20–35 mm diametro, foliis 8–10 anguste-lanceolatis, caule usitate eramoso c. 3 mm diametro base spica, floribus albis violaceis notatis, tubo perianthii 45–50 mm longo, tepalis subaequalibus 25–28 mm longis, 10–12 mm latis, filamentis inclusis vel 1–2 mm exsertis, antheris c. 6 mm longis, purpureis.

DESCRIPTION

Plants 80–110 cm high. CORM obovoid, 20–35 mm in diameter, tunics of papery to matted fibrous layers, fragmenting into vertical sections below. CATAPHYLLS membranous, the upper brown or becoming greenish above the ground. LEAVES 8–10, the lower four or five more or less basal and largest, reaching to about the middle of the spike, narrowly lanceolate to linear, 5–8.5 mm wide, the midrib and margins not or only lightly thickened and hyaline, upper leaves cauline, decreasing in size above, the blades almost filiform, imbricate and sheathing the stem up to the base of the spike. STEM unbranched or with one short branch, lightly flexed below the spike, c. 3 mm in diameter at the base of the spike.

SPIKE 14- to 18-flowered, lightly inclined; BRACTS narrowly lanceolate, attenuate apically, apparently green below, membranous above and becoming dry and light brown toward the apices, 15–30 mm long, the inner half to two-thirds as long as the outer, bilobed apically. FLOWERS white, the lower lateral tepals each with a small violet median mark; PERIANTH TUBE narrowly funnel-shaped, widening gradually and curving outward above, 45–50 mm long; TEPALS subequal, oval-elliptic, 25–28 mm long, 10–12 mm wide, the dorsal largest, inclined over the stamens, the lower slightly inclined toward the ground. FILAMENTS arcuate, included in the tube or exserted 1–2 mm from the tube; ANTHERS c. 6 mm long, apices without appendages, dark purple. OVARY narrowly obovoid, 3.5–4.5 mm long; STYLE arching over the filaments, dividing at the

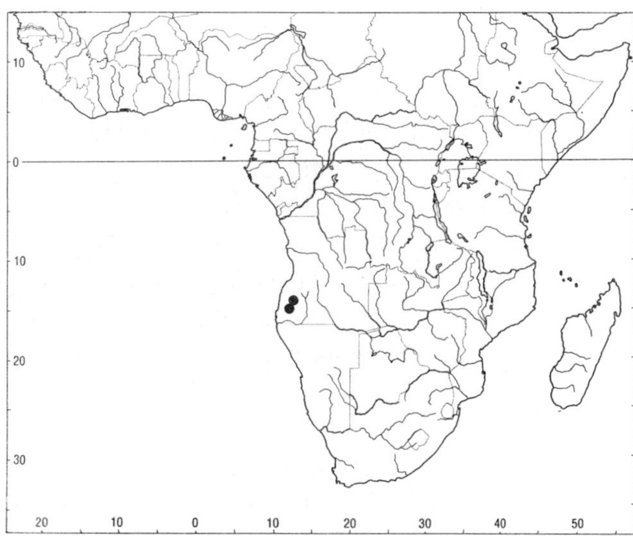

MAP 46. *Gladiolus stenosiphon.*

apex of the anthers, the branches c. 3.5 mm long, spreading beyond the anthers, expanded above. CAPSULES obovoid-ellipsoid, 10–15 mm long; SEEDS oval, c. 6 mm × 5 mm, broadly winged.

FLOWERING TIME. April and May, probably coming into bloom in March.

DISTRIBUTION & HABITAT

Restricted to southwestern Angola, *Gladiolus stenosiphon* is evidently an endemic species of recent origin, restricted to the Chela Mountains of Huila Province. Although reported to be common locally, it has been collected only rarely, first in 1937 by A. W. Exell and F. A. Mendonça, and then in 1968 and 1973. All collections are from the Chela Mountains near the road from Lubango (Sá da Bandeira) to Vila Arriaga, growing in savanna woodland among rocks. The parallel development of a white long-tubed flower in *G. harmsianus,* also from Angola but related to *G. gregarius,* and its significance are discussed under *G. harmsianus.*

DIAGNOSIS & RELATIONSHIPS

The long, slender perianth tube, 45–50 mm long, and almost entirely white flower are distinctive among tropical African *Gladiolus,* and these features combined with the narrow apically attenuate bracts distinguish *G. stenosiphon.* The western Angolan *G. harmsianus* has comparably long-tubed white flowers but it has rigid, apically obtuse overlapping bracts that tightly clasp the spike. The bracts of *G. harmsianus* indicate a relationship with the almost pan-tropical African *G. gregarius,* whereas the narrow bracts and slender, imbricate cauline leaves of *G. stenosiphon* correspond closely with *G. erectiflorus* and *G. verdickii* and it is to these two species that *G. stenosiphon* is related.

ADDITIONAL SPECIMENS

Angola. Huila: Lubango (Sá da Bandeira)–Namibe (Moçamedes) road, Chela Mountains, 22 Apr. 1968, *Kers 3361* (S); Chela Mountains, near summit, 1900 m, 17 May 1937, *Exell & Mendonça 2071* (BM).

46. *Gladiolus calcicola* Goldblatt, new species

MAP 47. Type: Ethiopia, Harerge, Mt. Hakim, 4 km south of Harar, rocky limestone slopes, c. 2200 m, 6 Oct. 1976, *de Wilde 7225* (ACD, holotype; BR, K, MO, WAG, isotypes).

EPONYMY

calcicola, "favoring calcium-rich soils," referring to the limestone-derived soils on which the species grows.

LATIN DIAGNOSIS

Plantae 30–70 altae, cormis 10–16 mm diametro, foliis 5–6 inferioribus longioribus ad basem spicae attingentibus linearibus (2–)3–5 mm latis, marginibus costatis leviter incrassatis, caule usitate eramoso, spica 2–4(–7) florum, bracteis 20–25(–30) mm longis, floribus pallide salmoneis, tubo perianthii c. 12 mm longo, tepalis 16–18 mm longis, filamentis c. 3 mm tubo exsertis, antheris c. 6 mm longis breviter apiculatis.

DESCRIPTION

Plants 30–70 cm high. CORM 10–16 mm in diameter, the tunics membranous, becoming fibrous, the fibers mostly vertical. CATAPHYLLS pale and membranous below, green to purple above ground. LEAVES five or six, the lower longest and reaching the base of the spike, after flowering reaching the spike apices, linear, 2–5 mm wide, the margins and midribs lightly thickened, upper leaves shorter and narrower than the lower, mostly sheathing but with blades developed, the sheaths imbricate and enveloping the stem almost to the first flower. STEM erect, rarely with one short branch, 2–3 mm in diameter at the base of the spike.

SPIKE 2- to 4(–7)-flowered, erect; BRACTS initially green, becoming membranous and dry above, 20–25(–30) mm long, apices attenuate, the inner about two-thirds as long as the outer.

FLOWERS pale salmon pink, the tepals darker in the midline; PERIANTH TUBE c. 12 mm long, curving outward and widening above; TEPALS subequal, 16–18 mm long, often shorter when dry, narrowly lanceolate, straight and directed forward. FILAMENTS 11–12 mm long, exserted c. 3 mm from the tube; ANTHERS c. 6 mm long, yellow, with short apiculi c. 0.5 mm long. OVARY ellipsoid, c. 4 mm long; STYLE arching over the stamens, dividing opposite the middle of the anthers, the branches c. 3 mm long, not reaching the anther apices. CAPSULES ellipsoid, 17–22 mm long; SEEDS c. 6 × 4 mm, broadly winged.

FLOWERING TIME. September to mid November.

DISTRIBUTION & HABITAT

Gladiolus calcicola is known from a small number of collections, all from the area immediately surrounding the city of Harar, Harerge Province, in southeastern Ethiopia. Plants grow in rocky situations in limestone at elevations of about 2000 m.

DIAGNOSIS & RELATIONSHIPS

A poorly known species, the relationships of *Gladiolus calcicola* are uncertain. The several long-bladed leaves that overlap one another and sheathe the stem resemble those of section *Tenuibracteus,* as do the shortly apiculate anthers, and it is in that section that the species is provisionally included. The other species of the section occur in southern tropical Africa.

ADDITIONAL SPECIMENS

Ethiopia. Harerge: fields around Awoday, near Alemaya, c. 2000 m, 16 Sept. 1976, *Jansen 7133* (WAG); 4 km from Bati (Alemaya College) toward Kombolcha, 2100 m, 19 Nov. 1974, *Bos & Getahun 9120* (flower and fruit) (ACD, WAG); foot of Hubeta, limestone slopes, 13 Oct. 1954, *Bally 10050* (EA).

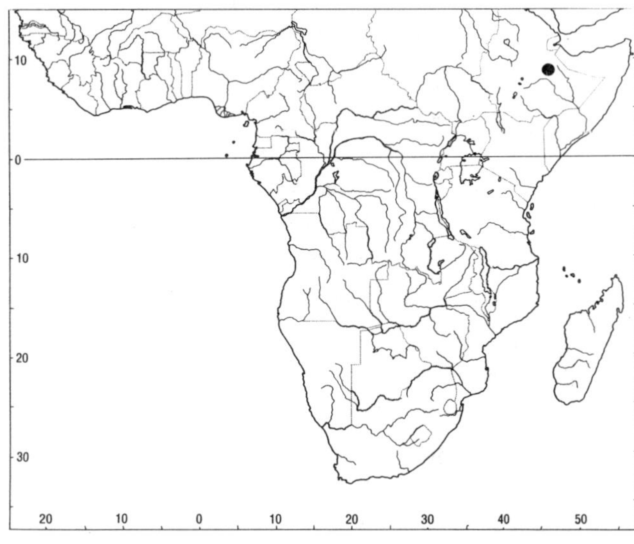

MAP 47. *Gladiolus calcicola.*

SECTION *OPHIOLYZA*

Synonymy

Gladiolus group *Dracocephali* Baker, J. Linn. Soc. Bot. 16: 176 (1878). Type: *G. dracocephalus* J. D. Hooker (= *G. dalenii* van Geel) (lectotype designated here).

Description

Plants usually large and stems usually unbranched. LEAVES lanceolate to linear (terete in *Gladiolus pallidus*), the upper smaller and sometimes entirely sheathing, usually contemporary with the flowers but borne on separate shoots after flowering in some species. SPIKES erect; BRACTS usually fairly long, (18–)20–70 mm long, generally slightly longer, or about as long as the perianth tubes. FLOWERS large to medium, (25–)35–95 mm long, the lower tepals with nectar guides often near the tepal bases, usually cream to yellow; PERIANTH TUBE (11–)20–45 mm long, typically slightly shorter to about as long as the bracts, in several species the tube much exceeding the bracts; TEPALS usually unequal, the dorsal largest, usually about as long as the tubes, occasionally slightly shorter, frequently hooded and horizontal, concealing the stamens, the lower three tepals less than half as long as the dorsal to about as long. ANTHERS with rounded apices.

Section *Ophiolyza* is widespread in Africa south of the Sahara, where there are more than 70 species; two species are shared with the Arabian Peninsula, and one of these also with Madagascar. The section is especially well represented in tropical Africa, but it is also diverse in the eastern half of southern Africa. There is a strong trend for the reduction in the size of the lower tepals in the section, occasionally to the degree that they are vestigial. This is exemplified particularly in *Gladiolus huillensis* and *G. magnificus* in southern tropical Africa, and in the *G. abyssinicus–G. longispathaceus* alliance in northeastern tropical Africa. Another notable specialization in the section is the temporal separation of the vegetative and reproductive phases of the growth cycle. In *G. dalenii* subspp. *andongensis* and *welwitschii, G. melleri,* and *G. roseolus,* the foliage leaves are produced on separate shoots in the wet season, after flowering is completed, often before the dry season is over.

47. *Gladiolus sericeovillosus* subsp. *calvatus* (Baker) Goldblatt

FIGURE 40, MAP 48. Goldblatt, Fl. Zambesiaca 12(4): 88 (1993).

SYNONYMY

Gladiolus ludwigii var. *calvatus* Baker, Fl. Capensis 6: 150 (1896). Phillips, Fl. Pl. S. Africa 4: pl. 125 (1924). *Gladiolus sericeovillosus* forma *calvatus* (Baker) Obermeyer, J. S. African Bot., Suppl. 10: 31 (1972). Type: South Africa, Transvaal, Umvoti Creek, Apr. 1890, *Galpin 925* (K, lectotype, designated by Lewis et al., 1972: 31; BOL, SAM, isolectotypes).

Gladiolus sericeovillosus var. *glabrescens* L. Bolus, S. African Gard. 18: 213 (1928). Type: South Africa, Transvaal, Waterval Boven (grown at Kirstenbosch Botanic Gardens in 1928), *Hutchinson s.n.* (BOL, as National Botanic Gardens 873/27, holotype).

Gladiolus dehnianus Merxmüller, Proc. & Trans. Rhodesia Sci. Assoc. 43: 151 (1951), Suessenguth & Merxmüller, Contrib. Fl. Marandellas District 76 (1951). Type: Zimbabwe, Marondera (Marandellas), dry veld, 30 Mar. 1941, *Dehn 14* (M, holotype, not seen; K, SRGH, isotypes).

Gladiolus elliotii sensu Lewis et al., J. S. African Bot., Suppl. 10: 27 (1972), in part, not Baker, J. Bot. 29: 70 (1891).

Gladiolus ochroleucus sensu Baker, Curtis's Bot. Mag. 103: pl. 6291 (1877), not sensu Baker (1876).

EPONYMY

sericeovillosus, "gray-haired," referring to the dense silvery pubescence on the leaves and bracts of typical subspecies *sericeovillosus*; *calvatus,* "bald," indicating the absence of pubescence.

DESCRIPTION

Plants 35–100 cm high. CORM globose, 2.5–3 cm in diameter, tunics coriaceous, decaying into vertical fibrous strips. CATAPHYLLS pale and membranous below the ground, the upper green above the ground, minutely pubescent. LEAVES five to seven, mostly basal, linear (narrowly lanceolate in South Africa), exceeding the spike sometimes by twice as much, 3–6(–8) mm wide, usually (microscopically) pubescent on the sheaths and often sparsely so on the lower parts of the leaves, the margins and midrib and sometimes other veins moderately to strongly thickened and hyaline, the upper one or two leaves cauline and much shorter than the basal. STEM occasionally branched, 4–5 mm in diameter at the base of the spike.

FIGURE 40. *Gladiolus sericeovillosus* subsp. *calvatus.* Corm, leaves, and flowering spike, × 0.5; single flower, full size; cross-section of leaf, × 2 (*Goldblatt 9055, 9083*).

SPIKE distichous, erect, 12- to 20(–40)-flowered; BRACTS pale green and soft-textured, becoming membranous or dry and pale above, 15–30 mm long, the outer acute-attenuate, usually slightly longer (or shorter) than the inner, the inner forked apically and with the margins united around the flower. FLOWERS pale gray-green to cream with the tepals densely speckled with minute dark red to maroon points, these often concentrated in the midline of each tepal, the lower lateral (and sometimes the lowermost) tepals yellow-green in the lower half (tepals sometimes in South Africa uniformly cream with green markings on the lower tepals); PERIANTH TUBE obliquely funnel-shaped, widening and curving outward near the apex, (6–)10–12 mm long, extended between the bracts; TEPALS unequal, the dorsal longest and hooded over the stamens, (18–) 20–22 mm long, often shorter when dry, the lower three c. 12 mm long, curving toward the ground, narrowed somewhat below but not abruptly expanded into limbs above. FILAMENTS arcuate, 10–12 mm long, exserted 6–8 mm from the tube; ANTHERS 8–10 mm long, reaching the upper third of the dorsal tepal. OVARY narrowly obovoid, c. 4 mm long; STYLE arching over the filaments, dividing near the apex of the anthers, the branches 4–5 mm long. CAPSULES obovoid, usually somewhat trilobed above, c. 12 mm long; SEEDS oval, c. 5 × 4 mm, strongly winged. CHROMOSOME NUMBER $2n = 30$.

FLOWERING TIME. Late summer to autumn, usually March and April.

DISTRIBUTION & HABITAT

Gladiolus sericeovillosus extends from the Transkei and eastern Cape in South Africa, northward through Natal and the Transvaal into Zimbabwe. Subspecies *calvatus* extends from central Transvaal north into Zimbabwe, where it occurs in the interior and eastern half of the country. It grows in open tall grassland or in light woodland in well-drained soils. It seems likely that subspecies *calvatus* also occurs in adjacent parts of Mozambique, but I have seen no records from there. Subspecies *sericeovillosus* occurs in the south of the range, from the southern Transvaal and Swaziland through Natal to the eastern Cape (Lewis et al., 1972).

DIAGNOSIS & RELATIONSHIPS

A strongly distichous, erect spike of moderate-sized, dull greenish flowers and linear leaves with strongly thickened margins and midribs that normally exceed the spikes (Figure 40) readily distinguish *Gladiolus sericeovillosus* subsp. *calvatus* in tropical Africa. Subspecies *sericeovillosus,* which occurs only in southern Africa, has broad leaves, 17–22 mm wide, with a dense, appressed pubescence, and the bracts and spike axis are covered with conspicuous whitish hairs. It is also generally more robust than subspecies *calvatus,* here raised from the rank of forma. Although there is a reasonably clear trend for the reduction of pubescence in populations of subspecies *sericeovillosus* from northern Natal, Swaziland, and the south-eastern Transvaal, it is almost always possible to assign plants to one or the other subspecies if leaf width and bract length are taken into account. Zimbabwean plants, which have leaves less than 7 mm wide, and usually no more than 5 mm, were described as *G. dehnianus* by Hermann Merxmüller in 1951 without reference to the glabrous and narrow-leaved Transvaal plant then known as *G. ludwigii* (a later synonym of *G. sericeovillosus*). In their revision of *Gladiolus* in southern Africa, Lewis et al. (1972) regarded *G. dehnianus* as conspecific with the southern African *G. elliotii*. Although *G. elliotii* is obviously closely allied to *G. sericeovillosus,* perhaps more so than Lewis et al. believed, they are separate species. *Gladiolus elliotii* can be distinguished mainly by its lanceolate leaves, shorter than the spike and with weakly developed midveins often no different from secondary veins of more or less equal thickness. *Gladiolus elliotii* also has somewhat larger flowers, usually speckled blue rather than dark red as in subspecies *calvatus* from Zimbabwe.

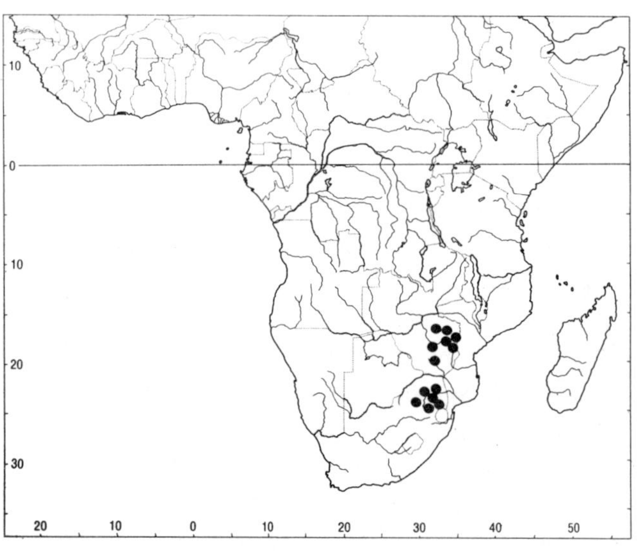

MAP 48. *Gladiolus sericeovillosus* subsp. *calvatus.*

SELECTED SPECIMENS

Zimbabwe. Manicaland: Mtare District, near Odzi, grassland near river, 11 May 1963, *Chase 8016* (K, P, SRGH); Mtare, east ridge of Nyamaganu Peak, 1200 m, 3 Jan. 1960, *Chase 7245* (K, SRGH). Nyanga District, 10 km west of Juliasdale on Rusape road, 31 Mar. 1991, *Goldblatt 9083* (MO). Mashonaland East: Harare (Salisbury), May 1906, *Flanagan 3252* (BOL); Cleveland Dam, granite sand, 15 May 1948, *Wild 2547* (K, SRGH); Marondera (Marandellas), 26 May 1943, *Dehn 14* (K, SRGH), 20 Nov. 1924, *Eyles 4374* (K, SRGH). Mashonaland West: Lomagundi District, east-facing slopes north of Great Dyke Pass, rocky grassland, 28 Apr. 1991, *Goldblatt 9055* (MO). Midlands: Mt. Beringwe, Apr. 1978, *Cannell 739* (SRGH); Jacobson farm, southeast of Gweru (Gwelo), 15 Apr. 1967, *Biegel 2070* (MO, SRGH).

48. *Gladiolus dalenii* van Geel

PLATES 27, 28, FIGURE 41, MAPS 49, 50, 51. Van Geel, Sert. Bot., Fasc. 28 (1829). Hilliard & Burtt, Notes Roy. Bot. Gard. Edinburgh 37: 297 (1979). Goldblatt, Bull. Mus. Hist. Nat., Paris, Sér 4, Sect. B, Adansonia 11: 251 (1989); Fl. Madagascar, Famille 45, Ed. 2, 42 (1991); Fl. Zambesiaca 12(4): 89 (1993). Type: South Africa, Natal, without precise locality, collector unknown, cultivated in Holland (illustration in Sert. Bot., Fasc. 28, lectotype designated by Hilliard & Burtt, 1979).

SYNONYMY

See under each of the three subspecies for the extensive synonymy of this species.

EPONYMY

dalenii, named for Cornelius Dalen, Dutch botanist who was to become director of the Rotterdam Botanic Garden. Dalen was responsible for the introduction of the species from Natal and its distribution to gardens in Europe; he also commissioned the painting that was published in the protologue, that now serves as the type of *Gladiolus dalenii.*

DESCRIPTION

Plants (50–)70–120(–150) cm high. CORM (15–) 20–30 mm in diameter, tunics of brittle membranous layers, the outer becoming irregularly broken, sometimes fibrous, reddish brown, usually bearing numerous tiny cormlets around the base. CATAPHYLLS firm-textured, the upper largest and up to 15 cm long, green or flushed purplish, occasionally densely short-pubescent. LEAVES either contemporary with the flowering stem (subsp. *dalenii*) and four to six, sometimes seven, or borne later on separate shoots (subspp. *andongensis* and *welwitschii*) and two to four on the flowering stem; when foliage leaves borne on the flowering stem at least the lower two basal or nearly so, long-laminate, narrowly lanceolate to more or less linear, (5–)10–20(–30) mm wide, about half as long as the spike, firm textured with moderately raised and thickened midribs and margins, the upper one or two leaves cauline and sheathing for at least half their length, sometimes entirely sheathing, often imbricate; when foliage leaves borne on separate shoots, then leaves of the flowering stem entirely sheathing or with blades up to 6 cm long, the foliage leaves of the separate shoots usually only two, lanceolate, or in subspecies *welwitschii* linear and the margins moderately thickened. STEM unbranched, 4–6 mm in diameter below the first flower.

SPIKE (2–)3- to 7(–14)-flowered; BRACTS green, (2.5–)4–7 cm long, sometimes dry and pale apically, the inner slightly shorter than the outer. FLOWERS either red to orange with a yellow mark on each of the three lower tepals or yellow to greenish and often with red to brown streaks on the upper tepals; PERIANTH TUBE (25–)35–45 mm long, nearly cylindric and curving outward in the upper half; TEPALS unequal, the three upper broadly elliptic-obovate, the dorsal largest, 35–50 mm long, 22–30 mm at the widest, horizontal to downcurved and concealing the stamens, the upper laterals about as long to c. 5 mm shorter than the dorsal, 20–30 mm wide, directed forward, often curving outward distally, the lower three tepals curving downward, 20–25(–30) mm long, 8–12 mm wide, the lowermost somewhat longer and narrower than the lower laterals. FILAMENTS c. 25 mm long, exserted 15–18 mm from the tube; ANTHERS 12–16 mm long, pale yellow. OVARY 6–8 mm long; STYLE arched over the stamens, dividing near the apex of the anthers, branches (4–)5–6 mm long. CAPSULES ellipsoid to ovoid, (18–) 25–35 mm long, 12–14 mm in diameter at the widest, SEEDS 8–12 × 5–9 mm, the wing well developed, lightly undulate, glossy light brown, seed body c. 2 mm in diameter. CHROMOSOME NUMBER $2n$ = 30, 60, 90 (45, 75).

FLOWERING TIME. Usually at least 3 weeks after the onset of the rainy season, (mid November to) December to May, rarely in June in southern Africa and eastern tropical Africa and Madagascar; May to August in western and central Africa; mostly August and September in Ethiopia; January to March in Eritrea and the Arabian Peninsula.

DISTRIBUTION & HABITAT

Gladiolus dalenii is by far the most widespread and common species of *Gladiolus.* It occurs virtually throughout the grasslands, savannas and woodlands of sub-Saharan Africa except the winter-rainfall region of southern Africa and the surrounding arid to semiarid Kalahari and Namib Deserts and the southern African Karoo. It also occurs in the highlands of southwestern Arabia and in Madagascar, where it is spread almost throughout the nonforested parts of the island except in the dry southwest. Thus *G. dalenii* may be found from Senegal in the west, across central Africa to Ethiopia, Eritrea, and western Saudi Arabia and Yemen. From there it extends southward through eastern Africa and Mozambique to Natal and the eastern Cape Province, South Africa, and across southern tropical Africa to the western Angolan highlands (Maps 49–51). *Gladiolus dalenii* favors moderately moist habitats and is thus most common in hill country and upland grassland, but it also occurs in fairly dry habitats with only a short wet season.

Across its range there are a number of variants, the most important of which is the one that flowers early in the season, 3–4 weeks after the first soaking rains of the wet season, and has reduced leaves on the flowering stem. Fully developed foliage leaves are produced from separate shoots on the same corm later in the growing season. Treated as separate species by some authors (*Gladiolus andongensis* Welwitsch ex Baker and *G. welwitschii* Baker from Angola; *G. goetzei* Harms from Tanzania; *G. pauciflorus* de Wildeman from Zaire; *G. mildbraedii* Vaupel from Rwanda), the reduced-leaved plants seem best regarded as two subspecies of *G. dalenii,* although they have not before been so treated. Plants with flowering stems lacking foliage leaves, subspecies *andongensis* and *welwitschii,* together have virtually the same distribution as subspecies *dalenii* except that they do not occur in South Africa, Madagascar, or Arabia, nor in Eritrea or the northern half of Ethiopia. In Kaffa Province, Ethiopia, some collections of an unusually slender form of subspecies *andongensis,* with flowers pale pink, may be hybrids with co-occurring *G. roseolus.*

ETHNOBOTANY

One of the few species of *Gladiolus* important to humans, *G. dalenii* is used in parts of West Africa in preparations to cure both constipation and severe dysentery, and is "highly esteemed in curing snake bites" (Irvine, 1930). At least in West Africa there are records that *G. dalenii* is cultivated on farms in the forest, where it was introduced from the savanna country to the north (Dalziel, 1937). How much of its remarkably wide distribution is due to deliberate human activity may never be known. Elsewhere in tropical Africa, Malaisse and Parent (1985: 62) report the use of corms of *G. dalenii* as food in upper Shaba Province, Zaire. The starchy corms are boiled and then leached in water for a week before consumption. In southern Africa, *G. dalenii* (reported under several synonyms) is a common component of the native herbalist's medicine horn, *lenaka.* Its use in treating cases of diarrhea and as a cold remedy are also documented (Watt & Breyer-Brandwijk, 1962; Jacot-Guillarmod, 1971).

A strikingly ornamental plant, *Gladiolus dalenii*

is widely cultivated. A late-summer-flowering form from southern Africa is perhaps the best known in horticulture. More important than its value as a wild species in garden displays is the role of *G. dalenii* as one of the parents in the original crosses that led to the development of the large-flowered *Gladiolus* hybrids, today one of the world's most important cut-flower crops.

DIAGNOSIS & RELATIONSHIPS

Gladiolus dalenii is readily recognized by the large flowers, 60–80 mm long, with the upper three tepals 35–50 mm long, much exceeding the recurved lower tepals, the strongly hooded dorsal tepal, well-exserted anthers, and fairly long perianth tube slightly exceeding the long floral bracts (Figure 41). The leaves are typically broad and sword-shaped, but as noted above, in subspecies *andongensis* and subspecies *welwitschii* the flowering stem lacks foliage leaves, and in some populations of subspecies *dalenii* the leaves may be narrow and no more than 5–8 mm wide. Whatever the width of the leaves, the midribs and margins are lightly thickened but not especially prominent.

Gladiolus dalenii may be confused with western Angolan endemic, *G. pallidus,* which has similar flowers though more often spotted than streaked with darker color (as is typical of *G. dalenii*) but linear leaves with very heavily thickened and raised margins. Also sometimes confused with *G. dalenii,* the southern tropical African *G. melleri* has small flowers with a tube 10–15 mm long, dorsal tepal 25–38 mm long, and bracts 15–23(–30) mm long. The flowering stem lacks foliage leaves, and filaments are included in the perianth tube. Treatment of *G. melleri* as a variety of *G. dalenii* (Geerinck, 1972) is unwarranted, although I have no doubt that the two species are closely related.

The type of *Gladiolus dalenii* is a watercolor illustration based on a plant from Natal and, like most collections of the species from southern Africa, has rather narrow flowers in which the upper tepals are closely appressed to the hooded dorsal tepal except toward the recurved apices. In most tropical African *G. dalenii,* the upper tepals are more loosely appressed to the dorsal and are sometimes flared outward for half their length, imparting a rather different appearance to the flowers. This does not seem to merit taxonomic recognition for it is not a consistent characteristic. Plants from Ethiopia may in fact sometimes be so similar to those from South Africa that it is virtually impossible to distinguish them, and taxonomic separation at any rank at all seems unworkable and unwarranted.

VARIATION

The most important variable in *Gladiolus dalenii* is the presence or absence of foliage leaves on the flowering stem. *Gladiolus dalenii* subsp. *dalenii* has well-developed sword-shaped leaves inserted on the lower part of the stem, whereas plants treated here as subspecies *andongensis* and *welwitschii* have only short, entirely sheathing, or short-bladed leaves on the flowering stem. In these two subspecies, fully developed sword-shaped (to linear) foliage leaves are produced on separate shoots on the same corm somewhat later in the growing season. It is not known whether the subspecies *andongensis* and *welwitschii* genotype can produce both leafless and leafy flowering shoots but this seems unlikely. These two subspecies more probably represent variants of the subspecies *dalenii* genotype that have evolved repeatedly across the range of the species. Although we lack data on the genetic changes that occur to allow the apparently very different developmental pattern and accompanying differences in flowering time that are present in plants with leaves borne on separate shoots, such as subspecies *andongensis* and *welwitschii,* it seems likely that they are profound. Moreover, the differences between subspecies *dal-*

enii, and subspecies *andongensis* and *welwitschii,* in the degree of development of the leaves on the flowering stem are usually clear. There are few true intermediates, although *G. pauciflorus* de Wildeman, evidently based on very similar plants from three different localities in Shaba Province, Zaire, seems to fall in this category. The flowering stems have two foliage leaves, each with blades 6–15 cm long. The discontinuity in leaf development and organization in *G. dalenii,* however, merits taxonomic recognition, and subspecific separation seems most acceptable. Geerinck's (1972) inclusion of plants with both leafy and leafless flowering stems in the same taxon, variety *natalensis,* reserving subspecies *melleri* for what I regard as the separate *G. melleri,* is not warranted.

Almost throughout its range, *Gladiolus dalenii* occurs in populations with flowers either pale yellow to greenish with or without red to brown streaking, or orange to red with bright yellow and sometimes green markings on the lower tepals. Only the common and typical orange-flowered form occurs in Arabia and Madagascar. Populations are almost always uniform for flower color but occasional intrapopulational variation has been recorded. Yellow-flowered plants, often more slender than the orange- or greenish-flowered forms, correspond to *G. primulinus,* and the red to orange to *G. dalenii, G. quartinianus,* and a large number of other named species now relegated to synonymy. Although sometimes treated as a separate species, yellow-flowered *G. dalenii* is certainly not always more slender. Comparisons made in Nigeria by Harper (1949) show the two cannot be distinguished by leaf dimensions. Most particularly dwarfed populations of *G. dalenii* do have yellow flowers, however, and its seems clear that regional and ecological differentiation of populations is occurring in the species. Thus the short-tubed yellow-flowered populations from western Zambia (e.g., *Robinson 5471,* with two or three foliage leaves and short spikes of two or three flowers) fit as awkwardly in *G. dalenii* as do collections from southern Tanzania and Malawi with purple flowers. All too little is known about this particular series of populations, always found in wetlands, and they may on further investigation be found to warrant taxonomic recognition. Consistent application of formal names to the numerous recognizable variants of *G. dalenii* would rapidly lead to the proliferation of some 10 or more taxa that would create an unmanageable taxonomy. Informal reference when necessary to important local populations is preferable.

Chromosome number is variable in *Gladiolus dalenii.* All African populations of subspecies *dalenii* so far examined are polyploid, thus having somatic numbers of either $2n = 60$ or 90, occasionally $2n = 45$ or 75 (Goldblatt et al., 1993), but Madagascan populations are diploid (Goldblatt, 1989). Variation in height of the plants and general vigor may be due at least in part to variation in ploidy level. The Madagascan plants are among the most slender in the species, and the flowers are slightly smaller than in continental African populations. The only counts for subspecies *andongensis,* from two Ugandan populations, are diploid, $2n = 30$.

HISTORY

Discovered first at the southern end of its range in Natal in the late 1820s, this common *Gladiolus* was given four names in the course of as many years after it was first described: *Watsonia natalensis* Ecklon (in 1827), *G. dalenii* van Geel (in 1829), *G. psittacinus* J. D. Hooker (in 1830), and *G. natalensis* Reinwardt ex J. D. Hooker (in 1831), the last an illegitimate renaming of *G. psittacinus.* The two first names were ignored for many years and in southern Africa the species came to be known either as *G. psittacinus,* which was based on a striking illustration in the accessible *Curtis's Botanical*

Magazine, or by the competing *G. natalensis,* which Hooker substituted for *G. psittacinus.* The name *G. natalensis* had been used for the species in Holland, and plants were distributed with this name by Professor C. G. C. Reinwardt at Leyden to growers in Europe. There is, however, no record of publication of the name *G. natalensis* prior to Hooker's substitution of it for *G. psittacinus* in 1831 (Hilliard & Burtt, 1979). Nevertheless, the confusion caused by Hooker's action has rippled through the nomenclature of *Gladiolus* until well into the 20th century, with authors favoring one or other name.

Just over a decade after its discovery in Natal, *Gladiolus dalenii* was collected in Ethiopia by Richard Quartin-Dillon, the botanist on the 1839 Lefebvre Expedition to that country. Collections from Ethiopia formed the basis for A. Richard's *G. quartinianus,* a name that continued to be used for the common tropical African form of *G. dalenii* for more than a century. Yellow-flowered *G. dalenii* was first collected in Mozambique by W. C. H. Peters (collected 1843–1847) and subsequently described by F. W. Klatt in 1864 as *G. luteolus.* In 1892, J. G. Baker described *G. primulinus,* by which name the yellow-flowered form of subspecies *dalenii* has most often been known when it was regarded as separate from typical orange-flowered plants. Subspecies *andongensis* and *welwitschii* were described first from Angola by J. G. Baker (1878a) as *G. andongensis* and *G. welwitschii,* both based on plants collected there in the 1850s by Friedrich Welwitsch. Later, nearly identical plants from Tanzania, then Tanganyika, were described as *G. goetzei,* and similar specimens from Rwanda were named *G. mildbraedii.* Slender, yellow-flowered plants with short-bladed leaves on the flowering stem were described by Emile de Wildeman (1913) from Shaba Province, Zaire, as *G. pauciflorus.* Whether these last plants also produce foliage leaves on separate shoots is unknown.

The numerous additional names applied to *Gladiolus dalenii* over the years arose in part because of poor communication and because of the parochial attitudes of botanists working in different parts of the continent. Variation in leaf shape and small differences in perianth size, often an artifact of preservation, were also responsible for some of the proliferation of names listed in synonymy below. Detailed enumeration of these names seems unnecessary. Baker, in his touchstone revision of the genus for tropical Africa (1898), reduced to synonymy under *G. quartinianus* many of the names until then in use for *G. dalenii.* He still recognized *G. quartinianus* as separate from any southern African plant, however, and he continued to maintain several species here regarded as conspecific with *G. dalenii.* Significantly, Baker did include *G. andongensis* and *G. welwitschii* in *G. quartinianus* (other synonyms of subspecies *andongensis* were only named after publication of Baker's work).

Only with the publication of the second edition of *Flora of West Tropical Africa* (Hepper, 1968) was it widely accepted that the tropical African *Gladiolus quartinianus* was conspecific with the southern African *G. psittacinus.* Both Lewis et al. (1972) and Geerinck (1972) accepted this conclusion but chose to use the name *G. natalensis* for the plant under the mistaken impression that it was a combination based on the synonym, *Watsonia natalensis* Ecklon, which predates *G. dalenii.* Hooker's *G. natalensis* is, however, not a combination but a new species (Hilliard & Burtt, 1979) and thus dates only from 1831. The existence of the name *G. natalensis* J. D. Hooker prevents the transfer to *Gladiolus* of Ecklon's *W. natalensis.* The independent choice of the epithet *natalensis* twice for the same species seems to have been coincidental.

Key to the Subspecies of *Gladiolus dalenii*

1 Leaves of the flowering stem with long well-developed blades (i.e., leaves and flowers contemporaneous) . . . . 48a. subspecies ***dalenii***

1′ Leaves of the flowering stem either entirely sheathing or with short blades to 10(–15) cm long and long-bladed foliage leaves produced on separate shoots later in the season (i.e., leaves and flowers not contemporaneous) 2

2(1′) Dorsal tepal as long as or slightly exceeding the upper lateral tepals; the base not surrounded by a neck of coarse fibers reaching ground level . . . . 48b. subspecies ***andongensis***

2′ Dorsal tepal 3–5 mm shorter than the upper lateral tepals; the base surrounded by a neck of coarse fibers reaching ground level . . . . 48c. subspecies ***welwitschii***

48a. *Gladiolus dalenii* van Geel subsp. *dalenii*

PLATES 27, 28, FIGURE 41, MAP 49.

SYNONYMY

Watsonia natalensis Ecklon, Topographishes Verzeichniss 34 (1827), not *Gladiolus natalensis* J. D. Hooker (1831). Type (according to Lewis et al., 1972: 46): South Africa, Natal coast (cultivated in Cape Town), *Ecklon s.n.* (S, holotype).

Gladiolus psittacinus J. D. Hooker, Curtis's Bot. Mag. 57: pl. 3032 (1830). Baker, Fl. Capensis 6: 158 (1896). Pole Evans, Fl. Pl. S. Africa 3: pl. 116 (1923). Hepper, Fl. West Trop. Africa, Ed. 2, 3(1): 141 (1968). *Gladiolus natalensis* Reinwardt ex J. D. Hooker, Curtis's Bot. Mag. 58: sub pl. 3084 (1831), as new name for *G. psittacinus*; Loddiges, Bot. Cab. 18: pl. 1756 (1831). Lewis et al., J. S. African Bot., Suppl. 10: 44–53 (1972), cited in error as (Ecklon) Reinwardt ex J. D. Hooker. Geerinck, Bull. Jard. Bot. Nat. Belgique 42: 281 (1972), excluding variety *melleri* (Baker) Geerinck. Wickens, Fl. Jebel Marra 158 (1976), an illegitimate name for *G. psittacinus*. Type: South Africa, Natal, without precise locality, illustration in Curtis's Bot. Mag. 57: pl. 3032.

Gladiolus quartinianus, A. Richard, Tent. Fl. Abyssinia 2: 306 (or 307?) (1851). Baker, Curtis's Bot. Mag. 110: pl. 6739 (1884), Fl. Trop. Africa 7: 371 (1898). Hutchinson & Dalziel, Fl. West Trop. Africa 2: 379 (1936). Type: Ethiopia, Tigre, damp sites in the Shire region, "locis humidis provinciae Shiré," 1839, *Quartin-Dillon s.n.* (P, lectotype designated here, the sheet with color illustration; BM, P, isolectotypes); near Adowa, September, *Quartin-Dillon s.n.* (not located, syntypes).

Gladiolus luteolus Klatt in Peter, Reise Mossambique, Bot. 2: 65 (or 515) (1864). Baker, Fl. Trop. Africa 7: 368 (1898). Type: Mozambique, lower Zambezi, woods near Boror, Apr. 1846, *Peters s.n.* (B, holotype).

Gladiolus garnieri Klatt, Linnaea 37: 511 (1872). Baker, Handbook Irideae 214 (1892). Perrier, Fl. Madagascar, Famille 45: 13 (1946). Type: Madagascar, without precise locality, 1866, *Garnier 68* (JE, holotype).

Gladiolus corneus Oliver, Trans. Linn. Soc. Bot. 29: 155, pl. 100 (1875). Baker, Handbook Irideae 222 (1892); Fl. Trop. Africa 7: 365 (1898). Type: Tanzania, mountains east of Lake Tanganyika, 1200 m, probably 1862, *Grant s.n.* (K, holotype).

Gladiolus saltatorum Baker, Trans. Linn. Soc. 29: 155 (1875). Type: Tanzania, Karangwe, 1862, *Speke & Grant s.n.* (K, holotype; plants are slender, probably yellow-flowered, and have leaves 6 mm wide).

Gladiolus ignescens Bojer ex Baker, J. Bot. 14: 334 (1876). Syntypes: Madagascar, without precise locality, "mont. prov. Emerina," *Hilsenberg & Bojer s.n.* (BM); "in montib. prov. Emirne," *Lyall 158* (K); *Pool s.n.* (not located).

Gladiolus newii Baker, J. Bot. 14: 334 (1876);

Handbook Irideae 214 (1892). Type: Tanzania, Mt. Kilimanjaro, at vegetation limit, Aug. 1871, *Rev. New s.n.* (K, holotype).

Gladiolus angolensis Welwitsch ex Baker, Trans. Linn. Soc. Bot., Ser. 2. 1: 269 (1878); Handbook Irideae 213 (1892). Type: Angola, Golungo Alto, Feb. 1856, *Welwitsch 1527* (BM, lectotype designated here; B, isolectotype); Golungo Alto, Mar. 1856 (fruit), *Welwitsch 1827a* (K, P, S, syntypes); Pungo Andongo, Apr. 1857, *Welwitsch 1530* (BM, K, P, syntypes).

Gladiolus sulphureus Baker, Trans. Linn. Soc. Bot., Ser. 2, 2: 350 (1887); Fl. Trop. Africa 7: 370 (1898), an illegitimate homonym, not *G. sulphureus* Jacquin (1786–1793) (= *Babiana stricta* var. *sulphurea* (Jacquin) Baker). Type: Tanzania, Kilimanjaro, 1500 m, without date, probably 1884, *Johnston s.n.* (K, holotype; plant with very short leaves).

Gladiolus primulinus Baker, Gard. Chron. 122 (1890); Fl. Trop. Africa 7: (1898). Hutchinson & Dalziel, Fl. West Trop. Africa 2: 379 (1936). Morton, W. African Lilies Orchids, fig. 45 (1961). Type: Tanzania, Dodoma, Usagara (as Sagara) Mountains, cultivated at Kew, June 1890, *Last s.n.* (K, holotype).

Gladiolus kilimandscharicus Pax in Engler, Hochgebirgsflora Tropischen Afrika 175 (1892). Baker, Handbook Irideae 214 (1892). Type: Tanzania, Mt. Kilimanjaro, 1889, *Meyer 278* (B, holotype; K, photo).

Gladiolus buettneri Pax, Bot. Jahrb. Syst. 15: 155 (1892). Baker, Fl. Trop. Africa 7: 372 (1898). Chevalier, Exploration Botanique Afrique 634 (1920). Type: Togo, Bismarckburg, July 1890, *Büttner 8* (B, lectotype designated here; K, isolectotype); Togo, May 1889, *Kling 22* (B, syntype; K, photo).

Gladiolus splendidus Rendle, J. Linn. Soc. Bot. 30: 406 (1895). Baker, Fl. Trop. Africa 7: 369 (1898). Type: Tanzania, Mt. Kilimanjaro, above Morang to 10,000 m, *Taylor s.n.* (BM, lectotype designated here, sheet 1, the better preserved; BM, isolectotype, sheet 2). Closely resembles the type of *G. quartinianus.*

Gladiolus taylorianus Rendle, J. Linn. Soc. Bot. 30: 405 (1895). Type: Kenya, Rabai Hills at Mwele (Mombasa district), Apr. 1886, *Taylor s.n.* (BM, holotype). Resembles *G. quartinianus.*

Gladiolus affinis de Wildeman, Pl. Nov. Hort. Then. 1: 161, pl. 35 (1901). Type: Mozambique, Morrumbala, Dec. 1900, *Luja 393b* (BR, holotype).

?*Gladiolus hockii* de Wildeman, Bull. Jard. Bot. Nat. Belgique 3: 264 (1911). Type: Zaire, Shaba, valley of the Little Luembe, Feb. 1910, *Hock s.n.* (BR, holotype, probably subspecies *dalenii* but without stem base).

Gladiolus luembensis de Wildeman, Bull. Jard. Bot. Nat. Belgique 3: 264 (1911). Type: Zaire, Shaba, Luembe Valley, Feb. 1910, *Hock s.n.* (BR, holotype; plants slender with short narrow leaves, yellow flowers).

Gladiolus garuanus Vaupel, Bot. Jahrb. Syst. 48: 541 (1913). Type: Cameroon, Garua, 320 m, 3 July 1909, *Ledermann 4475* (B, holotype; K, photo).

Gladiolus calothyrsus Vaupel, Bot. Jahrb. Syst. 48: 537 (1913). Type: Tanzania, between Gogoi and Mundo, in marshy places, c. 900 m, 26 July 1900, *Busse 172* (B, holotype; K, photo; plants tall, flowers reddish, spike with 14 flowers).

Gladiolus gallaensis Vaupel, Bot. Jahrb. Syst. 48: 533 (1913). Type: Ethiopia, Arusi, Galla highlands, Abinas Mountains, 2880 m, 10 July 1900, *Ellenbeck 1338* (B, holotype; K, photo).

Gladiolus boehmii Vaupel, Notizbl. Bot. Gart. Berlin-Dahlem 7: 32 (or 377) (1920). Type: Tanzania, near Gonda, Jan. 1882, *Böhm 60* (B, holotype; K, photo; Z, isotype; very like *G. quartinianus* but bracts rather long).

Gladiolus occidentalis A. Chevalier, Exploration Bot. Afrique 634 (1920), nomen nudum.

Gladiolus coccineus L. Bolus, S. African Gard. 18: 213 (1928). Type: Zimbabwe, Paulington, cultivated at Kirstenbosch, *Stokes s.n.* (BOL, holotype mounted on four sheets).

Gladiolus louwii L. Bolus, J. Bot. 67: 132 (1929). Type: Kenya, without precise locality, cultivated at Cape Town, *Louw s.n.* (BOL, holotype mounted on two sheets).

Gladiolus barnardii G. Lewis, S. African Gard. 22: 204 (1932). Type: Zimbabwe, near Harare, cultivated in Cape Town, *Barnard s.n.* (BOL, holotype mounted on three sheets).

Gladiolus caffensis Cufodontis, Österr. Akad. Wiss., Math.-Naturwiss. Kl., Sitzungsber. 156, 1: 471 (1947); Enum. Pl. Aethiopiae Sperm. 1590 (1974). Type: Ethiopia, Kaffa, without precise locality, *Bieber s.n.* (WU, holotype, not seen).

(Additional synonyms for *Gladiolus dalenii* based on plants from South Africa, including *G. dracocephalus* J. D. Hooker and many more, are cited by Lewis et al., 1972.)

DESCRIPTION

Plants (50–)70–120 cm high. CATAPHYLLS firm to coriaceous, often puberulent. LEAVES borne on the flowering stem, 4–6(–7), at least the lower two basal or nearly so, narrowly lanceolate to more or less linear, (5–)10–20(–30) mm wide, about half as long as the spike, firm-textured with moderately raised and thickened midrib and margins, the upper one or two leaves cauline and sheathing for at least half their length, sometimes entirely, often imbricate. SPIKE (2–)3- to 7(–14)-flowered; BRACTS herbaceous, (3.5–)4–7 cm long. FLOWERS with the tube 35–45 mm long, dorsal tepal (35–)40–50 mm long. FILAMENTS 25–30 mm long, exserted 15–20 mm from the tube; ANTHERS 12–16 mm long, pale yellow. CHROMOSOME NUMBER $2n$ = 30, 60, 90.

FLOWERING TIME. Usually at least 5–6 weeks after the onset of the rainy season, December to May (to June) in southern and east tropical Africa and Madagascar; late May to August in western and central Africa; mostly August and September in Ethiopia.

SELECTED SPECIMENS

Senegal. Kanéméré, 21 July 1965, *Fotius K257* (ALF, P).

Mali. Bamako, Aug. 1930, *Waterlot 1518* (P).

Guinea Bissau. Gabu, between Piché and Canquelifá, 12 Aug. 1950, *Espirito Santo 2729* (COI, LISC, NY).

Guinea. Near Beyla, June 1936, *Jacques-Felix 973* (P); Kourousa, July 1900, *Pobeguin 308* (P).

Ivory Coast: Upper Sassandra, 1909, *Chevalier 21461* (P); road from Bouna to Ferkessedougou near Comoé River, 4 Aug. 1967, *Geerling & Bokdam 450* (MO, WAG).

Ghana. Damongo, savanna, 13 July 1975, *Hall s.n.* (GC); Kpandai Leprosarium, 13 July 1969, *Enti s.n.* (GC 39544).

Togo. 12 km northwest of Sokodé, 9 Aug. 1965, *Davidson 20* (K); Adelé, 15 Nov. 1982, *Brunel 7758* (B).

Burkina Faso. Gourma, Fada, 20–26 July 1910, *Chevalier 24488* (COI, K, P); Banfora, 9 July 1958, *Aké Assi 5703* (K).

Benin. Agouagou, June 1910, *Annet 20* (P).

Nigeria. Bauchi: Tangale Waja, 6 km from Tula to Gombe, Nov. 1955, *Summerhayes 76* (BR, K, P); Bauchi Plateau, June 1930, *Lely 554* (K, MO). Kaduna: Mabiushi, 600 m, 14 July 1957, *Summerhayes 106* (K, P); Abinsi and vicinity, 3 Sept. 1912, *Dalziel 846* (K).

Cameroon. La Pastorale, 20 km north of Dschang, 1800 m, 25 July 1969, *de Wilde 5113* (P, WAG). Adamawa, Mambila Plateau, Maisamari, 14 June 1958, *Chapman 51* (BR, K, P).

Chad. Golé, 25 July 1969, *Fotius 1692* (P).

Central African Republic. Nana-Mambéré: 50 km east southeast of Bouar, 750 m, June, *Mildbraed 9620* (K). Haute Kotto: between Yalinga (95 km) and Ouadda, 13–20 June 1921, *le Testu 2835* (BR, P).

Sudan. Darfur: Jebel Marra, Nyertete, 1100 m, 13 Aug. 1964, *Wickens 2106* (K, MO). Equatoria: Near Khor Banza, road to Yei, 13 May 1939, *And-*

FIGURE 41. *Gladiolus dalenii* subsp. *dalenii,* × 0.67 (*Goldblatt s.n.*).

rews 1194 (K); Seriba Ghattas, territory of the Djur, 23 Jan. 1869, *Schweinfurth 1972* (E, K, P, WU, Z).

Ethiopia. Wollega: Near Didessa, 1220 m, 13 Aug. 1975, *Ash 3106* (K, MO, US). Tigre: Shiré, *Quartin-Dillon s.n.* (P). Shoa: Sire, between Robi and Ticcio, 10 Oct. 1971, *Ash 1258* (C, K, MO). Harerge: Alemaya, 15 km northwest of Harar, 10 June 1963, *Burger 2947* (K). Sidamo: Near Wolensu, 1250 m, 25 May 1982, *Fries, Tadesse, & Vollesen 3320* (C, K, ETH, UPS).

Eritrea: Amasen, Pianura Sabarguma, 2–10 Mar. 1902, *Pappi 3881* (BR, COI, EA, MO, P, S); Ghinda, 18 Jan. 1963, *Tekle Hagos 153* (ETH).

Saudi Arabia. Asir: Jebel Jihaf near Asdaf, northwest of Dhala, 2100 m, 14 Aug. 14 Aug. 1962, *Lavranos 1877* (K); Al Haddah near Taif, 2000 m, 26 Mar. 1979, *Collenette 1120* (K).

Yemen. Between Rubaisa and Turba in the Hoogariah, 1900 m, 26 Oct. 1974, *Wood 74/151* (K); the Sultana below Gabil, 2000 m, 1 Sept. 1977, *Radcliffe Smith & Henchie 4606* (K)

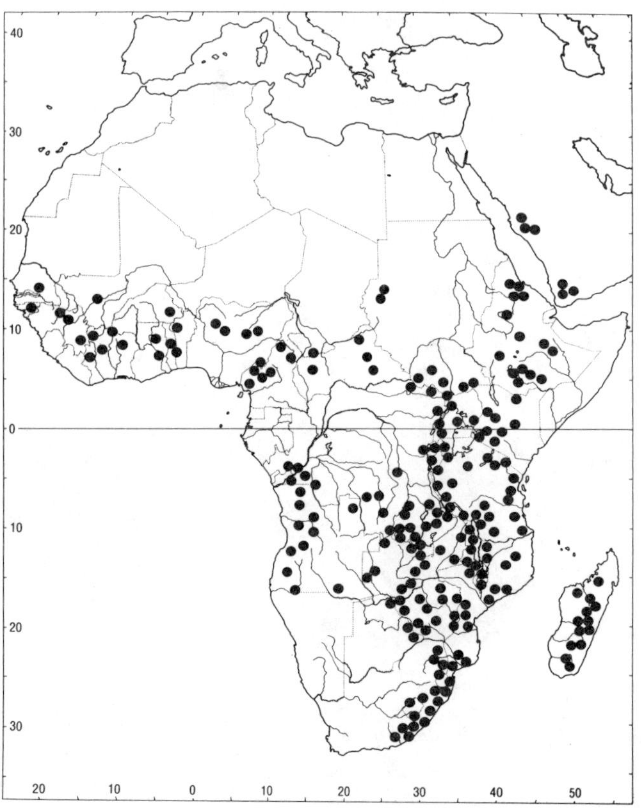

MAP 49. *Gladiolus dalenii* subsp. *dalenii*.

Uganda. Western: Kabalega (Murchison Falls) National Park, 12 June 1957, *Buechner 64* (EA). North Buganda: Kyagwe (Chagwe), Mabira Forest, without date, *Ussher 6* (K).

Kenya. Northern: Moyale, escarpment edge, 8 July 1952, *Gillett 13545* (BR, K, S). Central: Aberdares, Kinangop, 3600 m, 26 Oct. 1934, *Taylor 1259* (BM). Nyanza: Sotik, 1800 m, 16 June 1953, *Verdcourt 983* (BR, EA, K, MO, PRE); Eastern Mau Forest Reserve, 9 Sept. 1949, *Maas Gesteranus 6209* (BR, COI, G, MO, PRE).

Congo. Loudima plains, road to Brazzaville, Jan. 1891, *Thollon 4077* (P).

Zaire. Bas-Zaire: Mbanza–Ngungu, 3 Mar. 1987, *Billiet & Jadin 4270* (BR). Haut Zaire: Kagera National Park, Gabiro, 1500 m, Jan. 1938, *Lebrun 9582* (BR, P, S). Kivu: Katshuru, Jan. 1937, *Lebrun 9112* (BR). Kasai: Gandajika, 3 Jan. 1957, *Liben 2405* (BR, C, EA, K). Shaba: Upemba National Park, road from Lusinga to Mitwaba, 1760 m, 27 Jan. 1960, *Bamps 885* (BR).

Rwanda. Kitega. Kingota, 1650 m, 28 Jan. 1958, *van der Ben 1865* (BR).

Burundi. Buriri: Rumonge, 1200 m, 16 Jan. 1972, *Lewalle 1426* (BR, MO). Ruyigi: Gitwenge, 1750 m, 5 Jan. 1979, *Reekmans 7527* (BR, K, WAG).

Angola. Uige: 10 km west of Maquela do Zombo, 22 Jan. 1958, *Stanton 21* (BM). Bié: 6 km on road from Andulo to Tarala, c. 1650 m, 11 Nov. 1965, *Texeira et al. 9327* (LISC). Cuanza Sul: Malange, Jan. 1880, *von Mechow 397* (Z). Huambo: Haumbo (Nova Lisboa) District, Centro de Estudos at Chianga, 1750 m, 8 Nov. 1060, *da Silva 3341* (COI). Huila: between Cacula and Hoque, 1968, *Kers 3665* (S). Cunene: Cunene River, spray of Ruacana Falls, 9 June 1937, *Exell & Mendonça 2747* (BM). Lunda: Near Cossa, Oct.–Dec. 1962, *Cavaco 1352* (MO, P).

Zambia. Western: 10 km south of Senaga, 31 Jan. 1975, *Brummitt, Chisumpa, & Polhill 14192* (K, SRGH, WAG). Copperbelt: Kitwe, Garneton, 1 Jan. 1962, *Linley 240* (SRGH). Northern: Mbala District, Kali dambo, 1500 m, 15, Jan. 1955, *Richards 4102* (BR, K). Central: Serenje, Kundalila Falls, 17 Dec. 1967, *Simon & Williamson 1416* (K, SRGH). Eastern: Chipata (Fort Jameson), Nzamane, 6 Jan. 1959, 900 m, *Robson 1043* (BM, BR).

Tanzania. Kagera: Ngara, 13 Dec. 1960, *Tanner 5475* (BR, K). Arusha: Ngardota National Park, 6 June 1965, *Greenway & Kanuri 11905* (BR, K). Tanga: 6 km northeast of Korogwe, flats by the Lwengera River, 27 June 1953, *Drummond & Hemsley 3066* (B, BR, EA, K, LISC, S, SRGH). Morogoro: Uluguru Mountains, Lukwangule, 19 Feb. 1933, *Schlieben 3485* (B, BR, LISC, S, Z). Lindi: 90 km west of Lindi, Mandara, 300 m, 15 Feb. 1935, *Schlieben 5988* (BR, G, Z). Mbeya: Station Kyimbila, 1100 m, 9 Jan. 1911, *Stolz 579* (G, S, WAG). Rovuma: Songea, Matengo Hills, Lupembe Hill, 1860 m, 10 Jan. 1956, *Milne-Redhead & Taylor 8096* (B, BR, K, LISC). Rukwa: Ufipa, Malonje Farm, 13 Mar. 1957, *Richards 8678* (BR, K).

Malawi. Northern: Nyika Plateau, road to Karasamba View, 11 Apr. 1969, *Pawek 2123* (K). Central: Chongoni Forestry School, 17 Jan. 1967, *Salubeni 501* (MAL, SRGH). Southern: Zomba, Chitinje Dambo, 3 Jan. 1985, *Salubeni & Tawakali 3919* (MAL, MO).

Zimbabwe. Matabeleland North: Nyamandhlovu Pasture Station, 16 Jan. 1954, *Plowes 1660* (MO, PRE, SRGH); Victoria Falls, south bank, Dec. 1913, *Rogers 13040* (G). Mashonaland North: Mazowe, Apr. 1906, *Eyles 360* (BM). Manicaland: Nyanga, Honde Valley, 750 m, 10 June 1962, *Plowes 2252* (SRGH).

Mozambique. Niassa: Nampula, 17 Feb. 1937, *Torre 1215* (COI, LISC). Zambézia: Milanje, 550 m, 17 Feb. 1972, *Correia & Marques 2707* (BR, LMU). Maputo: Bokissa, 25 km from Maputo, 22 Apr. 1981, *Jansen & Macuácua 7697* (MO, WAG).

Botswana. Ngamiland: Without precise locality, 1930–1931, *Curson 15* (P, PRE); Tsantsarra Pan, Chobe National Park, 22 Jan. 1978, *Smith 2201* (PRE, SRGH).

48b. *Gladiolus dalenii* subsp. *andongensis* (Baker) Goldblatt

MAP 50. Goldblatt, Fl. Zambesiaca 12(4): 93 (1993).

SYNONYMY

Gladiolus andongensis Welwitsch ex Baker, Trans. Linn. Soc. Bot., Ser. 2, 1: 269 (1878); Handbook Irideae 221 (1892). Type: Angola, Pungo Andongo, 13 Dec. 1856, *Welwitsch 1529* (BM, lectotype designated here, well preserved and complete; G, K, P, isolectotypes).

Gladiolus goetzei Harms, Bot. Jahrb. Syst. 28: 365 (1900). Type: Tanzania, Uhehe, hilly plateau near Iringa, 1300 m, Jan. 1899, *Goetze 534* (B, holotype; K, photo).

Gladiolus pauciflorus de Wildeman, Feddes Repert. Spec. Nov. Regni Veg. 12: 297 (1913), an illegitimate homonym, not *G. pauciflorus* Baker (1886), from eastern Africa. Type: Zaire, Shaba, Upper Katanga, Bugege, Oct. 1911, *Hock s.n.* (BR, lectotype designated by Geerinck, 1972; BR, isolectotype); without precise locality, Oct. 1911, *Hock s.n.* (BR, syntype); Tsinshenda, Oct. 1911, *Hock s.n.* (BR, syntype).

Gladiolus mildbraedii Vaupel, Notizbl. Bot. Gart. Berlin-Dahlem 7: 33 (or 378) (1920). Type: Rwanda, Mt. Katandaganya, southeast of Lake Kivu, 1900–2000 m, 14 Aug. 1907, *Mildbraed 866* (B, holotype).

EPONYMY

andongensis, named for the Pungo Andongo region of northwestern Angola where the plant was first collected.

DESCRIPTION

Plants 60–90 cm high. CATAPHYLLS firm to coriaceous, often puberulent. LEAVES not contemporaneous with the flowers, those on the flowering stem short and entirely sheathing, two to four, 6–14 cm long, or sometimes with blades 2–3(–5) cm long and 6–12 mm wide, imbricate and sheathing the lower half of the stem; foliage leaves emerging from separate shoots later, usually at least two, narrowly lanceolate, 30–50 cm long, 4–16 mm wide. SPIKE (2–)3- to 9-flowered; BRACTS (25–)40–55 mm long. FLOWERS with the tube 25–33(–40) mm long; dorsal tepal 35–45 mm long, 22–25 mm at the widest. FILAMENTS c. 25 mm long, exserted 15–18 mm from the tube; ANTHERS 12–15 mm long. CHROMOSOME NUMBER $2n = 30$.

FLOWERING TIME. Early in the wet season, usually 3–4 weeks after the first rains, February to May in western Africa, November and December in eastern and central Africa.

DISTRIBUTION & HABITAT

Although the distribution of *Gladiolus dalenii* subsp. *andongensis* is extensive, it is much narrower than that of subspecies *dalenii.* It occurs entirely within the range of subspecies *dalenii* but it appears to be absent from large parts of tropical Africa. It occurs in the highlands of western Africa in Sierra Leone, Guinea, and Ivory Coast, and in Cameroon and adjacent western Central African Republic. In the eastern half of the continent it extends from western Ethiopia and southern Sudan through interior Tanzania to northern Malawi. It is relatively rare in Shaba Province of Zaire and northwestern Zambia. There is only a single record each for Angola and Zimbabwe, and only a few for the northern half of Mozambique. The subspecies is evidently distributed through the wetter highland areas of tropical Africa. Coastal records from the Shimba Hills of Kenya and Zambézia Province of Mozambique seem to be the exception. Plants favor seasonally wet sites and flower early in the wet season, within 2 to 4 weeks of the first rains. Subspecies *dalenii,* which is far more common, may occur sympatrically with subspecies *andongensis,* but in such places it always flowers 4–6 weeks later, well into the wet season.

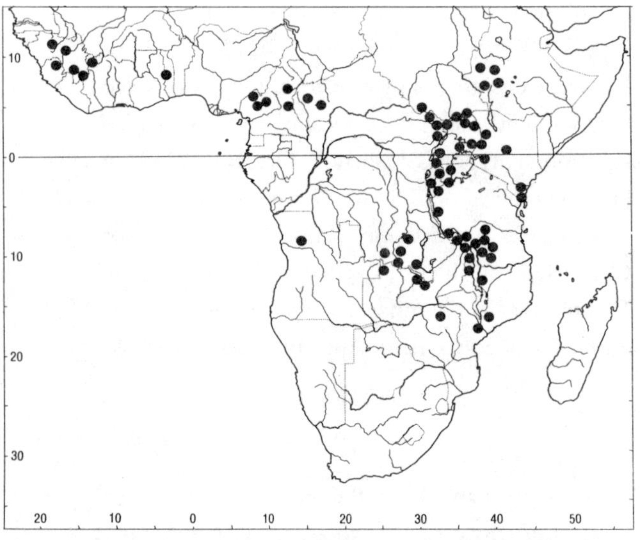

MAP 50. *Gladiolus dalenii* subsp. *andongensis.*

SELECTED SPECIMENS

Guinea. Macenta, Ténémadou, 3 Apr. 1949, *Adam 4224* (MO, P, SRGH). Nimba Mountains, Zouguépo Plateau, May 1957, *Schnell 24* (P).

Sierra Leone. Loma Mountains, high-altitude grassland, 1600 m, 17 Apr. 1966, *Jaeger 9867* (K, MO, P); Sulima Chiefdom, Bintumane Peak, plateau, in grass, 2 May 1949, *Deighton 5086* (K).

Ivory Coast. Upper Sassandra, summit of Mt. Dourou, 1123 m, 27 May 1909, *Chevalier 21725* (P); territory of the Dyolas, Mt. Boho near Zoanlé, 6 May 1909, *Chevalier 21468* (P).

Togo. Near Kuwde, west of Kpagouda, northwest of Farendé, 10 May 1976, *Hakki, Leuenberger, & Scheirs 903* (B).

Cameroon. La Pastorale, 20 km north of Dschang, 1800 m, 25 July 1969, *de Wilde 5113* (P, WAG). Batchingou Mountains, c. 20 km west of

Baganté, 4 May 1964, *de Wilde 2437* (B, EA, MO, P, Z).

Central African Republic. Bouar, 9 Apr. 1962, *Mosnier 859* (P); Cersé, May 1963, *Bille 1214* (P).

Sudan. Equatoria: Didinga, Mt. Lotuke, 1800 m, 19 Apr. 1939, *Myers 10932* (K); Imatong Mountains, 2300 m, 16 Feb. 1982, *Fries & Vollesen 917* (C, K).

Ethiopia. Kaffa: Gojeb River 67 km southwest of Jimma, 1300 m, 25 Feb. 1966, *de Wilde 10189* (BR, K, MO, PRE, WAG). Wollega: Near Dembidollo, 1600 m, 25 Feb. 1957, *Mooney 6811* (BR, ETH, K, S). Gamu Gofa: Bulki Mountain, 2100 m, 16 Apr. 1959, *Thesiger 1847* (BM).

Uganda. Northern: West Madi, Abeso, May 1948, *Eggeling 5774* (BR, K). Southern: Virunga Mountains, Muhavura, 3000 m, 18 Nov. 1934, *Taylor 1771* (BM). North Buganda: Kampala District, Kakenzi, 25 Apr. 1945, *Greenway & Thomas 7361* (K).

Kenya. Western: Kakamega Forest, west of Forest Station, 24 Apr. 1975, *Friis & Hansen 2597* (C). Rift Valley: Trans Nzoia, Cherangani, near Kitale, 20 May 1949, *Maas Geesteranus 4690* (K). Eastern: 9 km east of Nanyuki, 22 Feb. 1969, *Magor 23* (K). Coast: Shimba Hills, Longo Magandi, 30 Mar. 1968, *Magogo & Glover 581* (EA).

Zaire. Haut-Zaire: Nioka, Mt. Ri, Ineac Reserve, 24 Apr. 1952, *Troupin 640* (BR, K); Garamba National Park, valley of the Garamba, 30 Mar. 1952, *Troupin 640* (BR, K); Kivu: Kalehe, Mt. Kahuze, 2680 m, 30 Sept. 1971, *Ntakiyimana 176* (BR). Shaba Lubumbashi, dambo, 3 Jan. 1928, *Quarré 951* (BR, SRGH); Upemba National Park, near the source of the Louie, 29 Aug. 1949, *de Witte 7613* (BR).

Rwanda. Gikongoro, Nyungwe Forest, 30 Oct. 1982, *Troupin 16323* (BR, K).

Burundi. Bubanza: Randa, marsh, 850 m, 3 Dec. 1977, *Reekmans 6673* (BR, C, MO, P, WAG); Kihanga, Katunguru Valley, 900 m, 8 Nov. 1969, *Lewalle 4005* (BR, MO).

Tanzania. Kigoma: Mahari Peninsula near Kasala Peak, 6 Sept. 1959, *Harley 9519* (BR, K); Biharamulo, Kashasha, 10 Aug. 1960, *Tanner 5072* (BR, K). Rukwa: Ufipa, Chala Mt., 10 Dec. 1956, 2100 m, *Richards 7205* (B, BR, K). Iringa: Lupembe District, upper Ruhudje, 1931, *Schlieben 127* (B, BR, LISC). Morogoro: Uluguru Mountains, northern slopes, 450 m, 28 Jan. 1933, *Schlieben 3300* (B, BM, BR, G, LISC, Z). Rovuma: Songea, Matengo Hills, 29 Sept. 1956, *Semsei 2504* (K, PRE, SRGH).

Zambia. Western: Mwinilunga District, Ikelenge, 8 Nov. 1962, *Richards 16983* (K). Copperbelt: Ndola, 2 Jan 1954, *Fanshawe 622* (K).

Malawi. Northern: Nyika Plateau near Lake Kaulime, c. 2280 m, 8 Jan. 1974, *Pawek 7901* (MO, P); Rumphi, 8 Dec. 1965, *Banda 763* (MAL, SRGH).

Zimbabwe. Mashonaland North: Mazowe District, Mvurwe (Umvukwes), Ruorka Ranch, 1550 m, 16 Dec. 1952, *Wild 3915* (MO, SRGH).

Mozambique. Niassa: Lichinga Plateau, Dec. 1932, *Gomes e Sousa 1071* (COI). Zambézia: Maganja da Costa, 64 km from Vila de Maganja to Mocuba, c. 100 m, 9 Jan. 1968, *Torre & Correia 16983* (LISC).

48c. *Gladiolus dalenii* subsp. *welwitschii* (Baker) Goldblatt

MAP 51. Goldblatt in Stork & Lebrun, Enumeration Plantes Fleurs Afrique Tropicale 3: 90 (1995).

SYNONYMY

Gladiolus welwitschii Baker, Trans. Linn. Soc. Bot., Ser. 2, 1: 267 (1878); Handbook Irideae 213 (1892). Type: Angola, Huila, hillsides around Lopolo, Oct.–Nov. 1859, *Welwitsch 1541,* annotated as "*G. splendens*" (B, lectotype designated here, specimen complete with leaf shoot and corm tunics; BM, COI, K, LD, P, isolectotypes).

EPONYMY

welwitschii, named in honor of Friedrich Welwitsch, early botanical explorer in Angola.

DESCRIPTION

Plants 30–40 cm high. CORM with tunics of soft membranous layers, the outer becoming irregularly broken and somewhat fibrous, reddish brown. CATAPHYLLS membranous, hidden by a fibrous neck of dry leaf bases, the upper green, obscurely downy. LEAVES of the flowering stem usually three, short and almost or entirely sheathing, 5–7 cm long, imbricate and sheathing the lower half of the stem; foliage leaves produced from separate shoots later in the season, three, sometimes four, the blades linear, 45–60 mm long, 5–6(–9) mm wide, the midrib and margins thickened and hyaline, lightly pubescent on the sheaths. SPIKE 5- to 8-flowered; BRACTS 30–45 mm long. FLOWERS usually red, the tube 25–35 mm long; dorsal tepal 36–40 mm long, upper laterals 40–50 × 20–25 mm, the lower laterals c. 15 mm long, the lowermost 25 mm long. FILAMENTS 20–25 mm long, inserted in the middle of the tube, exserted c. 15 mm; ANTHERS 14–15 mm long, pale yellow.

FLOWERING TIME. Mostly September and October.

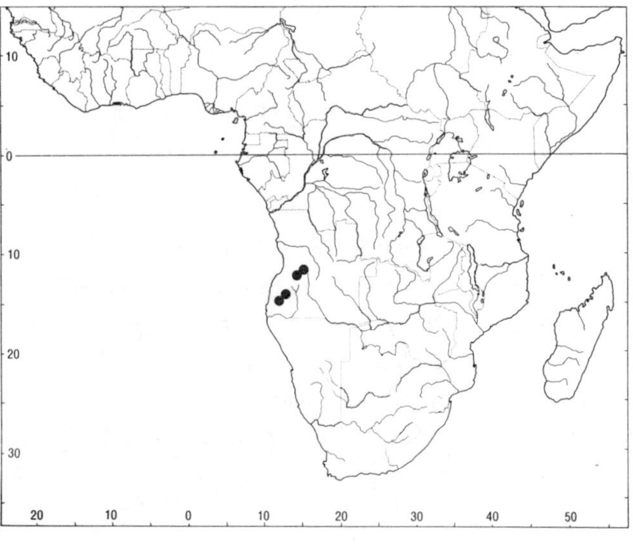

MAP 51. *Gladiolus dalenii* subsp. *welwitschii*.

DISTRIBUTION & HABITAT

Although poorly known, it seems all but certain that *Gladiolus dalenii* subsp. *welwitschii* is restricted to the highlands of southwestern Angola. Unlike the similar subspecies *andongensis,* which evidently has no preference for damp sites, subspecies *welwitschii* seems to occur only in permanently wet sites where it flowers in spring, in September and October, before the rains have begun. Fully developed foliage leaves are only known from plants grown in Paris (*Berthellot s.n.*), specimens of which have three to four long leaves on separate shoots produced after the flowers have faded. These leaves are consistent with dry and dead leaf remains attached to a wild-collected specimen (*Murta & Silva 712*) and confirm that plants produce long linear leaves during the growing season. That last collection shows that the flowering stem is actually produced from the same shoot as the leaves and that the flowering stems are actually the delayed terminal shoots of the leafy growth of the past season's vegetative cycle.

DIAGNOSIS & RELATIONSHIPS

The flowers of subspecies *welwitschii* correspond closely to those of subspecies *dalenii* and subspecies *andongensis,* except that the dorsal tepal seems to be consistently smaller by 5 to 10 mm than the upper laterals, and the perianth tube is somewhat shorter at 25 to 30 mm than is normal for those two subspecies, which have tubes usually 25–40 mm long. The neck of tough fibers sheathing the base of the stem, the short stature, and the three to four linear leaves, 40–60 cm long, combined with the slightly differently constructed flower make subspecies *welwitschii* easy to recognize. Subspecies *andongensis,* which also occurs in western Angola, although not on the Huila Plateau, does not have a fibrous neck around the stem base, and the foliage leaves are generally fewer, shorter, and broader than in subspecies *welwitschii.* Although the two subspecies are undoubtedly closely allied, the differences between them seem sufficient to merit their treatment as separate taxa.

SELECTED SPECIMENS

Angola. Cuanza Sul: Between Catenga and Vila Nora de Seles, 9 Sept. 1930, *Gossweiler 9362* (COI). Huambo: Vicinity of Huambo (Nova Lisboa), Centro de Estudos at Chianga, 1750 m, 29 Aug. 1970, *da Silva 3218* (COI). Huila: Huila Plateau, wet ground, Sept., *Berthelot 344* (P); between Lubango (Sá da Bandeira) and Chibia, 8 Oct. 1969, *Murta & Silva 712* (BM, COI, LISC); Chela Mountains, Tchivinguiro, 1750 m, 8 Oct. 1941, *Gossweiler 12735* (LISC).

49. *Gladiolus velutinus* de Wildeman

PLATE 29, FIGURE 42, MAP 52. De Wildeman, Feddes Repert. Spec. Nov. Regni Veg. 12: 297 (1913). Goldblatt, Fl. Zambesiaca 12(4): 93 (1993). Type: Zaire, Lubumbashi (Elisabethville), Dec. 1911, *Hock s.n.* (BR, lectotype designated here, specimen with leaves and cataphylls); Jan. 1912, *Hock 96* (BR, syntypes, with spikes only).

SYNONYMY

Gladiolus katubensis de Wildeman, Feddes Repert. Spec. Nov. Regni Veg. 12: 297 (1913). Type: Zambia, Katuba Stream, 30 Dec. 1907, *Kassner 2268* (BR, holotype; B, P, Z, isotypes).

Gladiolus vallidissimus Vaupel, Notizbl. Bot. Gart. Mus. Berlin-Dahlem 7: 34 (or 379) (1920). Type: Tanzania, "Njassa-Gebied, Station Katanda," 1900 m, 26 June 1909, *Münzner sub Exped. Fromm 262* (B, holotype).

Gladiolus xanthus L. Bolus, S. African Gard. 23: 47 (1933). Type: Zambia, Ndola West, in vlei, 25 Jan. 1932, *Duff 59/32* (BOL, lectotype designated by Goldblatt, 1993: 93); Nkana, cultivated in Cape Town, *Barnard s.n.* (BOL, syntype) (= *G. dalenii* subsp. *dalenii*).

EPONYMY

velutinus, "velvety," referring to the soft, dense pubescence of the leaf surfaces.

DESCRIPTION

Plants 1–1.5 m high. CORM depressed-globose, 30–50 mm in diameter, the tunics reddish and irregularly broken. CATAPHYLLS firm-textured, the upper green or flushed purple above the ground and then shortly and densely pubescent. LEAVES several, mostly basal or inserted in the lower part of the stem, about half as long as the spike, lanceolate, sometimes narrowly so, (13–)20–35 mm wide, the margins, midrib, and usually at least one other pair of veins thickened and hyaline, the surface densely and minutely papillose to short-pubescent. STEM unbranched, c. 7 mm in diameter at the base of the spike.

SPIKE 6- to 10-flowered, erect, the nodes 3–7 cm apart; BRACTS green, 4–6(–7) cm long, the inner 1–2 cm shorter than the outer. FLOWERS pale yellow with greenish highlights; PERIANTH TUBE 30–40 mm long, curving outward above, gradually expanding toward the apex; TEPALS unequal, the upper three c. 4 cm long, the dorsal hooded over the stamens, c. 2 cm wide, the upper laterals directed forward and curving outward only apically, the lower three tepals half to one-third as long as the upper, curving downward from the base. FILAMENTS 25–30 mm long, exserted at least 15 mm from the tube; ANTHERS 14–17 mm long, yellow. OVARY narrowly ovoid to obovoid, 7–9 mm long; STYLE arching over the stamens, dividing near the anther apices, the branches c. 5 mm long. CAPSULES and SEEDS unknown.

FLOWERING TIME. December and January in Zambia and Shaba Province, Zaire, April to June

FIGURE 42. *Gladiolus velutinus*, × 0.67 (*Lovett s.n.*, *Nash 15*).

in Tanzania and Uganda, thus from the middle to the end of the wet season.

DISTRIBUTION & HABITAT

Gladiolus velutinus has a relatively limited distribution across tropical Africa. It extends from southern Shaba Province in Zaire and adjacent northern central Zambia into western Tanzania and southwestern Uganda. The range is rather discontinuous, possibly a reflection of incomplete collecting, but *G. velutinus* seems consistently to occur in wetland sites such as marshes, dambos, and streambanks.

DIAGNOSIS & RELATIONSHIPS

Evidently closely related to *Gladiolus dalenii, G. velutinus* has a large flower with a hooded and horizontal dorsal tepal similar to larger-flowered forms of that species, but the flowers are always a pale lemon yellow with greenish highlights, and the flower is relatively narrow because the upper lateral tepals remain rather closely appressed to the dorsal tepal (Figure 42). The leaves are perhaps its most distinctive feature. Comparatively broad, with the basal 20–25 mm at the widest, the leaves generally have a second pair of thickened veins, one on either side of the midrib, and the entire surface of the blade is covered with a dense indumentum of papillose hairs. This leaf morphology and the height of the plants, 1.5–2 m, make it relatively easy to distinguish the species. No doubt *G. velutinus* is closely related to *G. dalenii,* but its wetland habitat combined with a fairly distinctive morphology provide a strong argument for its recognition.

HISTORY

Although first recorded in 1893 in northern Zambia by the Scottish naturalist, Alexander Carson, *Gladiolus velutinus* was only described in 1913 by Emile de Wildeman, who based his description on plants collected in 1911 and 1912 in Shaba Province, Zaire, by Adrien Hock. *Gladiolus velutinus* has seldom been accorded the recognition it deserves and has generally been regarded as conspecific with *G. dalenii* or one of its synonyms. It was so treated by Geerinck (1972) in the only revision of *Gladiolus* in any part of tropical Africa where *G. velutinus* occurs. Another early collection of *G. velutinus,* made by Theo Kassner in northern Zambia in 1907, is the type of *G. katubensis,* also described by de Wildeman in 1913. A collection from western Tanzania was named as *G. validissimus* by F. Vaupel in 1920, and plants from Ndola, Copperbelt Province, Zambia, were described as *G. xanthus* by H. M. L. Bolus in 1933. The plants described under these three later synonyms are identical with *G. velutinus* in all important taxonomic features, including the papillose to pubescent foliage, large yellow to yellow-green flowers, and preference for wet habitats. *Gladiolus xanthus* was actually based on two collections, one a yellow-flowered *G. dalenii.* The other, which has minutely pubescent leaves is chosen as the lectotype, as it accords well with the published description and matches exactly *G. velutinus.*

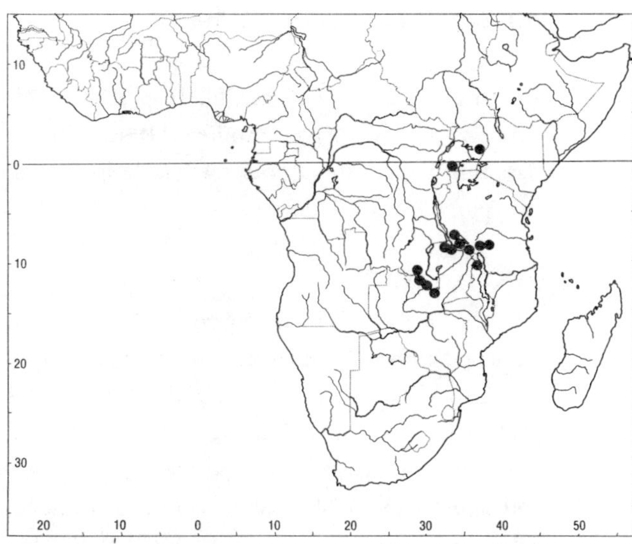

MAP 52. *Gladiolus velutinus.*

SELECTED SPECIMENS

Uganda. Bugisu: Between Buyaga and Siroko, 11 May 1970, *Katende & Lye 286A* (MO). North Buganda: Akole, 1600 m, without date, *Purseglove 502* (K).

Zaire. Shaba: Kipopo, 20 km northeast of Lubumbashi, dambo, 6 Dec. 1957, *Schmitz 6013* (BR, EA).

Zambia. Northern: Mbala (Abercorn), Kawimbe, wet dambo, Jan. 1954, *Nash 15* (BM, P, SRGH); Lumi marsh, Kawimbe, 29 Dec. 1959, *Richards 12021* (K); Fwambo, in 1893, *Carson 9/93* (K). Copperbelt: Ndola, dambo margins, 9 Jan. 1954, *Fanshawe 646* (K).

Tanzania. Iringa: Mufindi, Lake Kihanga, swampy stream bank, 1830 m, 11 Apr. 1987, *Lovett 1992* (K, MO, DSM). Mbeya: Mbosi, c. 1600 m, 8 Dec. 1935, *Horsbrugh-Porter s.n.* (BM). Rukwa: Sumbawanga, edge of Lwiche River, c. 2000 m, 21 June 1968, *Sanane 179* (K, P); 14 km northeast of Sumbawanga, Nzika stream, 2150 m, 21 June 1987, *Mwasumbi, Mohamed, & Kajula 13193* (K, MO).

Malawi. Northern: Nyika Plateau, Lake Kaulime, 4 Jan. 1969, *Richards 10446* (K).

50. ***Gladiolus pallidus*** Baker

FIGURE 43, MAP 53. Baker, Fl. Trop. Africa, Addenda 7: 577 (1898). Type: Angola: Huila, without date but before 1895, *Antunes 42* (K, holotype; B, isotype).

SYNONYMY

Gladiolus kubangensis Harms in Warburg, Kunene-Sambesi-Exped. 200 (1903). Type: Angola, Cuando Cubango, hills near the Kubango at Holombo, 1450 m, 8 May 1900, *Baum 898* (B, holotype; BM, COI, E, K, Z, isotypes).

Gladiolus baumii Harms in Warburg, Kunene-Sambesi-Exped. 200 (1903). Type: Angola, between Kulei and Kutsi, 1300 m, 30 Apr. 1900, *Baum 881* (B, holotype).

Gladiolus splendidus var. *tundensis* Chiovenda, Bull. Soc. Bot. Ital., 1924(2): 45 (1924). Type (not cited in the protologue): Angola, Namibe (Mossamedes), Tunda, 1923, *Mazocchi-Alemanni 161* (FI, lectotype designated here, photocopy seen, the most complete specimen); *123* (K, syntype); *147* (FI, syntype).

EPONYMY

pallidus, "pale," referring to the perianth color, cream to pale yellow with light pink to orange undertones.

DESCRIPTION

Plants 55–70 cm high. CORM 25–35 mm in diameter, the tunics coriaceous, decaying into irregularly shaped fragments and sometimes becoming more or less fibrous. CATAPHYLLS membranous, the inner usually reaching a short distance above the ground and then green. LEAVES six or seven, the lower three more or less basal, longest and usually somewhat exceeding the spike, the upper leaves largely sheathing, imbricate and usually concealing the stem to the base of the spike, glabrous, the blades linear, 2.5–3.5 mm wide, the margins and midrib heavily thickened, the margins sometimes raised and arched, partly covering the surface. STEM unbranched and more or less erect, c. 3 mm in diameter at the base of the spike.

SPIKE 3- to 5-flowered, erect; BRACTS green or flushed with purple, 35–55 mm long, about one internode long, the inner two-thirds to nearly as long as the outer. FLOWERS uniformly pale yellow or moderately to densely speckled with violet or red dots; PERIANTH TUBE 25–30 mm long, slender below, expanded gradually in the upper half and curving outward; TEPALS unequal, the dorsal largest and hooded over the stamens, (30–) 35–45 mm long, 18–20 mm wide, the upper laterals

FIGURE 43. *Gladiolus pallidus,* × 0.67 (*Bamps 4203, Murta 204*).

35–40 mm long, directed forward, perhaps ultimately curving outward, the lower three tepals 16–20 mm long, lanceolate, horizontal, curving downward distally, the lower laterals obovate and obtuse, the lowermost lanceolate and slightly longer. FILAMENTS arched below the dorsal tepal, c. 25 mm long, exserted 10–12 mm from the tube; ANTHERS 9–13 mm long, horizontal or tilted downward. OVARY oblong, c. 5 mm long; STYLE usually dividing just beyond the apex of the anthers, the branches 4–5 mm long. CAPSULES 20–32 mm long; SEEDS c. 9 × 6 mm, light brown.

FLOWERING TIME. Mainly April to June but also July and August and possibly even in November.

DISTRIBUTION & HABITAT

Known only from southwestern Angola, *Gladiolus pallidus* extends from the western highlands of Benguela Province through the Huila Plateau into the interior of Bié and Cuando-Cubango. Although most often collected in flower in May and June, it appears to have an unusually long flowering season, from April to August. A collection said

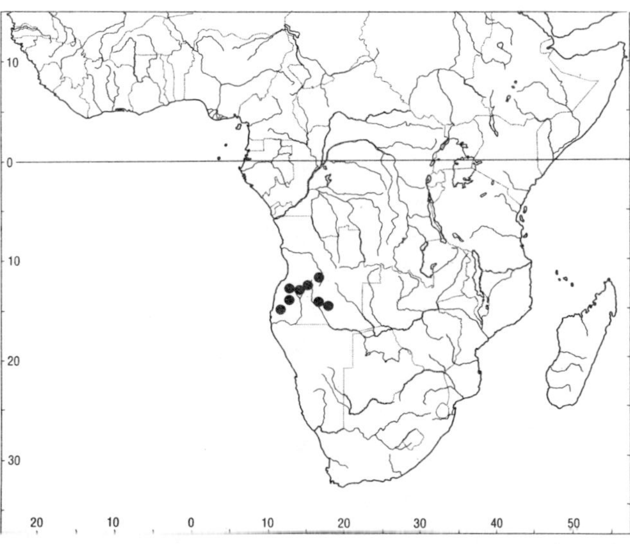

MAP 53. *Gladiolus pallidus.*

to have been made in November extends the flowering season considerably, if this is correct.

DIAGNOSIS & RELATIONSHIPS

Closely allied to the widespread tropical African *Gladiolus dalenii, G. pallidus* is almost as robust and large-flowered as that species but is easily distinguished by its unusual linear leaves, the blades of which have strongly thickened margins and midribs, usually raised over the unthickened part of the leaf surface. The blades are up to 5 mm wide but usually less than 3 mm. The flowers are somewhat smaller than is typical of *G. dalenii,* but the range of variation in flower size in the two species does overlap. *Gladiolus dalenii* also occurs in the Angolan highlands, where it flowers in January to March, and of course has a much wider distribution, extending throughout tropical and southeastern Africa.

HISTORY

Although collected fairly early for a tropical African species by Fr. Antunes in the 1890s and described by J. G. Baker in 1898, *Gladiolus pallidus* has had a rather confused history, possibly owing in part to the scanty type material. It was collected again in 1900 by Hermann Baum on the German Cunene–Zambezi Expedition, led by Otto Warburg, and described as *G. kubangensis* by Harms in 1903, who was evidently unaware of the similarities of his species with *G. pallidus.* A second Baum collection from the same expedition was also described by Harms in the report of the expedition, *G. baumii,* which seems to me to differ in no significant way from *G. kubangensis.* The Italian explorer, N. Mazocchi-Alemanni, who made a small collection of plants in Angola in 1922, also collected *G. pallidus.* These specimens were described by the Italian botanist, Achille Chiovenda, as *G. splendidus* var. *tundensis,* after the locality where they were found. Although *G. splendidus* is now regarded as conspecific with *G. dalenii,* the variety *tundensis* is unmistakably *G. pallidus.* There is no reason to regard any of these synonyms of *G. pallidus* as significant variants of the species. They seem simply to have been described under the assumption that they were new to science when discovered.

SELECTED SPECIMENS

Angola. Benguela: Near Chicala, c. 1900 m, 23 June, 37, *Exell & Mendonça 3036* (BM); Ganda, scattered in the grounds of the Centro de Estudos, 1730 m, 27 June 1963, *Texeira & Andrade 6950* (COI, LISC). Bié: Cuito (Silva Porto), Centro de Estudos de Ceilunga, c. 1700 m, 6 July 1963, *Murta 204* (COI, LISC). Cuando Cubango: Along the Kubango at Holombo, 8 May 1900, *Baum 898* (BM, COI, E, K, Z); near Menongue (Serpa Pinta), 19 Aug. 1926, *Pocock 591* (PRE). Huambo: Chianga, c. 1700 m, *Texeira & Sousa 6524* (COI, LISC); Mungo-Lunge, km 23, 11°58′ S, 16°14′ E, 1600 m, 16 Mar. 1973, *Bamps, Martins, & Maia 4203* (BR). Huila: Huila, rocky hills, 1740 m, June 1899 (flower and fruit), *DeKindt 639* (P); Lubango (Sá da Bandeira), Hoque, 2 June 1966, *Henriques 1014* (K, SRGH). Namibe: Tunda, in 1923, *Mazocchi-Alemanni 123* (K).

51. *Gladiolus nyasicus* Goldblatt

FIGURE 44, MAP 54. Goldblatt, Fl. Zambesiaca 12(4): 94 (1993). Type: Tanzania, Lumecha Bridge, c. 21 km north of Songea, 3 Jan. 1956, *Milne-Redhead & Taylor 8033* (K, holotype; B, SRGH, isotypes).

EPONYMY

nyasicus, named for the area in Tanzania and Malawi surrounding Lake Malawi, also known as Nyasa.

DESCRIPTION

Plants 40–60 cm high. CORM 15–20 mm in diameter, tunics of brittle membranous layers, the

FIGURE 44. *Gladiolus nyasicus*, × 0.67 (*Milne-Redhead & Taylor 8033, Robson & Jackson 1301*).

outer becoming irregularly broken or fibrous, light brown. CATAPHYLLS membranous, the upper largest and up to 8 cm long, green above ground. LEAVES borne on the flowering stem, four or five, the lower two or three basal or nearly so, laminate, narrowly lanceolate, 5–10 mm wide, about a third to half as long as the spike, firm-textured with moderately raised and thickened midrib and margins, the upper two or three leaves cauline and largely or entirely sheathing, often imbricate. STEM unbranched, 3–4 mm in diameter below the first flower.

SPIKE (2–)3- to 7(–10)-flowered; BRACTS pale green, fairly soft-textured, evidently becoming dry above, 24–30(–40) mm long, the inner somewhat shorter than the outer. FLOWERS pale yellow to cream, the dorsal tepal sometimes flushed with red; PERIANTH TUBE 18–25 mm long, nearly cylindric below, curving outward and expanded in the upper half; TEPALS unequal, the three upper broadly elliptic-ovate, the dorsal largest, (20–) 25–28 mm long, c. 16 mm at the widest, arched over the stamens, the upper laterals about as long to c. 5 mm shorter and c. 12 mm wide, directed forward and curving outward distally, the lower three tepals curving downward, 12–16 mm long, 7–10 mm wide, the lowermost somewhat longer than the lower laterals. FILAMENTS c. 16 mm long, exserted 4–6 mm from the mouth of the tube; ANTHERS 8–12 mm long, pale yellow. OVARY 4–7 mm long; STYLE arched over the stamens, dividing near the apex of the anthers, the branches c. 4.5 mm long. CAPSULES ellipsoid to ovoid, (18–)20–24 mm long, c. 12 mm in diameter at the widest; SEEDS, including the wing, c. 6.5 × 4 mm, lightly undulate, light glossy brown, seed body c. 1 mm in diameter.

FLOWERING TIME. Mid December to mid February.

DISTRIBUTION & HABITAT

Gladiolus nyasicus is restricted to southern tropical

Africa where it occurs in southern Tanzania and central and northern Malawi, forming an arc around the northern half of Lake Malawi. Relatively poorly known and often confused with *G. melleri, G. nyasicus* occurs in boggy grassland and dambos. That is also the typical habitat for *G. melleri,* which flowers earlier in the season from September to November, at least a month before the first *G. nyasicus* comes into bloom and always producing flowering stalks and foliage leaves on separate shoots.

DIAGNOSIS & RELATIONSHIPS

The relative modest stature and small pale yellow to cream flowers with smaller, downcurved lower tepals and a perianth tube 18–25 mm long distinguish *Gladiolus nyasicus* from the related *G. dalenii.* The latter, a much taller species, reaching 60–120 cm, has flowers with a perianth tube (25–)30–40 mm long, upper tepals 35–50 mm long, and altogether a more robust species. In addition to its smaller flowers, *G. nyasicus* can be distinguished from species of similar general appearance by its fairly short filaments, exserted 4–6 mm from the tube. In *G. dalenii* the filaments are exserted 15–20 mm. The short filaments recall *G. melleri* and the closely allied *G. oliganthus,* in both of which the filaments are included in the tube. In *G. melleri* a pink to red perianth and absence of foliage leaves on the flowering stem make confusion with *G. nyasicus* unlikely. *Gladiolus oliganthus* has a yellow perianth and short to long foliage leaves on the flowering stems, and care must be taken not to confuse *G. nyasicus* with that western Tanzanian endemic, which can be distinguished, in addition to its included filaments, by longer anthers and larger flowers with the dorsal tepal usually 30–35 mm long.

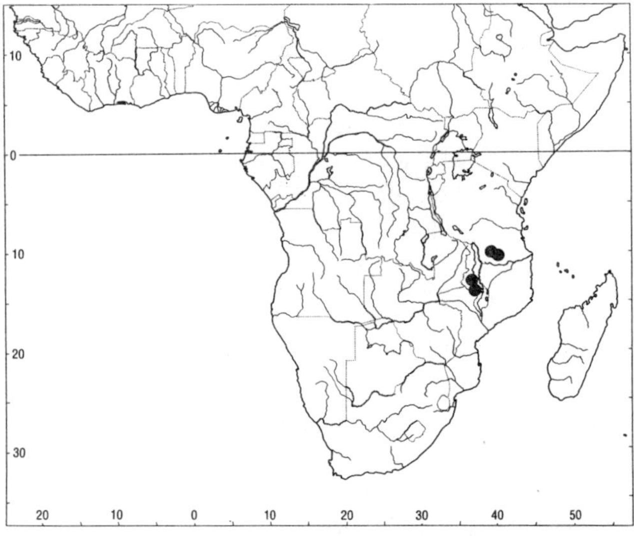

MAP 54. *Gladiolus nyasicus.*

SELECTED SPECIMENS

Tanzania. Rovuma: 12 km east of Songea by Nonganonga stream, 1050 m, 27 Dec. 1955, *Milne-Redhead & Taylor 7769* (B, EA, K).

Malawi. Central: Dedza, near Mphunzi, 1550 m, *Robson & Jackson 1301* (K, SRGH); Kachere village, 9 Jan. 1959, *Jackson 2274* (K, PRE, SRGH); base of Ciwao Hill, Chongoni Forestry School, 1650 m, 4 Feb. 1959, *Robson 1447* (K, LISC, PRE, SRGH).

52. *Gladiolus magnificus* (Harms) Goldblatt

PLATE 30, FIGURE 45, MAP 55. Goldblatt in Goldblatt & de Vos, Bull. Mus. Nat. Hist. Nat., Paris, Sér. 4, Sect. B, Adansonia 11: 426 (1989); Fl. Zambesiaca 12(4): 94 (1993). *Petamenes magnificus* (Harms) R. Foster, Contrib. Gray. Herb., New Ser. 114: 50 (1936). *Antholyza magnifica* Harms in Warburg, Kunene-Sambesi-Exped. 201 (1903). Type: Angola, on the Longa near Minnesera, 1200 m, 11 Jan. 1901, *Baum 651* (B, holotype; BM, COI, K, W, isotypes).

SYNONYMY

Antholyza zambesiaca Baker, Handbook Irideae 232 (1892); Fl. Trop. Africa 7: 374 (1898), not *Glad-*

iolus zambesiacus Baker (1892; from Central Africa). *Petamenes zambesiacus* (Baker) N. E. Brown, Trans. Roy. Soc. S. Africa 20: 227 (1932); Sölch, Prodr. Fl. Südwestafrika 155: 12 (1969). *Oenostachys zambesiacus* (Baker) Goldblatt, J. S. African Bot. 37: 443 (1971). Drummond & Plowes, Wild Fl. Rhodesia, pl. 44 (1976). Type: Botswana (or northwestern Zimbabwe), south of the Zambezi River, Leshumo Valley, in or before 1883, *Holub s.n.* (K, holotype).

Petamenes spectabilis (Schinz) Phillips, Bothalia 4: 44 (1941). *Chasmanthe spectabilis* (Schinz) N. E. Brown, Trans. Roy. Soc. S. Africa 20: 273 (1932). *Antholyza spectabilis* Schinz, Mém. Herb. Boissier 20: 13 (1900), not *Gladiolus spectabilis* Baker, Bull. Herb. Boissier, Sér. 2, 4: 1006 (1904) (= *G. saundersii* J. D. Hooker). Type: Namibia, Waterberg Plateau, 19 Apr. 1899, *Dinter s.n.* (possibly destroyed during World War II, evidently not at B, where expected).

EPONYMY

magnificus, "magnificent," alluding to the handsome appearance of the tall stems and spikes of large brightly colored flowers.

DESCRIPTION

Plants 80–140 cm high. CORM depressed globose, 2–3 cm in diameter, tunics of fine, wiry, straw-colored fibers. CATAPHYLLS firm-textured, pale below ground, green or flushed purple above ground, the inner lanceolate and sheathing the leaves for up to 10 cm. LEAVES five to seven, the lower four or five basal and longest, about half as long as the stem and reaching almost to the base of the spike, the blades narrowly lanceolate, 6–9 mm wide, midrib and margins lightly thickened, the margins hardly raised, the upper leaves progressively smaller and with shorter blades or the blades lacking and then entirely sheathing. STEM unbranched, c. 4 mm in diameter below the first flower.

FIGURE 45. *Gladiolus magnificus,* × 0.67 (*Plowes 1668*).

SPIKE 6- to 15-flowered, erect; BRACTS green, 2.5–3(–3.5) cm long, the inner somewhat shorter than the outer. FLOWER bright red with yellow markings on the lower tepals and in the throat, the lower three tepals at least sometimes each with a dark red-purple blotch in the midline; PERIANTH TUBE 25–30 mm long, narrow below, expanded near the top of the bracts and curving outward and nearly horizontal, c. 4 mm wide at the mouth; TEPALS very unequal, the dorsal largest, hooded and horizontal, 35–42 mm long, 18–22 mm at the widest, upper laterals broadly lanceolate, 13–15 mm long, directed forward, the apices curving outward, the lower three tepals narrowly lanceolate, to 10 mm long, more or less horizontal, or the lowermost recurving. FILAMENTS arched under the dorsal tepal, c. 35 mm long, exserted c. 20 mm from the tube; ANTHERS 12–17 mm long. OVARY ellipsoid, 5–6 mm long; STYLE dividing shortly below the apex of the anthers, the branches about 3 mm long. CAPSULES and SEEDS unknown.

FLOWERING TIME. January and February, occasionally in March.

DISTRIBUTION & HABITAT

Gladiolus magnificus is restricted to southern tropical Africa, where it occurs from southeastern Angola and northeastern Namibia through northern Botswana and western Zambia to eastern Zimbabwe. Although apparently rare in Angola and Zambia, *G. magnificus* is common in northwestern Zimbabwe and eastern Botswana, where it occurs in light woodland or forest clearings in sandy soil, often with tall grasses, an ecosystem known as Kalahari sandveld.

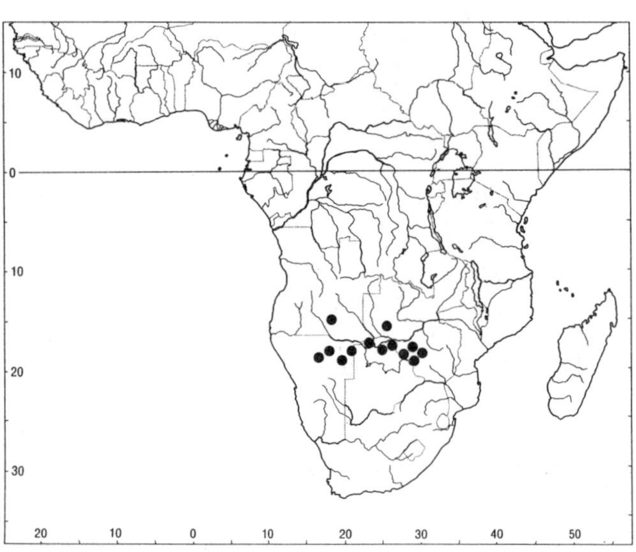

MAP 55. *Gladiolus magnificus.*

DIAGNOSIS & RELATIONSHIPS

Superficially resembling a tall slender variant of widespread *Gladiolus dalenii, G. magnificus* can readily be distinguished from its apparent close relative by its enlarged dorsal tepal, 35–42 mm long and about twice as long as the upper lateral tepals. The lower three tepals are much reduced in size and are generally somewhat less than a third as long as the dorsal. Apart from these striking floral differences the vegetative form of *G. magnificus,* including the unbranched stem and large floral bracts, is easily confused with *G. dalenii.* The structure of the flowers of *G. magnificus,* with their long perianth tube, exaggerated dorsal and almost vestigial lower tepals, conforms to the classic bird-pollinated flower of Iridaceae (Vogel, 1954; Goldblatt & de Vos, 1989). Field observations are needed to confirm the type of pollination in the species.

HISTORY

First collected in Botswana by the Czech naturalist, Emil Holub, in the early 1880s, and described by J. G. Baker as *Antholyza zambesiacus* in 1892, this striking species has had a surprisingly complex history. A second collection was made in 1899 by the German botanist, Kurt Dinter, in northern Namibia. These specimens were described by Hans Schinz in 1900 as *A. spectabilis.* Then a collection made in 1901 in southern Angola by Hermann Baum was described as *A. magnifica.* The confusion over the status of the genus *Antholyza* in the 1930s led to the species being transferred to both *Chasmanthe* and *Petamenes,* genera of Iridaceae based on Cape species, not closely related to *Gladiolus magnificus.* In 1972 the species was transferred to the tropical African segregate of *Gladiolus, Oenostachys.* Finally, in 1989 the species came to rest in *Gladiolus* (Goldblatt & de Vos, 1989), where it carries the epithet *magnificus* because the two earlier synonyms had already been used in *Gladiolus.*

SELECTED SPECIMENS

Angola. Cuando Cubango: On the Longa at Minnesera, forest edge, 1200 m, 11 Jan. 1901, *Baum 651* (B, BM, COI, K, W).

Zambia. Western: 30 km south of Mulobezi to Sesheke, 30 Jan. 1970, *Anton-Smith s.n.* (SRGH); Sesheke, Jan. 1925, *Borle s.n.* (PRE). Southern: Upper Zambezi, forest on banks of the Lufuba, without date, *Kiener s.n.* (P).

Namibia. Otjiwarongo: Foot of the Waterberg, Farm Okosongomingo, 16 Mar. 1955, *de Winter 2770* (K). Grootfontein: c. 45 km north of Gautscha Pan, 9 Feb. 1958, *Storey 6450* (PRE); bush at foot of Aha Mountains, 29 Jan. 1958, *Storey 6344* (PRE); Kaukoveld, 30 km south of Kano Vlei–Tshumkwe road, 13 Apr. 1967, *Giess 9816* (PRE). Okavango: 7 Mile Dune, c. 30 km south of Runtu, 9 Mar. 1956, *de Winter & Marais 5044* (PRE). Kaprivi: Singalamue, c. 90 km from Katima, 1000 m, 30 Dec. 1958, *Killick & Leistner 3198* (PRE).

Botswana: Northern: Seronga road, 0.4 km west of Masoko Pan, 22 Jan. 1978, *Smith 2204* (K, SRGH); c. 1.5 km east of footbridge on Namibia border along Geological Survey cutline, 25 Mar. 1980, *Smith 3342* (SRGH); Chobe District, Kasane, on farmlands, 900 m, Feb. 1967, *Mutakela 167* (SRGH).

Zimbabwe. Matabeleland North: Nyamandhlovu Pasture Station, open bush in Kalahari sandveld, 10 Jan. 1954, *Plowes 1668* (K, LISC, PRE, SRGH); Hwange (Wankie) Game Reserve, Gwaai corridor, *Baikiaea* woodland, 14 Feb. 1956, *Wild 4724* (COI, K, LISC, PRE, SRGH); Hwange, 3 km northwest of main camp, 30 Jan. 1969, *Rushworth 1485* (K, PRE, SRGH); 16 km northwest of Lupane, 20 Feb. 1972, *Westwater s.n.* (K, PRE, SRGH 215,228).

53. *Gladiolus huillensis* (Welwitsch ex Baker) Goldblatt

FIGURE 46, MAP 56. Goldblatt in Goldblatt & de Vos, Bull. Mus. Nat. Hist Nat., Paris, Sér. 4, Sect. B, Adansonia 11: 426 (1989); Fl. Zambesiaca 12(4): 99 (1993). *Antholyza huillensis* Welwitsch ex Baker, Trans. Linn Soc. Bot. ser. 2, 1: 270 (1878); Handbook Irideae 392 (1892); Fl. Trop. Africa 7: 374 (1898). *Petamenes huillensis* (Welwitsch) N. E. Brown, Trans. Roy. Soc. S. Africa 20: 276 (1932). *Oenostachys huillensis* (Welwitsch ex Baker) Goldblatt, J. S. African Bot. 37: 443 (1971). Geerinck, Bull. Soc. Roy. Bot. Nat. Belgique 105: 7 (1972). Type: Angola, Huila, between Lopolo and Humpata, Nov.–Dec. 1859, *Welwitsch 1539* (BM, lectotype; B, C, COI, G, K, P, isolectotypes).

SYNONYMY

Gladiolus subulatus Baker, Fl. Trop. Africa 7: 577 (1898). Type: Angola, Huila, without date but before 1895, *Antunes 149* (B, holotype not seen and possibly lost; LISC, isotype).

Antholyza degasparisiana Buscalioni & Muschler, Bot. Jahrb. Syst. 49: 463 (1913). *Petamenes degasparisiana* (Buscalioni & Muschler) N. E. Brown, Trans. Roy. Soc. S. Africa 20: 277 (1932). Type: Zambia, Kabwe (Broken Hill), date and collector's name unknown, *157* (location unknown).

Antholyza pubescens Vaupel, Notizbl. Bot. Gart. Berlin-Dahlem 7: 31 (or 376) (1920). Type: Zambia, "N'Yengeshi Spruit," 28 Dec. 1907, *Kassner 2211* (B, holotype; BM, HBG, K, P, isotypes).

Petamenes vaginifer Milne-Redhead, Hooker's Icon. Pl. 35 (Ser. 5, 5): pl. 3478 (1950). *Oenostachys vaginifer* (Milne-Redhead) Goldblatt, J. S. African Bot. 37: 443 (1971). Type: Zambia, Mwinilunga District, east of Dobeka Bridge, 5 Nov. 1937,

Milne-Redhead 3111 (K, lectotype designated by Goldblatt & de Vos, 1989; BR, K, isolectotypes).

EPONYMY

huillensis, named for the Huila Plateau in southwestern Angola where the species was first collected.

DESCRIPTION

Plants 30–100 cm high. CORM 16–20 mm in diameter, tunics firm-papery to coriaceous, fragmenting irregularly into mostly vertical strips. CATAPHYLLS pale and membranous, obtuse below, the upper green above the ground and acute, glabrous. LEAVES three to five, short and sheathing for most of their length, obscurely to obviously white pubescent, imbricate, usually at least two with a short blade to 8 cm long, the blade more or less linear, the margins and midrib thickened (well-developed foliage leaves produced from separate shoots after flowering are not known and probably not produced). STEM unbranched, sheathed except for a short distance below the spike, often with a short membranous sheathing leaf inserted on the upper third of the stem just above the uppermost sheathing leaf.

SPIKE 6- to 14(–20)-flowered; BRACTS green below, dry and light brown above, sometimes sparsely pubescent above, 13–20(–25) mm long, the inner about as long as or only slightly shorter than the outer. FLOWERS bright red, often marked with yellow on the lower tepals and in the throat; PERIANTH TUBE dimorphic, slender below and abruptly expanded into a cylindric upper part, the lower part 11–16 mm long, exserted from the bracts, c. 1 mm in diameter, the upper part 10–12 mm long, c. 3 mm in diameter, ascending to horizontal; TEPALS unequal, the dorsal largest and arched over the stamens, c. 15 mm long, spathulate, nearly 15 mm at the widest, upper lateral tepals deltoid, 5–6 mm long, joined to the dorsal for c. 3 mm, the lower three tepals lanceo-

FIGURE 46. *Gladiolus huillensis.* Corm and spike with leaves, × 0.5; portion of spike and capsules, full size (*Milne-Redhead 3111*).

late, directed forward, lower laterals c. 6 mm long, reaching the apices of the upper laterals, the lowermost shortest, c. 3 mm long. FILAMENTS c. 20 mm long, reaching to about the middle of the dorsal tepal; ANTHERS c. 7 mm long, reaching nearly to the apex of the dorsal tepal or barely exceeding it. OVARY c. 2.5 mm long; STYLE dividing opposite the middle or upper half of the anthers, the branches 3–3.5 mm long, not much expanded above, ultimately usually reaching the anther apices. CAPSULES broadly obovoid, 10–16 mm long; SEEDS c. 6 × 4 mm, broadly winged.

FLOWERING TIME. November and December.

DISTRIBUTION & HABITAT

Gladiolus huillensis appears to have two centers, one in the highlands of western Angola and another in northern and central Zambia and adjacent western Zaire. In Angola the species has been collected near Huambo and on the Huila Plateau but not elsewhere in the country. The few records from Zaire suggest that *G. huillensis* is rare and local there. Plants grow in light woodland, flowering early in the season before the leaf canopy is closed and the undergrowth thick and tall.

DIAGNOSIS & RELATIONSHIPS

Gladiolus huillensis is unmistakable in the bright red perianth and unusual form of its flowers, the upper tepal extended horizontally and much exceeding the others, the lateral tepals only about half as long, and the lower tepal tiny. This type of flower is one now associated with sunbird pollination (Goldblatt & de Vos, 1989). Other tropical African species with similar floral structure may not be immediately related to *G. huillensis* for they have larger corms and several basal leaves, all with well-developed and plane leaves. Their affinities are most likely with the *G. dalenii* complex. The smaller flowers and bracts of *G. huillensis* as well as its reduced leaf blades of linear form accord better with the *G. atropurpureus* group of section *Hebea*. *Gladiolus huillensis* cannot be distinguished from some species of section *Hebea* in fruit except for the sparse pubescence of the leaves, a feature occasionally encountered in the section. In *Gladiolus,* red flowers with the series of characteristics associated with bird pollination are known to have evolved independently several times (Goldblatt & de Vos, 1989), and floral form alone cannot therefore be used in assessing relationships.

Petamenes (*Oenostachys*) *vaginifer,* based on Zambian plants, was regarded as conspecific with *Gladiolus huillensis* (Goldblatt & de Vos, 1989). Differences between the types of the two are trivial and include the degree of pubescence and length of the leaf blades, often variable in *Gladiolus.* Some local variation is, however, evident in plants from Zambia and Zaire that tend to be more robust, have slightly larger flowers, and lack yellow pigmentation on the lower tepals. *Antholyza pubescens,* based on a northwestern Zambian collection, was separated from *G. huillensis* (as *Antholyza*) on the basis of its taller stems (to 1 m), leaves with longer and markedly pubescent blades, and more numerous flowers, but it matches exactly other Zambian specimens of *G. huillensis* in the fairly

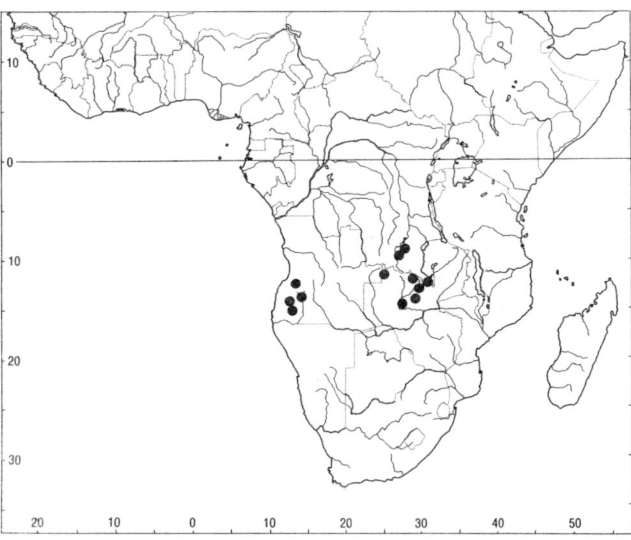

MAP 56. *Gladiolus huillensis.*

long bracts, to 2 cm, and the dark red lower tepals without any yellow marking.

The status of two collections of particularly small-flowered plants from western Angola that seem to belong to *Gladiolus huillensis* is puzzling. Plants from Huambo (*Raimundo et al. 1255*) and Huila (*Martins 74*) have tiny flowers with the upper tepal 8–13 mm long, fully half as long as in other collections of *G. huillensis*, and the stamens are very short. The filaments are only c. 10 mm long and reach to about the apex of the lateral tepals, and to about the middle of the dorsal tepal instead of to the apex. In *Martins 74* the dorsal tepal appears to arch downward. In other respects, except perhaps for moderate stature, the specimens accord with *G. huillensis*, including sparse lower leaf pubescence and fairly short leaf blades. The possibility that these are hybrids, perhaps with *G. atropurpureus*, makes taxonomic recognition seem unjustified until these plants can be investigated more thoroughly.

HISTORY

One of the most distinct of the species of *Gladiolus* discovered by Welwitsch during his exploration of Angola in the mid 19th century, *G. huillensis* was originally assigned to *Antholyza* because of its very zygomorphic bright red flower with the dorsal tepal much exceeding the length of the others. The species remained rather poorly understood, and Baker, who described *A. huillensis*, named a later collection of the plant *G. subulatus*. Based on a collection from the Huila Plateau, the type of *G. subulatus* matches exactly Welwitsch's collections of *G. huillensis*, which are also from the Huila Plateau. Plants from central Zambia collected by Theo Kassner and assigned to *G. pubescens* differ in no significant way from the western Angolan plants. I assume, from the description only, that *A. degasparisiana* is also conspecific with *G. huillensis*. The type, collected at Broken Hill, now Kabwe, in northern central Zambia, was thought to be the same as the northeastern African *G. schweinfurthii* (Cufodontis, 1974) but this seems unlikely in view of the type locality. The location of the type material, collected by Helena, Duchess of Aosta, is unknown.

SELECTED SPECIMENS

Angola. Huambo: Mt. Moco, 2050 m, 17 Dec. 1973, *Huntley, Roberts, & Ward 2* (PRE). Huila: Huila Plateau, Oct.–Jan., *Berthelot 345* (P); Huila, *Antunes 149* (LISC).

Zaire. Shaba: Biano–Lualaba, Manika Plateau, near Katentania, Nov. 1912, *Homblé 788* (BR); south of Biano Hotel, *Uapaca* woodland, 8 Dec. 1959, *Duvigneaud 4464* (BRLU).

Zambia Copperbelt: Luanshaya, plateau woodland, 19 Dec. 1954, *Fanshawe 1735* (K, NDO); Kataba, woodland, 12 Dec. 1960, *Fanshawe 5966* (K, NDO, SRGH); Kitwe, miombo, 17 Dec. 1964, *Fanshawe 9056* (NDO, SRGH); Kitwe, woodland, 19 Dec. 1960, *Linley 227* (K, MO, SRGH). Northwestern: Mwinilunga District, woodland just east of Dobeka bridge, 5 Nov. 1937, *Milne-Redhead 3111* (BR, K). Central: Mumbwa District, Kafue National Park, c. 2 km west of Chunga, 9 Dec. 1962, *Mitchell 15/84* (SRGH); Kafue National Park, Ngoma, 23 Dec. 1964, *Mitchell 25/87* (K, LISC).

54. *Gladiolus benguellensis* Baker

PLATE 31, FIGURE 47, MAP 57. Baker, Trans. Linn. Soc. Bot., Ser. 2, 1: 268 (1878); Handbook Irideae 221 (1892); Fl. Trop. Africa 7: 370 (1898). Goldblatt, Fl. Zambesiaca 12: 95 (1993). Type: Angola, Huila, dry pastures near Lopolo, Dec. 1859 and Feb. 1860, *Welwitsch 1540* (P, lectotype designated here, a well-preserved specimen; BM, B, C, COI, G, K, P, isolectotypes).

SYNONYMY

Tritonia tigrina Pax, Bot. Jahrb. Syst. 15: 152 (1893). Type: Angola, between Sanza and Malange, Oct. 1876, *Pogge 431* (B, holotype).

FIGURE 47. *Gladiolus benguellensis.* Corm, leaves, and flowering spike, × 0.5; single flower with bract and vertical section of flower, full size; cross-section of leaf, × 2 (*Richards 3808, Schaijes 3808*).

Gladiolus pubescens Pax, Bot. Jahrb. Syst. 15: 154 (1893). Baker, Fl. Trop. Africa 7: 264 (1898), an illegitimate homonym, not *G. pubescens* Lamarck (1791) (= *Babiana pubescens* (Lamarck) G. Lewis) and not *G. pubescens* Baker (1876) (= *G. pubigerus* G. Lewis). *Gladiolus paxii* Klatt in Durand & Schinz, Conspectus Fl. Africae 5: 222 (1895), as new name for *G. pubescens* Pax. *Gladiolus malangensis* Baker, Bull. Herb. Boiss., Sér. 2, 1: 867 (1901), an illegitimate name for *G. paxii* Klatt, based on the duplicate at Z of the type of *G. paxii. Gladiolus pubescifolius* G. Lewis, Ann. S. African Mus. 40: 132 (1954), an illegitimate name for *G. paxii* Klatt, but proposed in ignorance as a new name for *G. pubescens* Pax. Type: Angola, near Malange, Oct. 1879, *Teuscz sub von Mechow Expedition 280* (B, holotype; B, Z, isotypes).

Gladiolus macrophlebius Baker, Fl. Trop. Africa 7: 576 (1898). Type: Angola, Huila, before 1895, *Antunes 29* (B, holotype; K, drawing; COI, isotype).

Gladiolus longanus Harms in Warburg, Kunene-Sambesi-Expedition 201 (1903). Type: Angola, Chijija, banks of the Longa in marshy grassland, 1200 m, 5 Jan. 1900, *Baum 632* (B, holotype; BM, E, K, Z, isotypes).

EPONYMY

benguellensis, named after the Benguela region of western Angola where the species was first collected.

DESCRIPTION

Plants 35–60 cm high. CORM globose to horizontally elongate, often rhizomelike, dark red, 15–25 mm long, c. 10 mm in diameter, the old corms sometimes persistent and then attached horizontally, tunics membranous, reddish brown, fragmenting irregularly. CATAPHYLLS membranous, green above ground and glabrous or lightly to densely pubescent. LEAVES four to six, the basal longest, the blades usually well developed, reaching to about the base of the spike, narrowly lanceolate, 5–9 mm wide, the margins and midribs moderately to heavily thickened and hyaline, sometimes other veins also somewhat thickened, glabrous to pubescent, sometimes with a dense appressed short pubescence, the upper one or two cauline leaves usually entirely sheathing. STEM unbranched, erect.

SPIKE (3–)5- to 8-flowered, erect; BRACTS green, (20–)25–30 mm long, often somewhat attenuate and dry apically, the inner about two-thirds as long as the outer. FLOWERS scarlet red with the lower tepals each with a median white stripe, flowers occasionally entirely yellow; PERIANTH TUBE c. 15 mm long, expanding uniformly from base to apex, curving outward in the upper half; TEPALS unequal, the dorsal largest, 18–21 mm long, c. 12 mm wide, ascending to more or less horizontal, the upper laterals c. 20 mm long, the lower three tepals 16–18 mm long, c. 6 mm wide, more or less straight and held at 45° to the ground. FILAMENTS 5–8 mm long, unilateral and arcuate, included or barely exserted from the tube; ANTHERS 7–8 mm long. OVARY 3–4 mm long; STYLE dividing opposite the middle to upper third of the anthers, the branches 3–4 mm long, expanded gradually in the upper half. CAPSULES narrowly obovoid-ellipsoid, (14–)16–18(–22) mm long; SEEDS more or less oval, broadly winged, c. 7 × 4–5 mm.

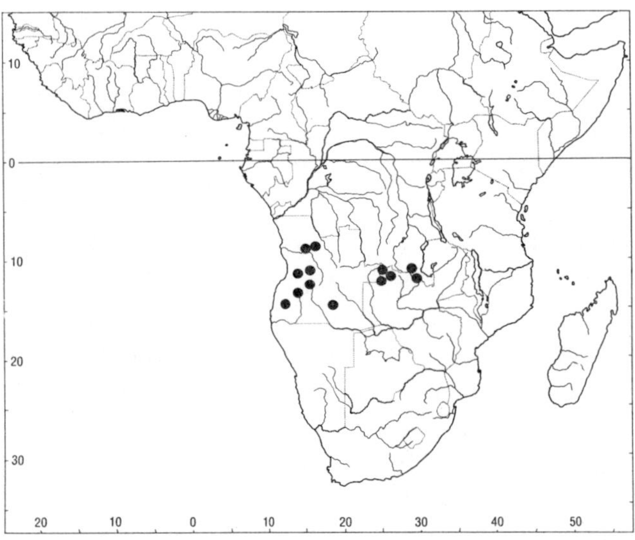

MAP 57. *Gladiolus benguellensis.*

FLOWERING TIME. Mainly December to February, but not infrequently in October and November.

DISTRIBUTION & HABITAT

Gladiolus benguellensis extends across southern central Africa from the western Angolan highlands to western Zambia and southern Zaire. It appears to be restricted to marshy sites, either permanently wet or drying out at the end of the dry season. Flowering occurs from the late dry season in September into the early part of the wet season in December and January. Over most of its range, *G. benguellensis* has bright red flowers, but populations with yellow flowers occur locally in southern Angola, where the type of yellow-flowered *G. longanus* was collected.

DIAGNOSIS & RELATIONSHIPS

Although somewhat variable in its foliage, *Gladiolus benguellensis* can always be recognized by its fairly small flowers with subequal tepals, short perianth tube, and unusually short stamens. The filaments, only 5–8 mm long, are usually included in the tube or just barely exserted. The structure of the flowers and bracts of *G. benguellensis* is similar to that of *G. dalenii* and *G. melleri* although they are about half as large as in those species, the bracts being 20–30 mm long and the upper tepals 18–21 mm long. The small flowers and their short stamens combined with normally heavily thickened leaf margins and midribs distinguish the species from its allies.

The leaves of the type of *Gladiolus benguellensis* are fairly broad, 10–14 mm wide, and quite glabrous and matched in this by the type of *Tritonia tigrina* and several other collections, including *Antunes 29* (type of *G. macrophlebius*), which is particularly robust. The leaves of the type collection of *G. paxii* differ substantially in being narrower and less prominently thickened on the margins, and the intercostal areas are covered by a fine dense appressed pubescence. Other collections from the Malange area of western Angola also have lightly pubescent leaves (e.g., *Young 775*). Plants from there sometimes also have newly emergent leaves with lightly thickened margins and an erect, globose, instead of horizontally elongate, corm. The only collection I have seen from the drier southern part of Angola, *Baum 632,* has narrow leaves with especially heavily thickened margins and pubescence only on the upper cataphyll.

White-flowered *Gladiolus manikaensis* from the Manika Plateau in western central Shaba Province, Zaire, is easily mistaken for *G. benguellensis* in the vegetative state, both in the peculiar rhizomelike corms with dark brown tunics and leaves with heavily thickened margins and oblique apices. In addition to perianth color, the flowers of the two species differ in size and in the orientation of the lower tepals. In *G. manikaensis* the upper tepal is c. 25 mm long, compared with 18–21 mm in *G. benguellensis,* and the lower three tepals curve downward, unlike the nearly straight tepals of *G. benguellensis.* Both species have short, included filaments, but in *G. manikaensis* the style divides near the anther apices, and the fairly long style branches exceed the anthers. In *G. benguellensis* the shorter style branches do not normally reach the anther apices.

HISTORY

Like the other Angolan collections of *Gladiolus* made by Friedrich Welwitsch in the period 1853–1861, *G. benguellensis* was only described after his death, in 1878. A subsequent collection, unusual in its lightly pubescent leaves, was made by P. Pogge in 1876. This was thought to belong in the genus *Tritonia* by F. W. Pax, who described it as *T. tigrina.* Specimens of *G. benguellensis* collected by H. Teucz on the German-sponsored expedition to Angola led by Alexander von Mechow in 1879 were also misunderstood and were described by Pax as *G. pubescens.* That name is a homonym twice-over, for combinations in southern African species of *Babiana* and *Gladiolus.* To remedy the situation, Klatt (1895) provided Pax's species with the new name *G. paxii.* In ignorance of this substitution, the South African botanist, G. J. Lewis (1954b), proposed the substitute name *G. pubescifolius* for the same plant. All these names fall into synonymy under *G. benguellensis.* There seems no reasonable explanation for this regrettable series of muddles. Good specimens of the type of *G. benguellensis* were available to Pax at Berlin, and it is difficult to see how he might have thought *G. pubescens* differed in any significant way. As if this were not enough, three more species were described in the course of 10 years from western Angolan specimens, virtually indistinguishable from the type of *G. benguellensis*: *G. macrophlebius* (Baker, 1898); *G. malangensis* (Baker, 1901), based on the same collection as *G. paxii*; and *G. longanus* (Harms, 1903). In the history of tropical African *Gladiolus,* this is one of the more unfortunate examples of confusion resulting from poor communication, and belief that new collections out of the inaccessible parts of Africa must be new to science.

SELECTED SPECIMENS

Angola. Malange: Malange, in vlei, Sept. 1932, *Young 775* (BM, COI, LISC, P, SRGH, W); Quipacaça, 1000 m, 8 Sept. 1922, *Gossweiller 8764* (K). Moxico: Jimbe River bridge, Zambia frontier, 1200 m, 10 Nov. 1962, *Richards 17153* (K, SRGH); west of Mujilezhi River in boggy grassland, 7 Jan. 1938, *Milne-Redhead 3962* (K). Benguela: Alto Catumbala, Ganda, 1400 m, Dec. 1940, *Faulkner A28* (K, PRE). Bié: Serpa Pinta, Munanque, marshes near Macuche River, 18 Nov. 1906, *Gossweiler 3538* (BM, COI). Cuando Cubango: Longa River, marshy land, Dec. 1921–Jan. 1922, *Dawe 344* (K). Huila: Huila Plateau, moist situations, in 1895, *Berthelot 349* (P); Cunene, 54.2 km from Cuvelai to Bambi, 16 Feb. 1973, *Menezes, Barosso, & Sousa 4668* (LISC).

Zaire. Shaba: Keyberg, Kisanga Valley 8 km from Lubumbashi, 11 Feb. 1949, *Schmitz 2234* (BR); Keyberg, Kimilolo Valley, 4 Jan. 1957, *Detilleux 375* (BR); pastures near Karavia, *Quarré 3650* (BR).

Zambia. Western: Mwinilunga District, swamp 3 km from Kabompo Gorge, 24 Nov. 1962, *Richards 17499* (K, SRGH); Lukela River, margin of wet riverine vlei, 27 Oct. 1966, *Leach & Williamson 13495* (K, SRGH). Copperbelt: Chingola, dambo, 7 Nov. 1968, *Mutimushi 2800* (K).

55. *Gladiolus manikaensis* Goldblatt, new species

FIGURE 48, MAP 58. Type: Zaire, Shaba, near Kolwezi, wet bog, 1425 m, 13 Oct. 1984, *Schaijes 2401* (K, holotype).

EPONYMY

manikaensis, from the Manika Plateau in western central Shaba Province of Zaire.

LATIN DIAGNOSIS

Plantae 35–60 cm altae, cormo elongato demum rhizomatiformi c. 10 mm in diametro, foliis 4–6 basalibus 3–4 laminatis marginibus costisque incrassatis 5–9 mm latis, spica (3–)5–8 florum, bracteis (20–)25–30 mm longis, floribus albis, tubo perianthii c. 15 mm longo, tepalo dorsale (21–)23–25 mm longo, inferioribus recurvatis, filamentis 7–8 mm longis inclusis, antheris 8.5–10 mm longis, ramis styli antheras excedentibus.

DESCRIPTION

Plants 35–60 cm high. CORM elongate, somewhat rhizomelike, dark red, 15–25 mm long, c. 10 mm in diameter, the old corms sometimes persistent and then attached horizontally, tunics membranous, reddish brown, fragmenting irregularly. CATAPHYLLS membranous, green above ground and then glabrous or lightly to densely pubescent. LEAVES four to six, the basal longest and laminate, the blade narrowly lanceolate, about half as long as the stem, 5–9 mm wide, the margins and midribs moderately to heavily thickened and hyaline, sometimes other veins also somewhat thickened, the upper one or two leaves cauline, usually entirely sheathing. STEM unbranched, erect, c. 3 mm in diameter at the base of the spike.

SPIKE (3–)5- to 8-flowered; BRACTS (20–) 25–30 mm long, green below, dry apically, more or less attenuate, the inner about two-thirds as long as the outer. FLOWERS white to pale lilac; PERIANTH TUBE c. 15 mm long, expanding uniformly from base to apex, curving outward in the upper half; TEPALS unequal, the dorsal largest, (21–) 23–25 mm long, c. 12 mm wide, more or less horizontal and hooded over the stamens, the upper laterals 20–22 mm long, the lower three tepals 16–18 mm long, c. 8 mm wide, recurving throughout and ultimately directed downward. FILAMENTS 7–8 mm long, unilateral and arcuate, included or barely exserted from the tube; ANTHERS 8.5–10 mm long. OVARY 3–4 mm long; STYLE dividing at or just beyond the anther apices, the branches 5–5.5 mm long, extending well past the anthers. CAPSULES and SEEDS unknown.

FLOWERING TIME. October to mid December, late in the dry season or early in the wet season.

DISTRIBUTION & HABITAT

Apparently restricted to western Shaba Province in Zaire, *Gladiolus manikaensis* occurs on the Manika Plateau; it has so far been recorded from Kolwezi, Tenke, and surroundings. It is probably restricted to marshy sites that either remain wet throughout the dry season or are seasonally dry. The persisting corms with weakly differentiated tunics suggest a marshy, permanently moist habitat.

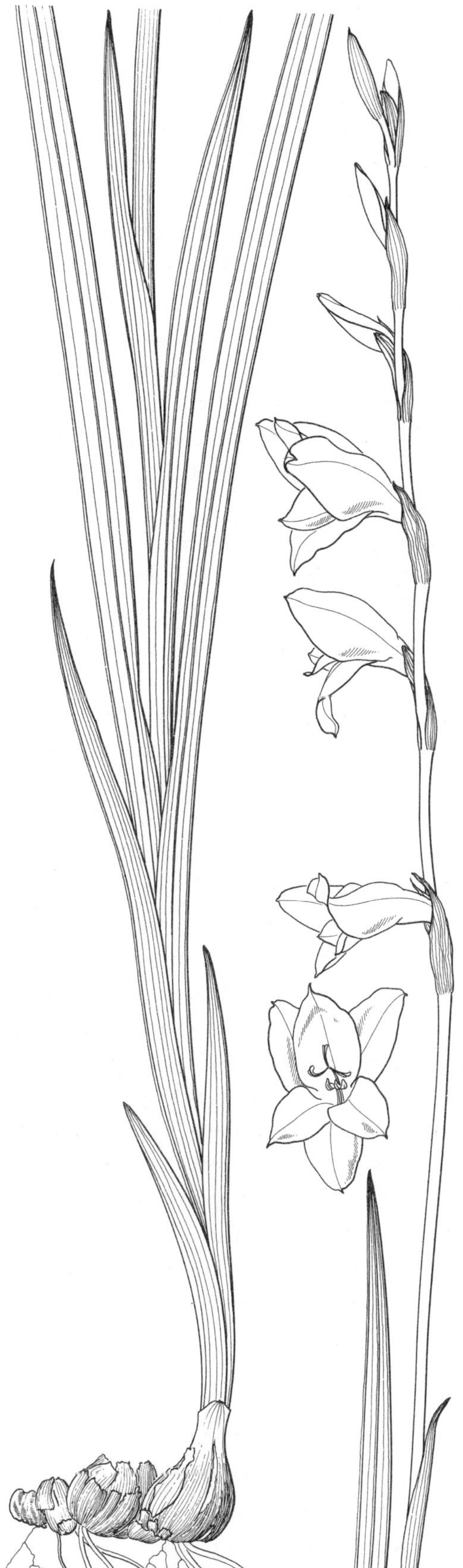

DIAGNOSIS & RELATIONSHIPS

Gladiolus manikaensis can be recognized by its tall, erect, unbranched stems, straight spikes of white flowers, and the distinctive persistent corms with poorly differentiated tunics. The moderately large flowers resemble most closely those of *G. melleri* and *G. benguellensis* in size, disposition of the tepals, and especially in the short filaments that are included in the perianth tube. Generally with scarlet or orange flowers, *G. melleri* has leaves produced from separate shoots after flowering has occurred, large globose corms, and usually shorter stems, and *G. manikaensis* is not likely to be confused with this species. It is more difficult to distinguish red-flowered *G. benguellensis* and *G. manikaensis*. Apart from perianth color, the two are remarkably alike in general aspect and especially in corm and leaf morphology. The major differences between them are in floral details. In *G. manikaensis* the style reaches the anther apices so the style branches extend beyond the anthers, the flowers are slightly larger so that the dorsal tepal is 21–25 mm long, and the lower tepals are strongly recurved. Smaller-flowered *G. benguellensis* has style branches that rarely exceed the anthers, the dorsal

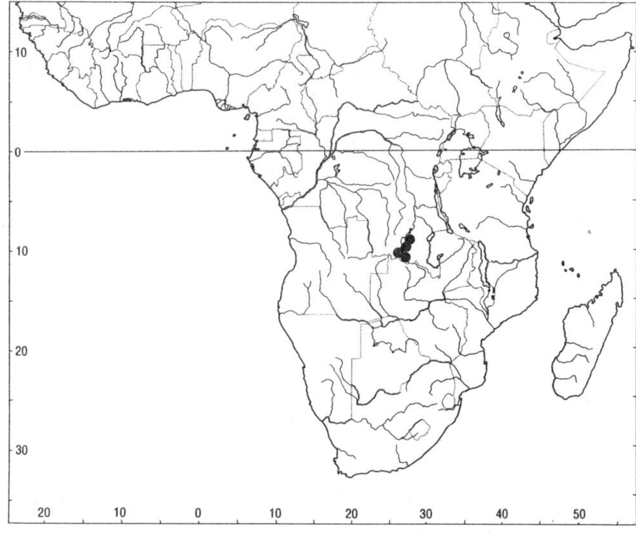

MAP 58. *Gladiolus manikaensis.*

FIGURE 48. *Gladiolus manikaensis.* Corm, leaves, and flowering spike, × 0.5; single flower, full size (*Schaijes 2704*).

tepal is 19–21 mm long, and the lower tepals are straight and held at 45° to the ground.

SELECTED SPECIMENS
Zaire. Shaba: Manika Plateau, environs of Katentania, Nov. 1912, *Homblé 821* (BR); near Kolwezi, wet bog, 1425 m, 13 Oct. 1984, *Schaijes 2401* (K); 17.5 km west-southwest of Kolwezi, Potopoto Valley, 1415 m, 6 Dec. 1983, *Schaijes 2067* (K). Kolwezi, Manika Plateau, Herman farm, marsh at river edge, 12 Dec. 1959, *Duvigneaud 4534G* (BRLU); Herman farm, steppe along the plateau beyond Musonoie River, 22 Jan. 1960, *Duvigneaud 5201* (BRLU); Dikuluwe, driest part of the marsh south of the copper deposit, 21 Jan. 1960, *Duvigneaud 5162* (BRLU); Dikuluwe, dry marsh, 13 Jan. 1959, *Duvigneaud 4130G* (BRLU)

56. *Gladiolus oliganthus* Baker

MAP 59. Baker, Fl. Trop. Africa 7: 576 (1898). Type: Tanzania, Tabora, Unyamwezi, Igonda (Gonda), moist meadows, 16 Feb. 1882, *Böhm 45* (B, holotype; B, Z, isotypes).

SYNONYMY
Gladiolus bussei Vaupel, Notizbl. Bot. Gart. Berlin-Dahlem 7: 32 (or 377) (1920). Type: Tanzania, near Songea, 14 Jan. 1901, *Busse 798* (B, holotype; B, isotype).

EPONYMY
oliganthus, "few-flowered," alluding to the spikes that have comparatively few flowers.

DESCRIPTION
Plants 30–60(–80) cm high. CORM 18–24 mm in diameter, tunics of brittle membranous layers, the outer irregularly broken or often fibrous, light brown. CATAPHYLLS membranous or papery, the upper largest and green, reaching 5–8 cm above the ground, hardly differing from the lower leaf. LEAVES contemporary with the flowers, as far as known, but sometimes very short, three to five, the lower two or three basal or nearly so, usually laminate, the blades narrowly lanceolate to nearly linear, 5–10 mm wide, about a third as long to nearly as long as the spike, firm-textured with moderately raised and thickened midrib and margins, the upper one to two, sometimes three, leaves cauline and largely or entirely sheathing, often imbricate; sometimes all the leaves with vestigial blades 2–3 cm long or blades entirely lacking. STEM unbranched, 3–6 mm in diameter below the first flower.

SPIKE (2–)3- to 7(–10)-flowered; BRACTS pale green, fairly soft-textured, evidently becoming dry above, 28–32(–35) mm long, the inner somewhat shorter than the outer. FLOWERS pale yellow to cream, without nectar guides; PERIANTH TUBE 18–25 mm long, nearly cylindric and curving outward in the upper half; TEPALS unequal, the three upper broadly elliptic to obovate, the dorsal largest, 30–35 mm long, to 15–18 mm at the widest, arched over the stamens, the upper laterals about as long, but somewhat narrower, directed forward, the lower three tepals curving downward, 20–25(–30) mm long, 8–12 mm wide, the lowermost about as long as the upper laterals and the lower laterals shortest. FILAMENTS 12–16 mm long, reaching the mouth of the tube or exserted 1–2 mm; ANTHERS 8–12 mm long, appressed to the upper tepal, pale yellow. OVARY oblong, 6–8 mm long; STYLE arched over the stamens, dividing near the apex of the anthers, branches c. 4.5 mm long. CAPSULES narrowly obovoid, (18–)20–24 mm long, c. 12 mm in diameter at the widest; SEEDS unknown.

FLOWERING TIME. Early December to late February, rarely into March.

DISTRIBUTION & HABITAT
Gladiolus oliganthus is restricted to southern tropical Africa, where it occurs in southern Tanzania, and possibly Malawi and eastern Zambia, although there are no records yet from either of

those two latter countries. Relatively poorly known and frequently confused with *G. melleri* or *G. dalenii, G. oliganthus* occurs in boggy grassland and dambos, a typical habitat for *G. melleri* and one in which *G. dalenii* also sometimes found. The foliage leaves are generally present by the time flowering takes place although they may be fairly short, and in some plants, even in the same population, leaves may have short blades or lack them entirely.

DIAGNOSIS & RELATIONSHIPS

A relatively modest stature but fairly large pale yellow flowers with downcurved lower tepals, the lower laterals of which are smallest, filaments included or barely exserted from the tube, and a perianth tube 18–25 mm long are characteristic of *Gladiolus oliganthus.* In addition, it usually has well-developed laminate foliage leaves on the flowering stem, although plants that flower early in the season, in December, may lack foliage leaves. *Gladiolus oliganthus* is most easily confused with *G. melleri,* which has a similar though somewhat smaller flower, but always a red to pink perianth and consistently included filaments. Of similar stature and general appearance, *G. melleri* has foliage leaves produced from separate shoots on the large corms later in the season, usually after the rains have begun. In *G. oliganthus* the leaves of the flowering stem may be fairly short or well developed, but rarely (e.g., *Bullock 1969*) do plants appear to produce additional leaves on separate shoots. Confusion with *G. dalenii* is possible but can be avoided if the flowers are carefully examined. In *G. dalenii* the filaments are always well exserted from the tube and the lowermost tepal is substantially shorter than the upper three, whereas in both *G. melleri* and *G. oliganthus* the lowermost tepal is about as long as the upper lateral tepals.

HISTORY

First collected in 1882 near Igonda, in the Tabora District of central Tanzania, by the German botanist, R. Böhm, *Gladiolus oliganthus* was described by J. G. Baker in 1898. A second collection, made in 1901 near Songea in southern Tanzania, was referred to the new *G. bussei* by F. Vaupel in 1920. Since then the two have seldom been associated and both have remained poorly understood. I have little doubt that they are conspecific, but the unusually large size of the type specimens of *G. bussei* and their poorly preserved flowers make it impossible to be certain.

SELECTED SPECIMENS

Tanzania. Mpanda: Ruigi River Forest Reserve, Biharamulo, 5 Dec. 1956, *Gane 94* (K); 30 km from Ikola to Mpanda, 1050 m, 8 Nov. 1959, *Richards 11753* (K, MO); c. 66 km south-southwest of Kibondo, 20 Nov. 1972, *Mutch 89* (EA). Iringa: Kilolo, Dabaga, 1700 m, 15 Jan. 1980, *Leedal 5842* (K). Mbeya: Mbozi, 8 km north of Tunduma, 1620 m, 10 Jan. 1975, *Brummitt & Polhill 13675* (K). Rukwa. Sumbawanga, Ufipa, dambo between Namanyere and Mkole, 4 Dec. 1986, *Moyer & Sanane 41* (MO, K); Ufipa, grass plain, 11 Dec. 1956, *Richards 7231* (K); Ufipa, Nkunde–Chapota road, c. 1800 m, 1 Dec. 1949, *Bullock 1969* (K).

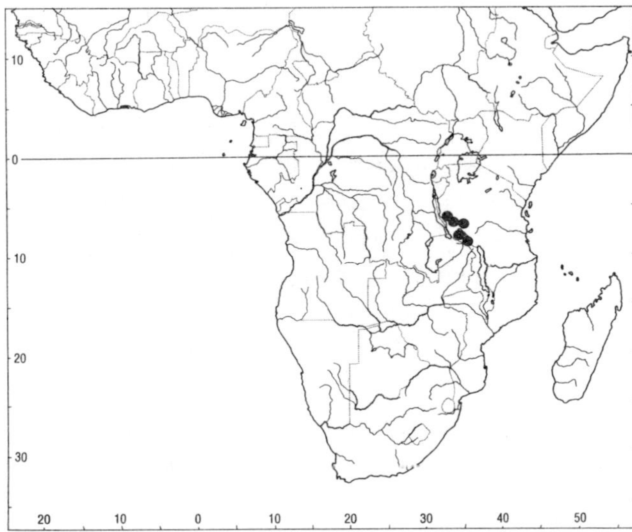

MAP 59. *Gladiolus oliganthus.*

57. *Gladiolus melleri* Baker

PLATE 32, FIGURE 49, MAP 60. Baker, J. Bot., New Ser. 5: 334 (1876); Handbook Irideae 212 (1892); Fl. Trop. Africa 7: 362 (1898). Pole Evans, Fl. Pl. S. Africa 20: pl. 768 (1940). Goldblatt, Fl. Zambesiaca 12: 97 (1993). *Gladiolus natalensis* var. *melleri* (Baker) Geerinck, Bull. Jard. Bot. Nat. Belgique 42: 285 (1972). *Gladiolus dalenii* var. *melleri* (Baker) Còrdova, Bull. Jard. Bot. Nat. Belgique 60: 326 (1990). Type: Malawi, Mpimbe Hill, Shire River, Oct. 1859, *Kirk s.n.* (K, lectotype designated by Goldblatt, 1993: 97); "Zambezi Highlands" (Manganja Hills in the protologue), Sept.–Nov. 1861, *Meller s.n.* (K, synypes).

SYNONYMY

Gladiolus brachyandrus Baker, Curtis's Bot. Mag. 105: pl. 6463 (1879); Handbook Irideae 214 (1892); Fl. Trop. Africa 7: 370 (1898). Type: Malawi, Shire Highlands, without date, cultivated at the Royal Botanic Garden, Edinburgh, *Buchanan s.n.* (figure in Curtis's Bot. Mag. 105: pl. 6463, no preserved specimens found at E or K).

Gladiolus brevispathus (Pax) Klatt in Durand & Schinz, Conspectus Fl. Africae 5: 214 (1895). *Gladiolus welwitschii* subsp. *brevispathus* Pax, Bot. Jahrb. Syst. 15: 155 (1893). Type: Angola, on the Cuango (Quango), 10°30′ S, Sept. 1876, *Pogge 426* (B, holotype; K, photo).

Gladiolus johnstonii Baker, Kew Bull. 283 (1897); Fl. Trop. Africa 7: 372 (1898). Type: Malawi, Mt. Zomba and vicinity, 2500–3500 ft, Dec. 1896, *Whyte s.n.* (K, lectotype designated by Goldblatt, 1993: 97); Mt. Malosa, 4000–5000 ft, Nov.–Dec. 1896, *Whyte s.n.* (B, K, Z, syntypes).

EPONYMY

melleri, named in honor of Charles Meller, naturalist and physician on David Livingstone's Zambezi Expedition.

DESCRIPTION

Plants 25–40(–60) cm high. CORM 2.5–3.5 cm in diameter, tunics coriaceous, fragmenting irregularly, straw-colored, the outer layer sometimes becoming fibrous. CATAPHYLLS often dry at flowering time, identical to the sheathing leaves except in position. LEAVES of the flowering stem usually all partly to entirely sheathing, rarely one, usually two or three, sometimes more, hardly differing from the cataphylls, sometimes with short blades 2–5(–10) cm long, often partly dry by anthesis; foliage leaves produced after flowering on separate shoots, evidently solitary, lanceolate, ultimately at least 30 cm long, c. 10 mm wide, the margins and midrib moderately thickened and hyaline; occasionally flowering stems with one or two long-bladed leaves. STEM unbranched, rarely branched, 3–4 mm in diameter at the base of the spike.

SPIKE 5- to 9(–12)-flowered; BRACTS green, often flushed reddish and becoming membranous above, the outer (15–)20–28(–35) mm long, the inner slightly shorter than the outer. FLOWERS orange-red to salmon, rarely whitish, without contrasting markings; PERIANTH TUBE (12–)18–20 (–30) mm long, widening evenly from the base, gently curving outward; TEPALS unequal, the upper three 25–38 mm long, 15–18 mm wide, strongly recurved distally, the dorsal horizontal or tilted downward, the upper laterals slightly narrower, directed forward, the lower tepals curving downward, the lower laterals c. 25 mm long, 12–15 mm wide, the lowermost 30–35 mm long, often about as long as the upper. FILAMENTS 8–10 mm long, included in the upper part of the tube; ANTHERS 10–12 mm long. OVARY 4–6 mm long; STYLE arched above the stamens, dividing 2–3 mm beyond the apex of the anthers, the branches 6–8 mm long. CAPSULES oblong to ellipsoid, 15–20 (–25) mm long; SEEDS oval to elliptic, c. 10 × 6–7 mm, broadly winged. CHROMOSOME NUMBER $2n = 30$.

FLOWERING TIME. August to November, occasionally into December, toward the end of the dry season.

DISTRIBUTION & HABITAT

Widespread across southern tropical Africa, *Gladiolus melleri* extends from northern Angola through southern Zaire and Zambia to western Tanzania, Malawi, central Mozambique, and Zimbabwe. Somewhat surprisingly, it does not seem to occur in the western Angolan highlands, where so many other species of tropical *Gladiolus* have their southwestern limit. *Gladiolus melleri* grows in seasonally wet sites, such as the margins of dambos and in poorly drained savanna, but blooms at the end of the dry season, from August to November, when such places are often quite dry. Its flowering is stimulated by fire, and it is common to see signs of burning on specimens.

DIAGNOSIS & RELATIONSHIPS

A plant with only sheathing leaves on the flowering stem and typically with brick-red to salmon flowers, *Gladiolus melleri* can be recognized in ad-

FIGURE 49. *Gladiolus melleri.* Corm with attached flowering spike, × 0.5; single flower, vertical floral section, and leafy shoot, full size (*Goldblatt et al. 8247*).

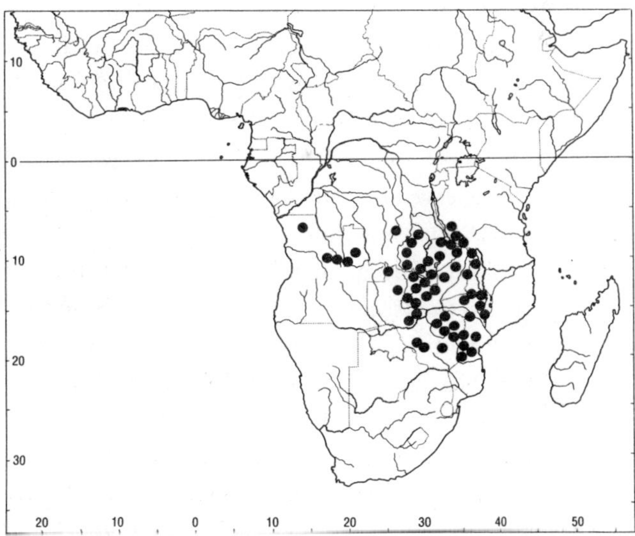

MAP 60. *Gladiolus melleri.*

dition by the relatively short bracts 15–28(–35) mm long, a perianth tube 12–20 mm long, shorter than the large upper tepals, and short filaments always included in the tube. *Gladiolus melleri* is most closely related to the southwestern Tanzanian *G. oliganthus,* which has somewhat larger flowers, yellow in color, generally has the leaves contemporaneous with the flowers, and always has short included filaments so characteristic of *G. melleri*. The flowers of *G. oliganthus* have a tube 20–30 mm long, beyond the normal range for *G. melleri,* but in other respects the flowers of the two species are similarly constructed. I doubt that *G. oliganthus* is simply a late-flowering plant and a form of *G. melleri* with leaves borne on the flowering stem, because the flowers differ in color, size, and in the length of the perianth tube. The occasional *G. melleri* that flowers out of season, or is grown in gardens, and has foliage leaves on the flowering stem is still recognizable as this species (e.g., *Gibbs 323* (BOL), cultivated in Cape Town). Although the flowers of *G. melleri* are typically brick-red to orange, paler-colored forms occur. In southwestern Tanzania the flowers are a pale salmon-pink, and on the Biano Plateau in southern Shaba, Zaire, plants have pale pink to nearly white flowers (*Schaijes 1956*). Those plants may be confused with the local Shaban endemic, *G. manikaensis,* which has pure white flowers, rather smaller in size than those of *G. melleri,* but in *G. manikaensis* the foliage leaves are always present on the flowering stem at flowering time.

Although to my mind a particularly distinctive species, *Gladiolus melleri* is also sometimes be confused with *G. dalenii* subsp. *andongensis,* which it resembles in the absence of foliage leaves on the flowering stem and, at least by *Gladiolus* standards, a relatively large flower. *Gladiolus melleri* can, however, always be recognized by its shorter perianth tube, normally 12–20 mm long, and smaller tepals and bracts, the latter typically 17–20 mm long. The three upper tepals of *G. melleri* are more or less equal and about as long as the lowermost, and only the two lower lateral tepals are substantially smaller. In the smallest-flowered forms of *G. dalenii* the perianth tube is 25–35 mm long, the dorsal tepal is largest, 35–50 mm long, and the bracts are 35–45(–70) mm long. The dorsal tepal is always arched and hooded over the stamens, and the lower tepals of the salmon- to red-flowered forms are usually marked with yellow. Also, the filaments are well exserted from the upper part of the perianth tube, whereas in *G. melleri* they are always included. In *G. melleri* the dorsal tepal is barely arched and instead extends forward more or less horizontally. The differences are more easily seen in fresh material, but there is really no reason for the two to be confused. Geerinck (1972) and Còrdova (1990) regarded *G. melleri* as no more than a variety of *G. dalenii,* but although the two are obviously related, differences in flower shape and size, including the short, included filaments, make it essential to treat them as separate species.

The western African and Ethiopian *Gladiolus roseolus,* which also has foliage leaves produced on separate shoots after flowering, is very like *G. melleri,* but it has somewhat smaller flowers, white to deep pink in color, cataphylls that are typically downy-pubescent, although not always so, and the filaments are exserted from the tube 4–5 mm. *Gladiolus melleri* can also be confused with the southwestern tropical African *G. benguellensis,* but although this species has flowers of almost exactly the same shape, they are smaller, thus the upper tepals are 21–24 mm long, and the anthers are 7–8 mm long. *Gladiolus benguellensis* also has the foliage leaves borne on the flowering stem rather than on separate shoots, and smaller corms and distinctively darker cataphylls. Another distinction between *G. melleri* and *G. benguellensis* is the styles, well developed in *G. melleri* and reaching at least to the apex of the anthers with branches 5–6 mm long, but shorter in *G. benguellensis,* not overtopping the anthers and with branches 3–4 mm long.

HISTORY

Discovered in 1860s in southern Malawi and described by J. G. Baker in 1876, *Gladiolus melleri* quickly became a relatively well-known southern tropical African plant. Two later gatherings, both closely resembling the type collections of *G. melleri,* were inexplicably described by Baker respectively as *G. brachyandrus* (in 1879) and *G. johnstonii* (in 1897). Plants from northern Angola, found in 1876, were not immediately compared with *G. melleri* but with the Angolan *G. welwitschii* and were described as subspecies *brevispathus* of that species by Pax in 1893, the taxon almost immediately elevated to species rank by Klatt as *G. brevispathus.* It differs in no taxonomically significant way from *G. melleri.* The reduction of *G. melleri* to varietal rank in *G. dalenii* (Geerinck, 1972; Còrdova, 1990), mentioned above, is not accepted here.

SELECTED SPECIMENS

Angola. Uige: 40 km north of Uige, 800 m, 9 Sept. 1958, *Stanton 79* (BM). Malange: Xa Sengue–Nova Gaia road, 4 Sept. 1932, *Young 764* (BM). Lunda Sul: Saurimo, Chicapa River, 4 Oct. 1932, *Young 441* (BM).

Zaire. Shaba: Kaniama, la Pastorale, Sept. 1931, *Quarré 2674* (BR, SRGH); Upemba National Park, Mukana, 1810 m, 26 Aug. 1947, *de Witte 2765* (BR, EA, K, PRE, SRGH); Biano Plateau, 26.5 km north northeast of Tenke, 1680 m, 17 July 1983, *Schaijes 1956* (BR).

Tanzania. Rukwa: Sumbawanga road, 30 km from Zambia border, 2100 m, 19 Oct. 1959, *Richards 11512* (K); Malonje Plateau, 21 km from Sumbawanga, 2100 m, 26 Oct. 1965, *Richards 20615* (B, K); Ishenta, Ulambya, 18 Oct. 1970, *Leedal 477* (EA).

Zambia. Luapula: Chishinga Ranch, 1450 m, 9 Sept. 1965, *Astle 3302* (K, SRGH). Northern: Kambole–Mbala (Abercorn) road, 15 Sept. 1960, *Richards 13263* (BR, K). Northwestern: 60 km south of Mwinilunga, 15 km west of Lunga River, 12 Aug. 1930, *Milne-Redhead 873* (K). Copperbelt: Ndola, woodland, 24 Sept. 1954, *Fanshawe 1566* (BR, K, NDO); 12 km north of Kitwe, 8 Sept. 1961, *Linley 181* (MO, S, SRGH). Central: Kabwe (Broken Hill), flood plain, 15 Sept. 1964, *Mutimushi 944* (NDO, SRGH). Southern: Pemba, Sept. 1909, *Rogers 8556* (BM, BOL, K, Z); Mumbwa District, south of Chinobi, 10 Sept. 1947, *Greenway & Brenan 8094* (BM, BOL, EA, K, Z).

Malawi. Northern: Mafinga foothills, 19 Dec. 1964, *Robinson 6310* (K, SRGH); Nyika Plateau, 20 km south of Chelinda, 19 Oct. 1975, *Pawek 10320* (EA, K, MO, PRE, SRGH). Central: Lilongwe District, Dzalanyama Forest Reserve, 3 Oct. 1962, *Banda 453* (MAL, SRGH); Dedza, near rifle range, 24, Sept. 1935, *Galpin 15055* (BOL, K, PRE). Southern: Bvumbwe, 1200 m, 14 Oct. 1982, *la Croix 349* (K).

Zimbabwe. Matabeleland South, Matopos, Research Station, 9 Oct. 1952, *Plowes 1491* (K, LISC, SRGH). Mashonaland North: Mazowe, Sept. 1906, *Eyles 417* (BM, BOL, SRGH); Chipolilo, Nyamunyeche Estate, 17 Oct. 1978, *Nyariri 422* (SRGH). Mashonaland East: 15 km north of Marondera (Marandellas), 4 Oct. 1953, *Wild 4141* (K, LISC, MO, PRE, SRGH). Manicaland: Mtare (Umtali) commonage, 9 Oct. 1953, *Chase 5113* (BM, SRGH); Nyanga, Pungwe Valley, 11 Nov. 1966, *Plowes 2828* (SRGH). Masvingo: Between Masvingo (Fort Victoria) and Ndanga, 20 Oct. 1930, *Fries et al. 2133* (BR, S).

Mozambique. Tete: Maravia, between Furancungo and Vila Gamito, 20 Oct. 1943, *Torre 6075* (LISC); Tete road 90 km south of Cotinho, 25 Sept. 1935, *Galpin 15066* (PRE). Manica e Sofala: Manica, km 4, Rotanda to Mavita, 30 Oct. 1965,

Correia 289 (LISC); Barué, Vila Goveia, Chôa Mountains, 17 Sept. 1942, *Mendonça 282* (LISC); Vila Pery, Oct. 1915, *Surcouf s.n.* (MO, P).

58. *Gladiolus roseolus* Chiovenda

PLATE 33, FIGURE 50, MAP 61. Chiovenda, Ann. Bot. Roma 9: 125–152 (1911). Type: Ethiopia, Semien, rocky meadow on the slopes of Lumalmo, 9 July 1909, *Chiovenda 778* (FI, lectotype designated here, photocopy seen). Tigre, Shiré, near Idaga (Edaga), 18 June 1909, *Chiovenda 578* (FI, syntype); Tzellemti, marshy meadow, near Buia, 3 July 1909, *Chiovenda 683* (FI, syntype).

SYNONYMY

Gladiolus heterolobus Vaupel, Bot. Jahrb Syst. 48: 535 (1913). Type: Cameroon, Sakje, "Posten Sagosche," 730 m, May 1909, *Ledermann 3839* (B, lectotype designated here, with flowers well preserved; K, photo); Cameroon, Tshape Pass, 1420 m, Feb. 1909, *Ledermann 2650* (syntype, not seen).

Gladiolus melleri sensu Hepper, Fl. West Tropical Africa, Ed. 2, 3: 144 (1968), not Baker.

EPONYMY

roseolus, "pale pink," the perianth color in some members of the species.

DESCRIPTION

Plants (40–)60–90 cm high. CORM 25–30 mm in diameter, tunics of firmly membranous layers, the outer more or less fibrous, red-brown to straw. CATAPHYLLS firm-membranous, usually densely short-pubescent to downy above the ground. LEAVES of the flowering stem two to four, rarely only one, short and almost entirely sheathing, sometimes with short blades, 6–14 cm long, imbricate and sheathing the lower half of the stem; laminate foliage leaves produced from one or more separate shoots, these narrowly lanceolate, c. 4 mm wide, the midrib and margins moderately thickened and hyaline, the sheaths sometimes lightly pubescent. STEM unbranched, 2–3 mm in diameter below the first flower.

SPIKE (2–)5- to 10-flowered; BRACTS green below at anthesis, becoming dry and straw-colored above, (20–)25–30 mm long, the inner somewhat shorter than the outer. FLOWERS whitish with a pink flush to pink, sometimes speckled with minute red dots, these densest near the base of the tepals, evidently without markings on the lower tepals; PERIANTH TUBE 18–22 mm long, cylindric, curving outward and widening above; TEPALS unequal, the upper three largest, ovate to elliptic, 26–30 mm long, 10–12 mm wide in the midline, the dorsal arched almost horizontally over the stamens, the lower three tepals lanceolate, curving downward, 20–24 mm long, 4–6 mm wide, lower laterals smallest. FILAMENTS c. 12 mm long, exserted 4–5 mm from the tube; ANTHERS 10–12 mm long, pale yellow. OVARY oblong, 5–6 mm long; STYLE arching over the stamens, dividing 2–4 mm beyond the apex of the anthers, branches 4–5 mm long. CAPSULES narrowly obovoid, 20–25 mm long; SEEDS broadly winged, c. 9 × 6 mm.

FLOWERING TIME. April to mid June.

DISTRIBUTION & HABITAT

Gladiolus roseolus extends across interior western Africa from Togo to central Cameroon and Nigeria, and it also occurs in Ethiopia. Plants grow in light woodland, often on rocky sites, and bloom early in the wet season. Blooming early in the season before the surrounding vegetation has grown tall, *G. roseolus* typically has flowering stems without foliage leaves. As in several other tropical African species of *Gladiolus,* notably *G. unguiculatus* in western Africa, the leaves are produced later in the season from separate shoots and reach their full size after the capsules ripen.

DIAGNOSIS & RELATIONSHIPS

Particularly distinctive among the northeastern and western African species of *Gladiolus*, *G. roseolus* stands out in its short-tubed, white to pink flowers and flowering stems that lack foliage leaves. The flowers have a strongly arched to almost hooded dorsal tepal and much smaller lower tepals that curve downward throughout. The relatively short filaments suggest a relationship with species such as the southern tropical African *G. melleri*, but in that species the filaments are shorter and reach only to the mouth of the tube. Although somewhat variable in color, the flowers of *G. melleri* are typically bright orange or brick-red, and the stem is typically shorter than in *G. roseolus*.

HISTORY

Although first collected by Georg Heinrich Schimper in Ethiopia in the first half of the 18th century, and re-collected by Georg Schweinfurth near the Ethiopia–Sudan frontier in 1862, these records were overlooked by contemporary botanists. The German, Ernst Steudel, annotated specimens of Schimper, "*Gladiolus minor*," but the name was never published. Only the much later

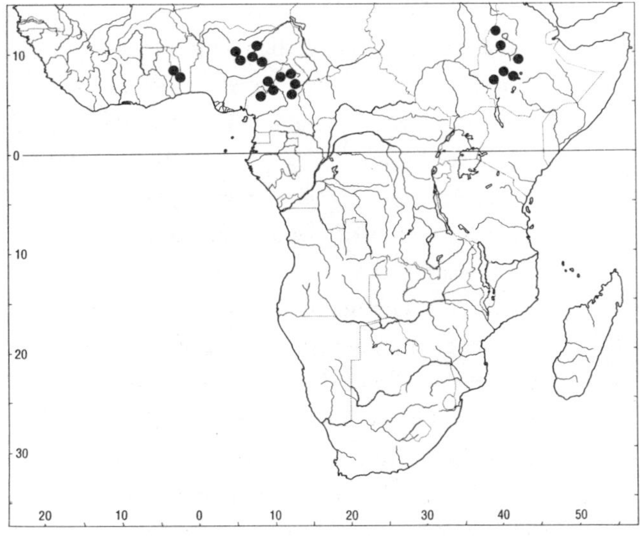

MAP 61. *Gladiolus roseolus*.

FIGURE 50. *Gladiolus roseolus*, full size (*Thulin & Hunde 4020*).

collections made by the Italian botanist, Achille Chiovenda, in 1909, formed the basis for *G. roseolus* (Chiovenda, 1911). Collections also made in 1909 in Cameroon by the German traveler, C. L. Ledermann, were described by F. Vaupel as *G. heterolobus* in 1913. Neither *G. roseolus* nor *G. heterolobus* were recognized in either edition of *Flora of West Tropical Africa,* but specimens of the species were cited there under the related *G. melleri.*

SELECTED SPECIMENS

Togo: 8 km north of Niamtougou, heights south of Défalé, 9 May 1978, *Hakki, Leuenberger, & Schiers 857* (flower and fruit) (B); Kuwde, west of Kpagouda, northwest of Farendé, hilly country 10 May 1978, *Hakki, Leuenberger, & Schiers 903* (B, K).

Nigeria. Gongola: Mambilla, Gashaka Game Reserve, Chappal Tali, 1600 m, 13 Apr. 1976, *Chapman 4310* (K); Bauchi: Bauchi Plateau, among rocks, May 1928, *Lely 284* (K, MO); top of Zaranda Mountain, 1700 m, 18 May 1921, *Lely 190* (K). Kaduna: Mile 3, Kachia–Zonkwa road, 850 m, 4 May 1957, *Summerhayes 92* (BR, K, LISC, P); 5 km south of Mando, woodland, 18 June 1950, *Keay s.n.* (FHI 25870 at K, P); Gwari, between Mando and Kwaga, 1 June 1950, *Keay s.n.* (FHI 25812 at K, P).

Cameroon. Benoué: Banglang, 60 km southwest of Tchamba, 3 Apr. 1976, *Fotius 2495* (P, YA); Mbéré–Meiganga, June 1939, *Jacques-Felix 4284* (P); Gotel Mountains, 30 km north-northeast of Banyo, Apr. 1967, *Letouzey 8617, 8617bis* (P); Wum, "Vom Rest House," Benue Plateau, 19 Apr. 1972, *Wit 1310* (K);

Ethiopia. Gonder: Gallabat, on the Gendoa, 10 June 1862, *Schweinfurth 479,* annotated "*Gladiolus matammensis* Schfth." (B). Kaffa: west slopes of the Omo River Valley, 149 km from Jimma to Addis Ababa, 10 July 1969, *de Wilde 5441* (BR, WAG); c. 140 km southwest of Addis Ababa on the road to Jimma, 2200 m, 9 Sept. 1965, *de Wilde et al. 7476* (BR, K, MO, WAG); 65 km from to Jimma to Bonga, 28 May 1971, *Gilbert 2024* (EA, K). Gojjam: 94 km on the new road to Giba from south of Injibara, 1200 m, 29 May 1980, *Thulin & Hunde 4020* (K, UPS). Shoa: Guder, riverbank around the waterfall, 10 May 1975, *Bos & Jansen 10210* (WAG); 30 km northeast of Addis Ababa, c. 2200 m, 6 May 1966, *de Wilde 10927* (WAG); north side of Gibie Gorge, heavy black clay, 1400–1800 m, *Edwards, Kelbessa, & Bekele 2289–2379* (ETH).

59. *Gladiolus pauciflorus* Baker

PLATE 34, FIGURE 51, MAP 62. Baker in Johnston, Kilimanjaro Exped. 346 (1886), name without description; Trans. Linn. Soc. Bot., Ser. 2, 2: 350 (1887); Handbook Irideae 206 (1892); Fl. Trop. Africa 7: 368 (1898). Type: Tanzania, Kilimanjaro, 600–1500 m, *Johnston s.n.* (K, holotype; BM, possible isotype, only upper part of the spike).

EPONYMY

pauciflorus, "few-flowered," referring to the relatively small number of flowers per spike, sometimes only two to four, although usually more.

DESCRIPTION

Plants 80–105 cm high. CORMS 15–22 cm in diameter, the tunics of matted fibers, dark brown. CATAPHYLLS green or often purple, the two upper reaching well above the ground. LEAVES occasionally three, usually four or five, the lower two to four basal and largest, narrowly lanceolate, (6–) 8–15 mm wide, reaching to about the base of the spike; the upper one or two cauline, with blades usually shorter than the sheaths, the uppermost of these almost entirely sheathing. STEM erect and unbranched, 3–3.5 mm diameter at the base of the spike, sometimes sheathed almost to the first flower, 2–4 mm in diameter at the base of the spike.

SPIKE (2–)4- to 8(–10)-flowered; BRACTS green, the outer (3.5–)4–6 cm long, the inner half to two-

FIGURE 51. *Gladiolus pauciflorus,* × 0.67 (*Gilbert 3903*).

thirds as long as the outer, shortly forked apically. FLOWERS cream to yellowish green, sometimes pink to reddish, or flushed orange, the lower three tepals often each with a dark purple median streak; PERIANTH TUBE (20–)35–45 mm long, cylindric below, widening toward the apex; TEPALS broadly or narrowly lanceolate, the upper three largest, 30–45 × 18–24 mm, the lowermost nearly as long as the upper, the lower laterals substantially smaller. FILAMENTS 22–24 mm long, exserted 10–14 mm from the tube; ANTHERS 8–11 mm long, acute, not apiculate, yellow. OVARY c. 5 mm long; STYLE arched over the stamens, dividing just below the anther apices, the branches 4–7 mm long, ultimately exceeding the anthers. CAPSULES obovoid, 15–20 mm long; SEEDS more or less oblong, 6–7 × c. 4 mm.

FLOWERING TIME. Mainly mid April to mid June.

DISTRIBUTION & HABITAT

Gladiolus pauciflorus is restricted to eastern Africa, where it ranges from the lower slopes of Kilimanjaro in northern Tanzania in the south, across Kenya to southern and eastern Ethiopia in the north.

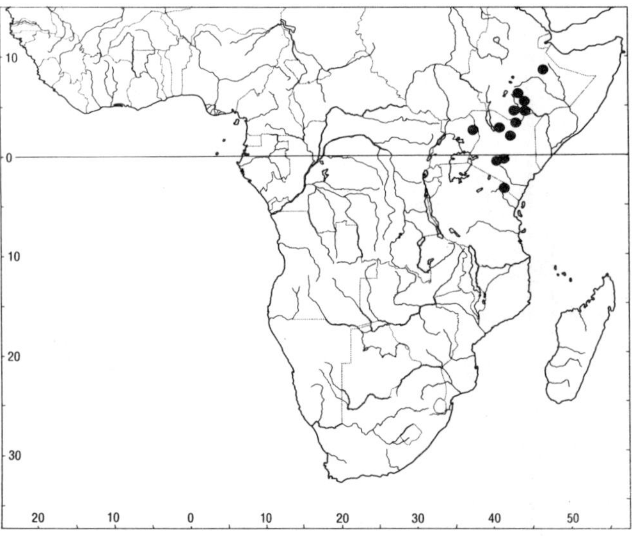

MAP 62. *Gladiolus pauciflorus.*

It occurs in highlands, most often above 1000 m, in open grassland or savanna. Most common in Ethiopia, *G. pauciflorus* occurs in Sidamo, Bale, and Harerge Provinces, which are relatively poorly known botanically. There are relatively few records of the species from Kenya, but it has been recorded from both Mt. Kulal and Mt. Marsabit, isolated massifs in the dry north of the country, as well as in the central Kenyan highlands. The only record from Tanzania that I have seen is the type, collected by H. H. Johnston in 1883 on Mt. Kilimanjaro. This specimen matches plants from Kenya and eastern Ethiopia in all important respects.

DIAGNOSIS & RELATIONSHIPS

One of the larger-flowered species of the genus, *Gladiolus pauciflorus* can be recognized by its robust form and fairly large flowers, with a perianth tube 35–40(–45) mm long and tube of about the same length. The flowers are typically pale-colored, either cream or yellowish green, often with a dark purple stripe along the midline of the lower three tepals. The lowermost tepal is usually about a long as the upper or only slightly shorter, but the lower laterals are substantially shorter than the upper tepals. The species as understood here is fairly variable. Plants from Mt. Kulal in northern Kenya (*Oteke 70*) and collections from Sidamo Province, Ethiopia, appear to have either pink (or red now faded to pink) flowers (*Sandford sub Mooney 7406*). The flowers of most other collections from Ethiopia are cream, but they may also be flushed with orange (*Gilbert et al. 8074, 8077*), or pale yellow throughout in the same populations. The dark median streak on the lower tepals is most pronounced in Ethiopian populations, especially those from Bale Province (e.g., *Gilbert & Sebsebe D., 8610*), as illustrated in Figure 51. One population from Sidamo (*Gilbert 3870*) is especially puzzling. The flowers are described as rose with the lower laterals yellow with blackish streaks, and these have the shortest perianth tubes of all specimens assigned to the species, 25–30 mm long. I considered the possibility that they represent a separate species, but they seem connected by a series of intermediates from the same area and until more is known about them it seems prudent to assume they are merely depauperate *G. pauciflorus.* These plants are rather slender and also have only one or two flowers per spike. Plants from other populations may have five to seven or even 10 flowers per spike, but this character is one that is normally very variable and not normally taxonomically significant.

I also include a pink-flowered plant from Mt. Lorosuk in Uganda here. This is at best provisional, for the single collection comprises fairly short plants, c. 20 cm high, with the long, rather broad perianth tube typical of *Gladiolus pauciflorus,* but the flowers themselves are rather small for the species, as are the bracts, 25–30 mm long, and the narrow leaves, c. 2 mm wide. The specimens lack corms that may provide a clue to their correct identity. Until more specimens of this Ugandan plant come to light it seems best to place it in *G. pauciflorus,* but there remains the possibility that this is a new species.

ADDITIONAL SPECIMENS

Ethiopia. Harerge: 10 km northwest of Jijiga on the road to Harar, c. 1800 m, 24 Apr. 1970, *Gilbert & Gilbert 1839* (C, EA, ETH); c. 90 km from Harer to Jijiga, c. 1300 m, 18 May 1969, *de Wilde & Tadesse Ebba 5040* (ACD, K, WAG). Bale: 8 km from Bidre on the road to Dello Mena (Maslo), 1550 m, 2 June 1988, *Gilbert & Sebsebe D. 8610* (K); 13 km north of Sidambale Bridge on the track to Biddera and Mena, 1400 m, 20 May 1975, *Gilbert 3903* (K). Sidamo: 3 km north of Bittata on the Waddera–Negeli road, 1575 m, 8 June 1986, *Gilbert, Sebsebe D., & Vollesen 8074* (K, UPS); 12 km north of Negeli to Waddera, 8 June 1986, *Gilbert, Sebsebe D., & Vollesen 8077* (K, UPS); Moyale, 13 May 1959, *Thesiger 2054* (BM).

Uganda. Karamoja: Mt. Lorosuk, Karasule, 2440 m, 15 Aug. 1965, *Wilson 1679* (EA).

Kenya. Southern: Ulu, Joyce Estate, 5700 m, 20 June 1967, *Churches s.n.* (EA 13784, K, P). Kirinyanga District, Mwea Plains, 1160 m, 19 Dec. 1972, *Robertson 1806* (EA). Eastern: Mt. Kulal, Narangani, 8 June 1960, *Oteke 70* (K); Marsabit, 5 June 1960, *Oteke 39* (K); Marsabit, 1400 m, 14 Jan. 1972, *Bally & Smith 14790* (K); 5 June 1960, *Oteke 39* (K).

Tanzania. None seen from here except the type.

60. *Gladiolus gunnisii* (Rendle) Marais

MAP 63. Marais, Kew Bull. 28: 313 (1973).

SYNONYMY

Acidanthera gunnisii Rendle, J. Bot. 36: 31 (1898). Baker, Fl. Trop. Africa 7: 360 (1898). Type: Somalia, Toghdeer, top of Wagga Mountain, 2000 m, Feb. 1895, *Lort Phillips s.n.* (BM, lectotype designated here, in good condition and nearly complete; K, isolectotype); Diamoleh, 1895, *Gillett & Aylmer s.n.* (BM, K, syntypes).

Acidanthera nelloi Chiovenda, Nuov. Giorn. Bot. Italia, Ser. 2, 26: 109 (1919). Type: Eritrea, Ghinda, 1905, *Beccari 266* (FI, holotype, photocopy seen).

EPONYMY

gunnisii, named in honor of F. G. Gunnis, member of a small expedition that collected in the mountains south of Berbera, in former British Somaliland, in the winter of 1895.

DESCRIPTION

Plants 25–35(–45) cm high. CORM globose-conic, 11–14 mm in diameter, the tunics of fine-to medium-textured, compacted fibers. CATAPHYLLS pale and membranous, the upper reaching shortly above the ground. LEAVES three to five, the lower two or three basal, about a third as long as the stem, the blades linear, 2–3(–4.5) mm wide, the midribs and sometimes the margins lightly thickened, the upper leaves cauline and progressively shorter, sometimes the uppermost entirely sheathing. STEM unbranched, c. 1.5 mm in diameter at the base of the spike.

SPIKE (1–)2- to 3-flowered; BRACTS green, 25–40(–45) mm long, the inner nearly equal or 2–4 mm shorter than the outer. FLOWERS white or sometimes pale yellow, strongly fragrant; PERIANTH TUBE slender, 80–120 mm long, expanding in the upper 10 mm; TEPALS evidently subequal, nearly elliptic, the dorsal probably horizontal, remaining tepals spreading, 25–30 mm long. FILAMENTS c. 9 mm long, included in the tube or barely exserted for c. 1 mm; ANTHERS 8–9 mm long, the bases sometimes within the tube. OVARY 4–5 mm long; STYLE dividing c. 5 mm beyond the anther apices, the branches c. 5 mm long. CAPSULES ellipsoid, apically emarginate, 15–20 mm long; SEEDS oblong, 5–6 × 2.5 mm, light brown, broadly winged at the distal and proximal ends.

FLOWERING TIME. Mostly December and January in Somalia and Eritrea, May and June in Ethiopia.

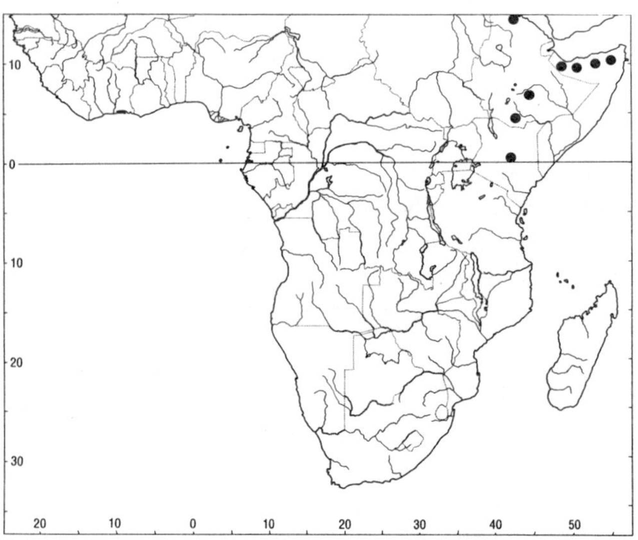

MAP 63. *Gladiolus gunnisii.*

DISTRIBUTION & HABITAT

Gladiolus gunnisii is restricted to the mountains of the Horn of Africa and has been recorded mostly from northern Somalia, but there are also three records from Eritrea and two from southern Ethiopia. In Somalia and Eritrea, plants flower mainly in the wetter months, October to February, and favor sheltered situations on north-facing slopes, often among rocks. In southern Ethiopia, plants flower in May and June. The difference in flowering time suggests that the Ethiopian plants may represent a separate species, but they do not seem to differ in any significant way from those from Somalia and Eritrea.

DIAGNOSIS & RELATIONSHIPS

The white to cream flower and long perianth tube, 80–120 mm long, included filaments, and narrow, grasslike leaves are the characteristic features of *Gladiolus gunnisii.* Although the long tube, white flower, and subequal tepals suggest a relationship with section *Acidanthera,* and perhaps most closely with the eastern African and southern Arabian *G. candidus, G. gunnisii* does not have the apiculate anthers that characterize the section. Moreover, the relatively short and nearly equal inner and outer bracts and the smaller tepals are also inconsistent with section *Acidanthera.* It seems unlikely that *G. gunnisii* belongs in section *Acidanthera,* but its relationships are otherwise uncertain. The species is provisionally associated with the central Tanzanian endemic, *G. richardsiae,* and the two are placed in section *Ophiolyza.*

As noted above, there are two collections of *Gladiolus gunnisii* from southern Ethiopia, and the geographical disjunction combined with the difference in flowering time from the Somalian and Eritrean plants suggest the possibility that they represent a separate species. The Ethiopian specimens seem more robust and have broader leaves, up to 5.5 mm wide, and bracts 4.5 cm long (2.5–3 cm in wild-collected plants from Somalia). These features seem significant until cultivated plants of Somalian origin are compared with the Ethiopian specimens. The Ethiopian plants also have broader leaves and longer bracts. This makes it seem likely that variation between Ethiopian and Somalian plants is the result of differences in growing conditions.

SELECTED SPECIMENS

Eritrea. Ghinda, Dec. 1902, *Tellini 7* (FI); Ghinda Plateau, Dec. 1902, *Tellini 266* (FI).

Ethiopia. Bale: Web Gorge, 15 km west of Goro on road from Ginir to Robe, 2250 m, 1 June 1983, *Gilbert, Ensermu, & Vollesen 8024* (C, ETH, K). Sidamo: 33 km on road from Negeli to Filtu, 1500 m, 20 June 1982, *Friis, Tadesse, & Vollesen 3153* (C, ETH, K, ETH).

Somalia. Sanaag: 64 km south-southwest of Bosaso, eastern end of Al Madu Range, foot of limestone escarpment in grassland, 10 Jan. 1973, *Bally & Melville 15706* (EA, G, K, MO); Al Madu Range, Geldin, c. 1500 m, 15 Oct. 1956, *Bally 11132* (G, K); north and west of Galgallo, Jan. 1973, *Lavranos & Horwood 10243* (E, K).

Kenya. Northern: c. 18 km from Isiolo to Meru, 4 Dec. 1955, *Hales 307* (EA).

61. *Gladiolus richardsiae* Goldblatt, new species

MAP 64. Type: Tanzania, Singida, Itigi District, road from Chunya to Itigi, 1500 m, swamp in thick clay mud, 26 Jan. 1968, *Richards 22961* (K, holotype; BR, isotype).

EPONYMY

richardsiae, named in honor of Mary Richards, British plant collector whose work in Zambia and Tanzania made possible a substantial increase in the botanical knowledge of that part of Africa.

LATIN DIAGNOSIS

Plantae 45–55 cm altae, cormis ignotis, foliis 5 inferioribus 3 basalibus anguste lanceolatis 5–8 mm latis marginibus costisque leviter incrassatis, caule eramoso c. 2.3 mm diametro infra spicem, spica 2–3 florum, bracteis (28–)40–47 mm longis manifeste siccis, floribus albis, tubo perianthii 80–85 mm longis, tepalis superioribus 40–45 × 16–19 mm longis, inferioribus brevioribus, filamentis in tubo inclusis, antheris c. 12 mm longis apicibus obtusis, stylo diviso ultra apices antheris.

DESCRIPTION

Plants 45–55 cm high. CORMS unknown. CATAPHYLLS membranous below ground, firm-textured and purplish or becoming dry and brown above the ground. LEAVES five, the lower three basal, the blades narrowly lanceolate, one-third to half as long as the stem, 5–8 mm wide, fairly firm-textured, the margins and midribs moderately thickened, the two upper leaves inserted on the stem, mostly or entirely sheathing. STEM unbranched, c. 2.3 mm in diameter at the base of the spike.

SPIKE straight, 2- to 3-flowered; BRACTS (28–)40–47 mm long, subequal or the inner slightly shorter, evidently dry at least above at flowering time. FLOWERS white, the tepals evidently without markings; PERIANTH TUBE 80–85 mm long, arching gently outward; TEPALS unequal, the upper three largest, 40–45 × 16–19 mm, the lower laterals smallest, c. 33 mm long, the median lower tepal 35–40 mm long. FILAMENTS c. 25 mm long, included in the tube and reaching to just below the mouth of the tube; ANTHERS c. 12 mm long, evidently unilateral, the apices with minute obtuse appendages. OVARY ovoid, 6–8 mm long; STYLE extending over the anthers, dividing 2–10 mm beyond the anther apices, the branches, 6–7 mm long. CAPSULES and SEEDS unknown.

FLOWERING TIME. January.

DISTRIBUTION & HABITAT

Gladiolus richardsiae is known from only a single gathering comprising four specimens, collected in 1968 in the Itigi District of Singida Region in central Tanzania. According to the information provided with the collection, the species occurs in heavy clay mud in a marshy habitat. Although the species is probably rare, Singida Region is poorly explored botanically, and additional collecting will almost certainly yield more populations. It is interesting to note that *G. canaliculatus,* also from Singida, is known from a single collection, and there are few records even of common and widespread species of *Gladiolus* from the Region.

DIAGNOSIS & RELATIONSHIPS

Although *Gladiolus richardsiae* closely resembles species of section *Acidanthera* in its long-tubed white flower with apparently subequal tepals, it seems likely that it does not belong in the section. Except for *G. candidus,* species of section *Acidanthera* have flowers with dark purple markings on the lower tepals, and all members of the section have anthers with conspicuous long-attenuate apical appendages. Additionally, the species of section *Acidanthera* have green floral bracts, the inner of which is usually much smaller than the outer and concealed within it. The anthers of *G. richardsiae* lack acute apiculate appendages, and the bracts are nearly equal and apparently dry at flowering. Its seems more likely that *G. richardsiae* is most closely related to tropical African species of section *Ophiolyza*. Of the tropical African members of that section, *G. richardsiae* seems most closely related to the northeastern African *G. gunnisii,* but that species is characterized by narrow, rather soft-textured leaves, flowers with a perianth tube 80–120 mm long, and somewhat shorter tepals, 25–30 mm long. It also grows in rocky habitats, not notably moist even in the wet season.

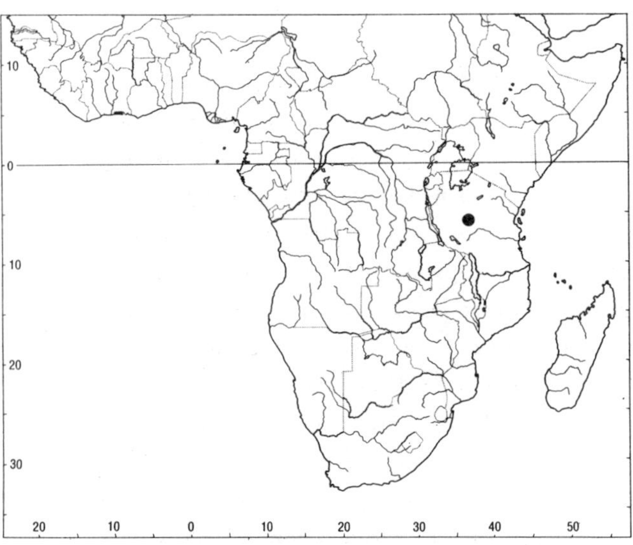

MAP 64. *Gladiolus richardsiae.*

SPECIMENS

Known only from the type, cited above.

62. *Gladiolus watsonioides* Baker

FIGURE 52, MAP 65. Baker, J. Linn. Soc. 21: 405 ("1885," i.e., 1886); Curtis's Bot. Mag. 113: pl. 6919 (1887); Handbook Irideae 226 (1892). *Antholyza watsonioides* (Baker) Baker, Fl. Trop. Africa 7: 376 (1898). *Homoglossum watsonioides* (Baker) N. E. Brown, Trans. Roy. Soc. S. Africa 20: 278 (1932). Type: Tanzania, Kilimanjaro, probably 1884, *Thomson & Johnston s.n.* (K, holotype).

SYNONYMY

Gladiolus watsonioides var. *minor* Baker, Handbook Irideae 227 (1892). Type: Tanzania, Kilimanjaro, "up to 13,000 ft," 1884, *Johnson 124* (K, holotype; BM, isotype).

Antholyza gracilis Pax in Engler, Hochgebirgsflora tropischen Afrika 177 (1892). *Homoglossum gracile* (Pax) N. E. Brown, Trans. Roy. Soc. S. Africa 20: 278 (1932), not *Gladiolus gracilis* Jacquin. Type: Tanzania, Kilimanjaro, vegetation limit on Kibo, 2800–4800 m, Nov. 1889, *Meyer 216* (B, lectotype designated here); Kilimanjaro, 3200 m, July 1887, *Meyer 130* (B?, not seen, location unknown).

Gladiolus mackinderi J. D. Hooker, Curtis's Bot. Mag. 128: pl. 7860 (1902). Type: Kenya, Mt. Kenya, 3000 m, 1899?, cultivated at Kew Gardens in 1901, *Mackinder s.n.* (figure in Curtis's Bot. Mag. 128: pl. 7860; apparently there is no preserved material, but *Mackinder s.n.* in 1899 at BM probably represents wild-collected plants from the same source as the type figure).

Gladiolus aberdaricus N. E. Brown, Trans. Roy. Soc. S. Africa 20: 267 (1932), new name for *Antholyza speciosa* C. H. Wright, Kew Bull. 338 (1914), not *G. speciosus* Thunberg (= *G. alatus* var. *speciosus* (Thunberg) G. Lewis). Type: Kenya, Aberdare Mountains, 3050 m, without date, *Battiscombe 838* (K lectotype designated by Brown, sheet 1, the most complete; K, isolectotypes).

EPONYMY

watsonioides, "*Watsonia*-like," alluding to the similarity to species of the southern African genus *Watsonia,* which also has tall spikes with long-tubed, bright red flowers.

DESCRIPTION

Plants 55–100 cm high or more. CORM 15–20 mm in diameter, the tunics firm to soft-membranous, fragmenting irregularly, red-brown. CATAPHYLLS coriaceous or more or less fibrous, green or purplish above the ground. LEAVES five to seven, mostly basal, linear to linear-lanceolate, (2–)5–14 mm wide, usually reaching to the base of the spike, sometimes slightly longer, the upper two or three leaves inserted on the stem and shorter that the basal. STEM sometimes with one, rarely two, branches, (2–)4 mm in diameter at the base of the spike.

SPIKE (3–)6- to 14-flowered; BRACTS green or more often flushed dark purple, especially those on the lower part of the spike, often very large, (18–)40–70 mm long, those on the upper part of the spike 15–50 mm. FLOWERS scarlet, with the throat and sometimes the base of the lower three tepals yellow, the tube externally red above, bright green below, often whitish or red on drying; PERIANTH TUBE (18–)30–40 mm long, the lower part erect, slender and cylindric, enclosed in the bracts, (7–)15–20 mm long, fairly abruptly expanded into a horizontal, cylindrical upper part (10–) 15–20 mm long, (3.5–)5–7 mm in diameter; TEPALS subequal or the dorsal slightly longer, ovate, (12–) 26–35 mm long, (4–)12–14 mm wide, the dorsal extended horizontally, other tepals spreading at more or less right angles to the tube. FILAMENTS (12–)18–25 mm long, exserted for (4–)10–12 mm from the tube, extended horizontally; ANTHERS (5–)10–12 mm long, yellow. OVARY (4–)6–7 mm long; STYLE usually exceeding the anthers and dividing beyond them (in small-flowered plants sometimes dividing opposite the anthers), branches 3.5–5 mm long. CAPSULES obovate to ellipsoid, (12–)25–30 mm long; SEEDS (4–)6–8 × 3–6 mm, the wing broad. CHROMOSOME NUMBER $2n = 30$.

FLOWERING TIME. Mostly August to October but apparently throughout the year; small-flowered plants mostly found in bloom September to November.

DISTRIBUTION & HABITAT

Gladiolus watsonioides is restricted to the higher mountains of Kenya and northern Tanzania. In Kenya it is common on Mt. Kenya, in the Aberdare Mountains, and in the Cherangani Hills, all of which reach above 3000 m. Localities in Tanzania include Kilimanjaro, Mt. Meru, Loolmalasin Peak, and Mt. Hanang, all also 3000 m or higher, and the slightly lower Ngorongoro Crater. Nearly all collections are from above 3000 m but occa-

FIGURE 52. *Gladiolus watsonioides.* Corm, leaves, and flowering spike, × 0.5; single flower and vertical section, full size (*D'Arcy 7318, Schlieben 4475*).

sionally as low as 2500 m. Plants grow on open slopes in the heath (Ericaceae) zone above the forest, sometimes in lava rubble, and in glades in juniper forest at lower elevations.

Smaller-flowered plants, sometimes treated as variety *minor*, are recorded only from the Kilimanjaro Massif in Tanzania and the Aberdare Mountains in central Kenya, where they occur in similar habitats. The small size may be related to local or seasonal conditions. Available evidence does not support their recognition as a separate taxon.

DIAGNOSIS & RELATIONSHIPS

The flower of the typical form of *Gladiolus watsonioides* bears a remarkable resemblance to that of some southern African species referred in the past to the genus *Homoglossum* (a minor segregate of *Gladiolus*), and also to some species of *Watsonia*, in the deep red color and long perianth tube with a slender lower part and a wide cylindrical upper part. Its several broad, plane leaves and firm-membranous corm tunics are quite different, and there can be no doubt that they are not closely related. The similarity of their flowers is due to convergence, presumably for bird pollination.

Although sometimes formally recognized, either as a separate species, *Antholyza gracilis*, or *Gladiolus watsonioides* var. *minor*, the few small-flowered and short-bracted plants collected on the Aberdare Mountains and on Kilimanjaro, are included here in *G. watsonioides*. Those plants have narrow leaves, flowers no more than 32–38 mm long, and tepals 10–14 mm long. This is smaller by more than half than the dimensions of the smallest-flowered specimens of typical *G. watsonioides*, in which the tepals are 30–35 mm long and the whole flower 65–75 mm long. Accommodation of so great a range of perianth size in a single species is unusual, especially as there is no evidence of clinal variation in *G. watsonioides*, or of a continuous range of variation between the larger- and smaller-flowered variants, as claimed by Hedberg (1957). Provisionally, I assume that the small-flowered plants are depauperate as a result of unusual local seasonal or soil conditions. Their true significance remains to be more fully investigated.

HISTORY

Gladiolus watsonioides was described by J. G. Baker in 1886 from plants collected by Joseph Thomson and Henry H. Johnson on Mt. Kilimanjaro in 1884. A second collection of the species, from Mt. Kenya, named *G. mackinderi* by J. D. Hooker in 1902, and a third, from the Aberdare Mountains in central Kenya, named *Antholyza speciosa* by C. H. Wright in 1914, differ hardly at all from the type of *G. watsonioides*. Those plants also have spikes of up to 12 bright red, long-tubed flowers, and the long perianth tube, which is fairly narrow below and abruptly widened 15–20 mm above the base into a broad cylindrical upper portion. The resemblance of the tube to that of *Antholyza* and *Homoglossum* (= *Petamenes*) resulted in the confusing transfer of *G. watsonioides* first to *Antholyza* (Baker, 1898) and then to *Homoglossum* (Brown, 1932). *Antholyza* is a synonym of *Babiana* (Goldblatt, 1990b), and *Homoglossum* is regarded as congeneric with *Gladiolus*

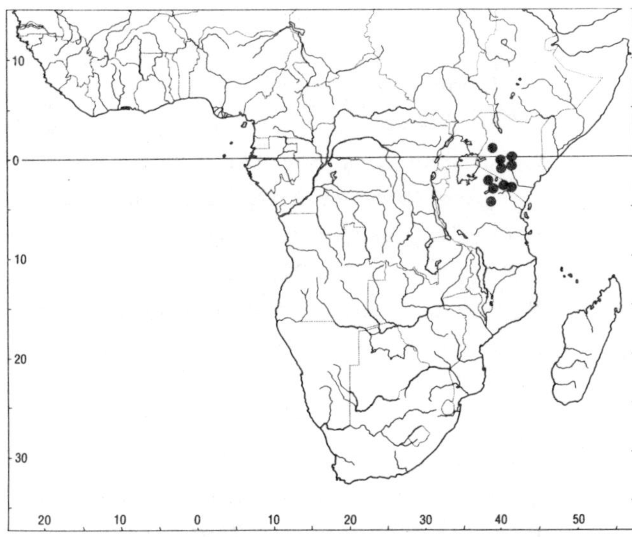

MAP 65. *Gladiolus watsonioides.*

(Goldblatt & de Vos, 1989). *Gladiolus watsonioides* is almost certainly not related to the southern African species that were at one time included in *Homoglossum* or *Petamenes.* The species was transferred back to *Gladiolus* by Hedberg in 1957, where it has since remained. The small-flowered variant of *G. watsonioides* was described independently a few months apart in 1892 as *G. watsonioides* var. *minor* by J. G. Baker and as *Antholyza gracilis* by Ferdinand Pax.

SELECTED SPECIMENS

Kenya. Eastern: Mt. Kenya, Sagana Valley, marshy moorland, 3200 m, 23 Aug. 1949, *Schelpe 2781* (BM, MO, S); Mt. Kenya, ridge south of Teleki valley, rock crevice, 3900 m, 12 Aug. 1948, *Hedberg 1891* (K, S, UPS); Central: Aberdare Mountains, Kinangop, Jan. 1932, *Dale 2730* (K), 17 July 1948, *Hedberg 1638* (S, UPS); Cherangani Hills, Embobub forest, 3500 m, Jan. 1971, *Tweedie 3884* (K); Gilgil, c. 2000 m, Nov. 1933, *Mainwaring 6017* (K); western Aberdare Mountains, glades on Mt. Kipipiri, Kinangop Plateau, 2700 m, Sept. 1934, *Gardner 3232* (K); Marangu (Fort Hall) District, The Elephant, 3300 m, 5 Sept. 1965, *Gillett 16900* (K).

Tanzania. Kilimanjaro: Mt. Kilimanjaro, 19 Feb. 1934, *Schlieben 4808* (BM, BR, LISC, PRE, S, Z); Rongai Caves, 3500 m, 16 Mar. 1962, *Gower 42* (EA, K); Kilimanjaro, Marungu, 2700 m, Nov. 1893, *Volkens 1308* (G, K, PRE). Arusha: Ngorongoro Conservation Area, Empakaai Crater, c. 3200 m, Sept. 1977, *Raynal 19208* (P); Meru Crater, 2520 m, among lava boulders, Dec. 1965, *Richards 20887* (BR, K, P); Mt. Meru, Najuka Camp, 4 Feb. 1968, *Richards 23058* (MO); Mt. Meru, northeast rim of crater in *Erica* heath, Oct. 1977 (flower and fruit), *Raynal 19435* (BR, K, P); eastern slopes of Mt. Meru, 3300 m, upper ericaceous belt, 27 Oct. 1948, *Hedberg 2305* (S, UPS); Mt. Hanang, 3300 m, 2 Sept. 1932, *Burtt 4009* (K); Monduli, Loolmalasin Peak, 3500 m, 17 Sept. 1932, *Burtt 4212* (K).

63. *Gladiolus longispathaceus* Cufodontis

FIGURE 53, MAP 66. Cufodontis, Senck. Biol. 1: 248 (1969); Enum. Pl. Aethiopiae Sperm. 1590 (1974). Type: Ethiopia, Gamu Gofa, Mt. Dita, 3000 m, 22 Aug. 1955, *Kuls 720* (FR, holotype).

EPONYMY

longispathaceus, "with long spathes," alluding to unusually large, red-flushed floral bracts, sometimes up to 10 cm long.

DESCRIPTION

Plants (45–)60–90 cm high. CORM 15–30 mm in diameter, tunics softly membranous, breaking irregularly into thin strips, rarely becoming almost fibrous, straw-colored, usually with numerous small cormlets clustered around the base. CATAPHYLLS often purple, the upper reaching 10–15 cm above the ground and then usually green. LEAVES five or six, the lower four or five more or less basal and largest, the upper one or two cauline and reduced, narrowly lanceolate to nearly linear, plane, reaching at least to the base of the spike, sometimes slightly exceeding it, 7–15 mm at the widest, fairly firm-textured but the margins and midribs not thickened. STEM erect, unbranched, 3–4 mm in diameter at the base of the spike.

SPIKE 8- to 12-flowered; BRACTS very large, green or flushed red above or almost entirely reddish, the outer 6–8(–10.5) cm long, the inner about two-thirds as long as the outer. FLOWERS bright red, the lower three tepals yellow; PERIANTH TUBE 35–40 mm long, slender and erect in the lower 20–25 mm, expanding and curved outward into a cylindrical, more or less horizontal upper part c. 15 mm long; TEPALS unequal, the dorsal largest, extended nearly horizontally, 32–35 mm long, broadly ovate, 20–22 mm wide, the upper laterals also broadly ovate, c. 28 mm long, the lower tepals reduced, the lower laterals lanceolate,

FIGURE 53. *Gladiolus longispathaceus*, × 0.5 (*Friis, Gilbert, & Vollesen 3809*).

c. 15 mm long, the lowermost ovate, 18–23 mm long. FILAMENTS 27–35 mm long, exserted 12–15 mm from the tube; ANTHERS 7–10 mm long. OVARY c. 5 mm long; STYLE arched over the stamens, dividing just beyond the anther apices of the anthers, the branches c. 4 mm long, strongly expanded above when unfolded. CAPSULES broadly ovoid, 10–14 mm long; SEEDS angular with reduced winglike extensions at one or both ends, 3–4 mm long.

FLOWERING TIME. Mainly in August, according to scanty records, and also in early November, probably from July to mid November.

DISTRIBUTION & HABITAT

Gladiolus longispathaceus is restricted to the high mountains of southern Ethiopia above 2400 m. Most collections are from the Mendebo Mountains in northwestern Bale Province, but the type collection and one other record are from the Gamu highlands to the west in Gamu Gofa. Plants grow in moist sites, often along streams, and typically at elevations above 2400 m, although occasionally somewhat lower. The Gamu Gofa collections are from higher elevations, 3000 m on Mt. Dita, and 4200 m on Mt. Guge (Mt. Tola), which is close to the summit of this, the highest peak in the province.

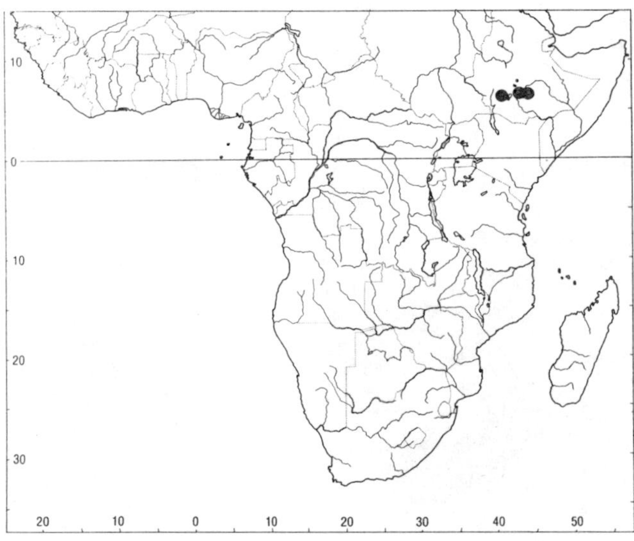

MAP 66. *Gladiolus longispathaceus.*

DIAGNOSIS & RELATIONSHIPS

In general aspect *Gladiolus longispathaceus* closely resembles a somewhat large-bracted form of the central and northern Ethiopian *G. abyssinicus.* Not only are the plants generally more robust and the bracts consistently longer, but the structure of the perianth differs considerably. The three upper tepals are about equal in length, 28–35 mm long and 14–15 mm wide (to 20 mm wide in the plants from Mt. Guge), and the lower tepals are 15–18 mm long with the lowermost longer than the lower laterals except in specimens from Mt. Tola. In *G. abyssinicus* the upper lateral tepals are about half as long as the dorsal and thus 12–20 mm long, and the lower lateral tepals are 8–15 mm long, always somewhat longer than the lowermost, which is often reduced to a short linear cusp.

Aspects of *Gladiolus longispathaceus* also recall the Ugandan *G. dichrous,* both in the broad leaves, the large reddish floral bracts, and the small broadly ovoid capsules. The flowers of *G. longispathaceus* are much larger than in that species. The three large upper tepals resemble those of another eastern African high-mountain species, *G. watsonioides,* but the comparatively small lower tepals, about half as large as the upper laterals, are bright yellow. No doubt *G. longispathaceus* is closely related to both *G. abyssinicus* and *G. watsonioides,* and even in a sense intermediate between them. This whole group of species, including *G. dichrous,* seems to have diversified in the northeastern African high mountains from common ancestral stock, possibly most closely resembling *G. watsonioides.*

SELECTED SPECIMENS

Ethiopia. Bale: Mendoyu Awraja, Harenna Forest, c. 3 km south of Rira on road to Dello

Mena, 2800 m, 10 Aug. 1986, *Mesfin T. 5108–5127* (ETH); Dello Awraja, Harenna Forest, c. 45 km on Dello Mena–Goba road, 2400 m, 3 Aug. 1986, *Mesfin T. 4819–4866* (ETH); Togona, Micha basin, 3.8 km south-southwest of Goba, Aug., *Herbert 69* (K); between Debbivan and Wolchefit Pass, 3000 m, 7 Nov. 1952, *Scott 253* (K); 50 km north of Dolo Menna on Goba road, 2750 m, 5 Nov. 1984, *Friis, Gilbert, & Vollesen 3809* (K); Dinshu, c. 160 km east of Shashamane on Goba road, c. 3200 m, 25 July 1970, *de Wilde 6817* (BR, MO, WAG). Gamu-Gofa: Gamu highlands, Mt. Guge (Tola), within 6 m of the summit, 4200 m, in 1968, *Mulvaney 23* (K).

64. *Gladiolus abyssinicus* (Brongniart ex Lemaire) Goldblatt & de Vos

PLATE 35, FIGURE 54, MAP 67. Goldblatt & de Vos, Bull. Mus. Nat. Hist. Nat., Sér. 4, Sect. B, Adansonia 11: 425 (1989). *Antholyza abyssinica* Brongniart ex Lemaire, Fl. Universel 4: 207 (1845); A. Richard, Tent. Fl. Abyss. 2: 306 (1851). Baker, Handbook Irideae 231 (1892); Fl. Trop. Africa 7: 375 (1898). *Oenostachys abyssinica* (Brongniart ex Lemaire) N. E. Brown, Trans. Roy. Soc. S. Africa 20: 280 (1932). Cufodontis, Enum. Pl. Ethiopiae Sperm. 1591 (1974). Collenette, Fl. Saudi Arabia 264, fig. (1985). Type: Ethiopia, Mt. Selleuda (Sholoda), Oct. 1939, *Quartin-Dillon s.n.* (P, lectotype designated by Goldblatt & de Vos, 1989; BR, P, isolectotypes); near Adwa (Adoua), *Quartin-Dillon s.n.* (P, syntype); "Prov. Shiré," Aug. 1839, *Quartin-Dillon s.n.* (P, syntype).

SYNONYMY

Petamenes latifolius N. E. Brown, Trans. Roy. Soc. S. Africa 20: 277 (1932), new name for *Antholyza abyssinica* sensu Baker, Fl. Trop. Africa 7: 375 (1898). Type: Ethiopia, Hedja (Mt. Hedscha, 2800 m), date unknown, *Schimper s.n.* (K, lectotype designated here); Gaffat, 10 Oct. 1863 (or without locality or date), *Schimper 1206* (E, K, syntypes).

EPONYMY

abyssinicus, "from Abyssinia," the old name for Ethiopia.

DESCRIPTION

Plants 45–65 cm high. CORM 15–25 mm in diameter, tunics soft-membranous, breaking irregularly into thin strips, rarely becoming almost fibrous, straw-colored, often with numerous small cormlets concealed around the base. CATAPHYLLS pale and membranous, the upper reaching above the ground and then green. LEAVES five or six, the lower four or five more or less basal and largest, the upper one or two cauline and reduced, narrowly lanceolate to nearly linear, plane, reaching at least to the base of the spike, sometimes slightly exceeding it, 7–15 mm at the widest, fairly firm-textured but the margins and midribs not thickened. STEM sometimes with one branch, usually 3–4 mm in diameter at the base of the spike.

SPIKE 8- to 12-flowered; BRACTS usually very large, firm, green or more often flushed red above or almost entirely, (35–)45–60(–70) mm long, the outer twisted to lie between axis and flower, glossy within, the inner about half as long or less. FLOWERS red on the upper three tepals, greenish and tipped with yellow on the lower, the throat and tube yellowish, in life sometimes only the upper tepals exposed; PERIANTH TUBE 27–32 mm long, the lower part slender and erect, c. 15 mm long, expanding and gradually curved outward into a cylindrical, more or less horizontal upper part, 12–16 mm long; TEPALS very unequal, the dorsal largest, extended nearly horizontally, (20–)24–35 (–40) mm long, up to 14 mm wide, the upper laterals directed forward, lanceolate, 12–20 × 12 mm, the lower tepals reduced, the laterals lanceolate, 8–15 mm long, the lowermost nearly linear 6–12 mm long. FILAMENTS 25–30 mm long, exserted for up to 15 mm; ANTHERS 8–12 mm long. OVARY c. 5 mm long; STYLE dividing near to or slightly beyond the apices of the anthers, the branches c. 4 mm long, much expanded in the upper half. CAPSULES obovoid-ellipsoid, 10–12 mm long; SEEDS angular with reduced winglike extensions at one or both ends, c. 4 mm long.

FLOWERING TIME. August to October in Ethiopia, following the rainy season of July to November; March and April in Arabia, following the winter rains of November to February.

DISTRIBUTION & HABITAT

Gladiolus abyssinicus has its center of distribution in the highlands of central Ethiopia, where it extends from the western edge of the Rift Valley in Shoa Province through Welo, Gojam, and Tigre, to the Semien Mountains of northern Gonder. Outlying populations have also been recorded in eastern Ethiopia and western Saudi Arabia. In eastern Ethiopia, records are from the Gara Mullata Mountains in Harerge Province. The presence of the species in Saudi Arabia was only established in 1974 by the English botanist, Sheila Collenette. She has subsequently found the species at several sites in the southwest of the country in the mountainous escarpment southeast of Jiddah. There are no reports of *G. abyssinicus* from the southern extension of this escarpment in western Yemen, al-

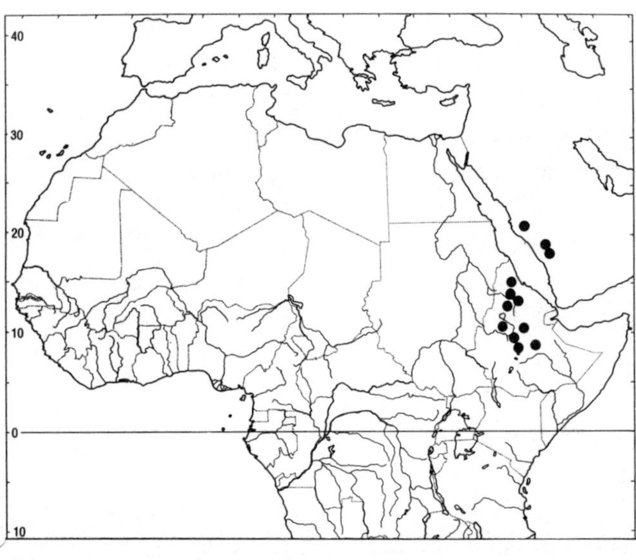

MAP 67. *Gladiolus abyssinicus.*

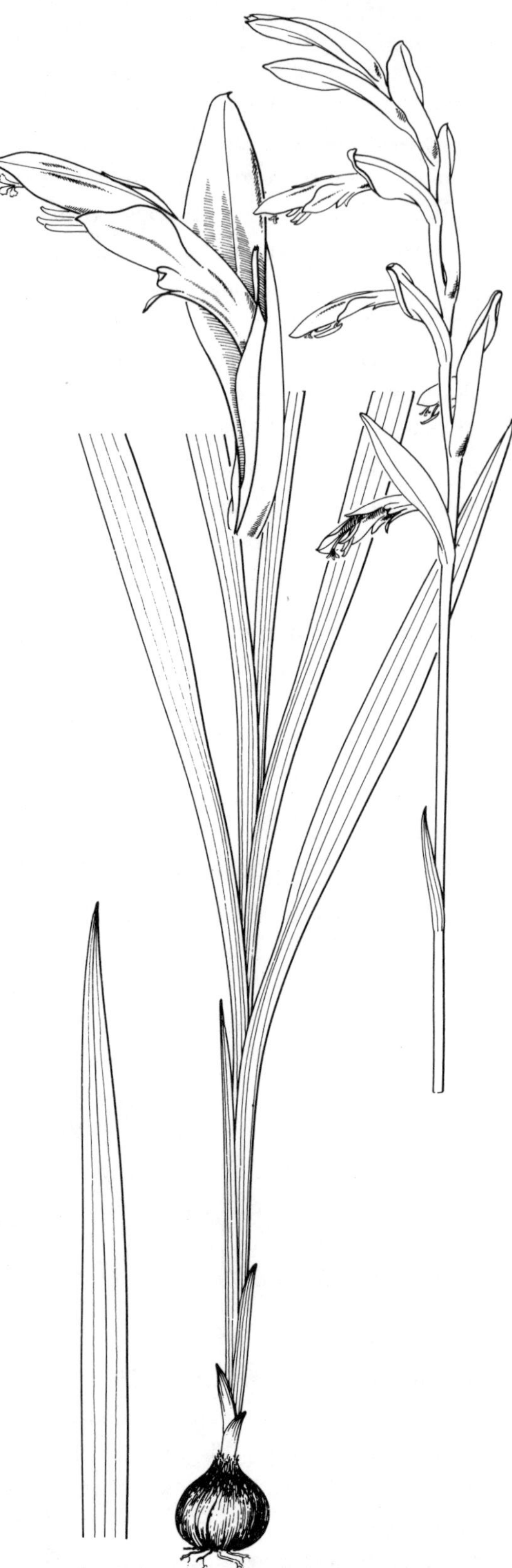

FIGURE 54. *Gladiolus abyssinicus.* Corm, leaves, and flowering spike, × 0.5; single flower and bract, full size (*Collenette 3492*).

though its presence there would not be surprising. Plants favor damp rocky sites where the ground remains wet for a longer time than in the open.

DIAGNOSIS & RELATIONSHIPS

A strongly zygomorphic flower with the dorsal tepal about twice as long as the upper lateral tepals and the lower three tepals reduced to short cusps, in combination with the long floral bracts, distinguish *Gladiolus abyssinicus* from most other species of the genus. Most closely related to *G. abyssinicus* and easily confused with it is *G. schweinfurthii*. The two have similarly shaped flowers, much smaller in the latter. Flowers vary considerably in size and may be 70–80 mm long in robust plants from the north of its range (e.g., *Schimper 767, 842*; *de Wilde & Gilbert 86*), but most collections from Shoa to the south have flowers 50–60 mm long. Depauperate plants or those with the flowers not fully open may appear even smaller, 40–50 mm long, and then resemble those of the related *G. schweinfurthii* (e.g., *de Wilde 8159*). It is sometimes difficult to distinguish the smaller-flowered plants of *G. abyssinicus* from *G. schweinfurthii*, and it is possible that the two are not separable at species rank. Most collections that can confidently be referred to *G. schweinfurthii* are from relatively low elevations either in Eritrea or Somalia, below 2000 m, and have small flowers 25–36 mm long, dorsal tepals 12–18(–22) mm long and floral bracts 18–24 (–28) mm long. Typical *G. abyssinicus* usually occurs above 2000 m, either in the Ethiopian highlands or in western Saudi Arabia, and has flowers at least 40 mm long, dorsal tepals 20–24 mm long, and bracts 35–70 mm long. Additional field and laboratory study is required before the status of *G. schweinfurthii* can be settled, but provisionally I continue to recognize the species.

HISTORY

Gladiolus abyssinicus, called *azerazaei* or *enzerezei* in Amharic, was first discovered by Europeans in 1770 when the explorer, James Bruce, and the artist who accompanied him, Luigi Balugani, found it growing in the northern Ethiopian highlands. Balugani's drawing was brought back to Europe by Bruce, but unlike some of Balugani's other paintings and drawings, it did not receive any attention. Its existence was for all purposes unknown until 1991 when Balugani's works were finally published (Hulton et al., 1991). The second collection, by the French explorers, Quartin-Dillon and Petit, was made in 1839. Plants that flowered later in Paris constituted the type for Brongniart's *Antholyza abyssinica*. That name was only validly published in 1845 by Charles Lemaire, by which time additional collections of the species had been made by Georg Schimper. One of Schimper's collections later became the type of N. E. Brown's *Petamenes latifolius*. Understanding of *G. abyssinicus* was still so poor in 1932 that Brown considered what is a single species here to be two species in two different genera. Brown transferred *Antholyza abyssinica* to *Oenostachys*, a genus distinguished from *Gladiolus* only by the very large and reddish flower bracts.

SELECTED SPECIMENS

Ethiopia. Gonder: near Adi Arca, edge of Semien Mountains, 2400–3000 m, 29 Aug. 1952, *Ferguson s.n.* (K); Semien Mountains, valley off main Geech–Ambaras Valley, 3360 m, 19 Sept. 1969, *de Wilde & Gilbert 86* (ACD, MO, UPS, WAG). Gojam: basalt slopes, 2550 m, 1 Oct. 1973, *Tewolde & Getahun 2469* (ETH). Tigre: Debr-Eski, toward Woina, 2000–2700 m, 4 Oct. 1852, *Schimper 767* (P); Saona, toward Atséga, 2700–3000 m, 5 Sept. 1852, *Schimper 842* (BM, G, K, P). Shoa: between Maigoigoi and Dobre Sina, probably in 1839, *Quartin-Dillon & Petit 114 (240b)* (P); British Embassy hill, Addis Ababa, 2500 m, 25 Sept. 1954, *Mooney 5807* (BR, ETH, K); Mt. Zuquala, c. 60 km south of Addis Ababa, c. 2900 m, 30 Oct. 1965, *de Wilde 8590* (BR, C, ETH, K, MO, P, WAG); Mt. Entoto, c. 25 km northwest of Addis Ababa on Blue Nile road, c. 2500 m, 9 Oct. 1965, *de Wilde 8159*

(BR, C, K, MO, P, UPS, WAG). Welo: Dessie Pool near Addis Ababa, 15 Oct. 1950, *Archer 8058* (K). Harerge: south face of Gara Mullata Mountain, c. 50 km west of Harar, 24 Sept 1961, *Burger 1024* (ACD, F, US).

Eritrea. Dolgollo (Donkollo), Ghinda, 1000 m, 10 Feb. 1891, *Schweinfurth 126* (G, K).

Saudi Arabia: 2 km north of An Numas, in 1974, *Collenette 74/278* (K); Jabal Ibrahim, off the Taif–Abha road, damp crevices, 2100 m, 27 Mar. 1982, *Collenette 3492* (K); Tanuma, c. 1600 m, 17 Mar. 1980, *Lavranos & Collenette 18386* (MO); South Hejaz, Jabal Ibrahim, 2300 m, Apr. 1982, *Lavranos & Collenette 20357* (MO).

65. *Gladiolus schweinfurthii* (Baker) Goldblatt & de Vos

MAP 68. Goldblatt & de Vos, Bull. Mus. Nat. Hist. Nat., Sér. 4, Sect. B, Adansonia 11: 425 (1989). *Antholyza schweinfurthii* Baker, Gard. Chron., Ser. 3, 15: 588 (1894); Fl. Trop. Africa 7: 375 (1898). J. D. Hooker, Curtis's Bot. Mag. 126: pl. 7709 (1900). *Petamenes schweinfurthii* (Baker) N. E. Brown, Trans. Roy. Soc. S. Africa 20: 277 (1932). *Homoglossum schweinfurthii* (Baker) Cufodontis, Bull. Jard. Bot. Nat. Belgique 42(3), Suppl. (1972), Enum. Pl. Aethiopiae Sperm. 1591 (1974). Type: Eritrea, without precise locality, 1890, *Schweinfurth s.n.* (K, presumed lectotype so designated by Cufodontis, 1972: 1591). A lectotype evidently designated by Diels in 1915, *Schweinfurth 143,* was collected at Ambelaco near Maldi, 14–18 Feb. 1894, and is not the material collected in 1890 and available to Baker when he drew up the description. The specimen, presumed to be at B but not explicitly so stated, is evidently no longer extant.

EPONYMY

schweinfurthii, commemorating Georg Schweinfurth, the 19th century botanist and explorer in northeastern Africa.

DESCRIPTION

Plants (30–)50–75 cm high. CORM 8–15 mm in diameter, tunics soft-membranous, fragmenting irregularly into narrow vertical strips, rarely becoming almost fibrous, light brown. CATAPHYLLS pale and membranous, green above the ground. LEAVES usually four or five, rarely three, at least the lower two basal and largest, half to two-thirds as long as the stem, not reaching the base of the spike, the blades plane, lanceolate to nearly linear, 4–12(–20) mm at the widest, usually rather soft-textured (e.g., *Pappi 1677*), the margins and midribs not thickened, the upper one or two leaves cauline and much shorter than the basal. STEM simple or with one or two branches, c. 3 mm in diameter at the base of the main spike.

SPIKE 2- to 7(–12)-flowered; BRACTS pale green or flushed red to purple, the outer 18–24(–28) mm long, the inner nearly or slightly less than half as long as the outer. FLOWERS bright red to orange-red on the upper tepals, greenish fading to yellow on the lower tepals, the throat, and tube, in life the tube included in the bracts; PERIANTH TUBE 11–16 mm long, the lower part slender and erect, 5–8 mm long, fairly abruptly expanded and curved outward into a cylindrical, more or less horizontal upper part, 6–8 mm long; TEPALS very unequal, the dorsal largest, extended horizontally 12–18 (–22) mm long, the upper laterals directed forward, lanceolate, 8–12(–14) mm long, the lower tepals reduced, the laterals narrowly lanceolate, 6–8 mm long, the lowermost a linear cusp 3–6 mm long. FILAMENTS 16–20 mm long, exserted 5–8 mm from the tube; ANTHERS 4.5–8 mm long, reaching 1–2 mm below the apex of the upper tepal. OVARY 3–4 mm long; STYLE ultimately reaching near to the apices of the anthers, the branches 3–4 mm long, extended beyond the anthers and much expanded above. CAPSULES globose-ovate, (7–)9–12 mm long; SEEDS c. 2.5 mm long, somewhat angular, with vestigial wings at the longer ends.

FLOWERING TIME. Mainly August to October.

DISTRIBUTION & HABITAT

Gladiolus schweinfurthii is restricted to northeastern Africa. It is most common in Eritrea, but there are records from Ethiopia, northern Somalia, and western Kenya. In Eritrea, *G. schweinfurthii* occurs at elevations from 750 to 2600 m. A record from Arusi Province, Ethiopia, rather distant from other localities for the species, may simply be depauperate *G. abyssinicus,* but it seems to match *G. schweinfurthii* closely and is provisionally included here. Somalian records are from the Al Hills inland from the north coast. There are only two Kenyan records, one from the Cherangani Hills and the other from Gilgil, both a considerable distance from other stations. Flower size and structure in the Kenyan specimens are completely consistent with other *G. schweinfurthii,* although the bracts are somewhat larger than normal for the species. The flowers on the plants from Kenya are poorly preserved, and that makes it difficult to say more about these populations.

DIAGNOSIS & RELATIONSHIPS

The small, strongly zygomorphic flowers, 25–36 mm long, and relatively short floral bracts, 18–24 (–28) mm long, usually reddish in color, are generally diagnostic for *Gladiolus schweinfurthii.* The prominent dorsal tepal is 12–18(–22) mm long and nearly twice as long as the upper lateral tepals, whereas the lower tepals are 3–8 mm long, the median lower being particularly short and cusplike. The species is closely related to the larger-flowered *G. abyssinicus* and is sometimes confused with it because of the similar structure of the flowers. In fact, it is not entirely clear whether the two are actually separate species. There is, however, an apparent morphological discontinuity between the plants assigned to the two, and in the absence of evidence to the contrary I continue to recognize *G. schweinfurthii. Gladiolus abyssinicus* can almost always be distinguished by its larger flowers with the dorsal tepals 20–24 mm long (shrinkage in dry flowers may be 40%, making floral dimensions sometimes an unreliable guide to distinguishing the two species) and the lower tepals 6–15 mm long, the lower slightly smaller than the lower laterals but narrowly lanceolate rather than cusplike as in *G. schweinfurthii.* The bracts of *G. abyssinicus* are also larger, usually 50–70 mm long, although in depauperate plants sometimes as short as 30 mm, and the outer bracts are twisted in life so that they lie between the flower and the spike axis (Figure 54), a feature apparently not found in *G. schweinfurthii. Gladiolus schweinfurthii* is generally a more slender plant, often with more softly textured leaves and the stem rarely branched, whereas the robust *G. abyssinicus* often has a branched stem, and broader and stiffer leaves. Even what appear to be robust specimens of *G. schweinfurthii* (e.g., *Schweinfurth & Riva 1510,* which has a three-branched stem) have small bracts and flowers.

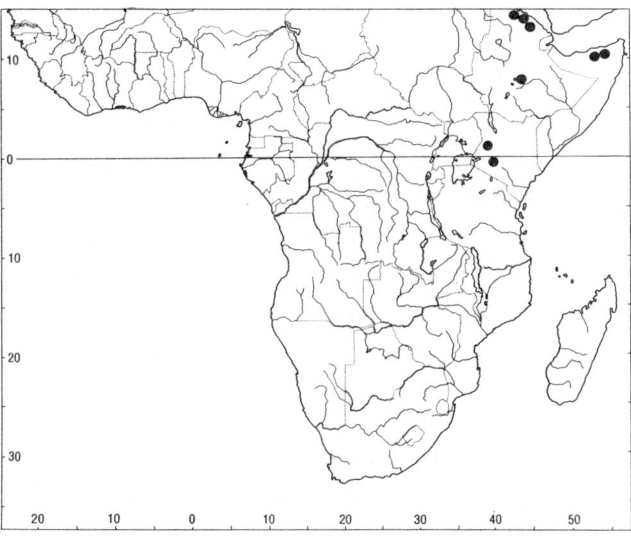

MAP 68. *Gladiolus schweinfurthii.*

HISTORY

First collected in 1890 by Georg Schweinfurth, the species was assigned to the genus *Antholyza* by J. G. Baker, who named the species in Schweinfurth's honor. For a long time the plant was poorly known, but the strongly zygomorphic flowers prompted its transfer to *Petamenes* by Brown (1932) when *Antholyza* was reserved for a South African species related to *Babiana*. After *Petamenes* was shown to be congeneric with the earlier *Homoglossum* (Goldblatt, 1971), Cufodontis (1974) transferred the species to *Homoglossum*. For reasons that are puzzling, Cufodontis treated the closely related *Gladiolus abyssinicus* as a member of the genus *Oenostachys*.

SELECTED SPECIMENS

Eritrea: Ocule Cusai, near Halai, c. 2600 m, 2 Sept. 1902, *Pappi 1677* (G, MO, P); c. 59 km on road from Massawa to Asmara, 750 m, 10 Nov. 1969, *de Wilde 4679* (BR, ETH, MO, WAG); Dongollo, near Ghinda, 12 Mar. 1902, *Pappi 4219* (BR, COI); 16–17 km northeast of Ghinda, 870 m, 9 Feb. 1986, *Mesfin & Sebsebe 3833* (ETH); "Dengergera," north slopes of Mt. Bizen, 2000 m, 10 May 1892, *Schweinfurth & Riva 1865* (fruit) (G, K); Mt. Kube, east of Bizen, 1850 m, 7 Apr. 1892, *Schweinfurth & Riva 1510* (G, K).

Ethiopia. Arusi: 7 km south of Asala, 2182 m, 12 Oct. 1964, *Selassie 400* (ETH). Gojam: Choke Mountains, upper Ghiedeb Valley, 4 Aug. 1957, *Ballantine 406* (K).

Somalia. Bari: escarpment of Bunder Murrayha, Buraha Dhaxsi, crevices in limestone, c. 1150 m, 16–17 Nov. 1986, *Thulin & Warfa 5856* (UPS). Sanaag: Dalo Forest near Erigavo (Ceerigaabo), 4 July 1945, *Glover & Gilliland 1092* (BM, K); Sugli, Al Hills, 1500 m, 13 Nov. 1929, *Collenette 270* (K); Bari, escarpment south of Gunder Muragha, crevices in limestone, 16–17 Nov. 1986, *Thulin & Warfa 5556* (K); Gatun Libeh Forest Reserve, cleared juniper forest on limestone cliff face, 1 Feb. 1973, *Bally & Melville 16215* (K).

Kenya. Rift Valley: Cherangani, 2300 m, Sept. 1935, *Dale 3449* (BR, K, PRE). Central: Gilgil, 1800 m, 18 Sept. 1932, *Hoppe sub Napier 2294* (K).

66. *Gladiolus dichrous* (Bullock) Goldblatt

Map 69. Goldblatt, Bull. Mus. Nat. Hist. Nat., Sér. 4, Sect. B, Adansonia 11: 426 (1989).

SYNONYMY

Oenostachys dichroa Bullock, Kew Bull. 465 (1930). Type: Uganda, Mt. Elgon, 2700–3000 m, 23 Aug. 1921, *Lankester s.n.* (K, holotype).

EPONYMY

dichrous, "bicolored," referring to color of the flowers, which have the three larger upper tepals white and the three smaller lower ones green.

DESCRIPTION

Plants 45–90 cm high. CORM 25–30 mm in diameter, the tunics membranous or more or less fibrous, CATAPHYLLS more or less membranous, or dry. LEAVES about five, mostly basal, at least the uppermost inserted well above the ground, and sheathing for about half its length, the lower leaves reaching to about the base of the spike, narrowly lanceolate, 9–15(–20) mm wide, neither the midrib not the margins noticeably thickened. STEM normally unbranched, occasionally with up to three lateral branches, 4–6 mm in diameter at the base of the spike.

SPIKE (6–)9- to 15-flowered, straight below, arched outward above; BRACTS 35–55(–70) mm long, imbricate, usually red or reddish purple (uniformly white in the only known population from Kenya). FLOWERS either predominantly white, but the lower three tepals greenish at least apically, the distal parts of the upper tepals sometimes flushed reddish, or in plants from northern Uganda and Sudan the tepals bright red to orange; PERIANTH TUBE c. 35 mm long, the lower part straight and slender, the upper part curved at base and horizontal or ascending, c. 20 mm long, also cylindric; TEPALS unequal, the dorsal largest, (12–)14–18 mm long, extending horizontally, upper laterals 8–10 mm long, the lower tepals 5–7 mm long, nearly linear and extending forward. FILAMENTS c. 40 mm long, exserted c. 10 mm from the tube; ANTHERS 7–8 mm long, yellow, ultimately extending beyond the dorsal tepal. OVARY broadly ovoid, c. 5 mm long; STYLE dividing opposite the middle of the anthers, the branches 5–7 mm long and ultimately the apices exceeding the anthers. CAPSULES ovoid, c. 15 mm long; SEEDS discoid, 5–7 × 4–5 mm, broadly winged.

FLOWERING TIME. June to August.

DISTRIBUTION & HABITAT

Although best known from Mt. Elgon in southeastern Uganda where it was first collected, *Gladiolus dichrous* extends through the higher mountains of eastern Uganda into southern Sudan where it occurs in the Imatong Mountains. There is also a record from the Cherangani Hills in western Kenya, where plants seem to match *G. dichrous* except that both the bracts and flowers are white with green highlights. *Gladiolus dichrous* occurs in mountainous habitats, usually above 2500 m. Habitats are usually rocky, and at least sometimes the plants grow on steep cliffs.

DIAGNOSIS & RELATIONSHIPS

The form of the flowers of *Gladiolus dichrous,* with their relatively longer upper tepals and reduced and almost vestigial lower tepals, is exactly like that of two other northeastern African species, *G. abyssinicus* and *G. schweinfurthii,* and it is to those two species that this odd species is most closely related. Both also have relatively large, sometimes very large, floral bracts, usually partly or entirely red or purple, but in neither are the bracts so large that they conceal most of the flower. In some individuals of *G. dichrous,* notable in the type and other specimens from the southern end of its range in the Mt. Elgon area, the bracts are so enlarged that they obscure the flowers and were responsible for the species being referred to a separate genus, *Oenostachys.* The corms, capsules, and seeds are typical of *Gladiolus,* and there seems no doubt that the species correctly belongs in this genus and may be very closely related to *G. abyssinicus.* Interestingly, the two species occupy similar habitats, high mountain sites in fairly wet situations, often in seeps, along streams, or on wet shady cliffs.

Flower color varies across the range of *Gladiolus dichrous.* Plants from the southern part of its range have white flowers, with the upper tepal flushed red above and the lower tepals greenish, but in specimens from southern Sudan all the tepals are entirely red to orange. A population from Kenya, the only record of the species from that country and referred here with hesitation, has pure white flowers and, in addition, white floral bracts. Unfortunately, the white-bracted form is known from only one gathering. More information about that variant is needed before its significance can be properly assessed.

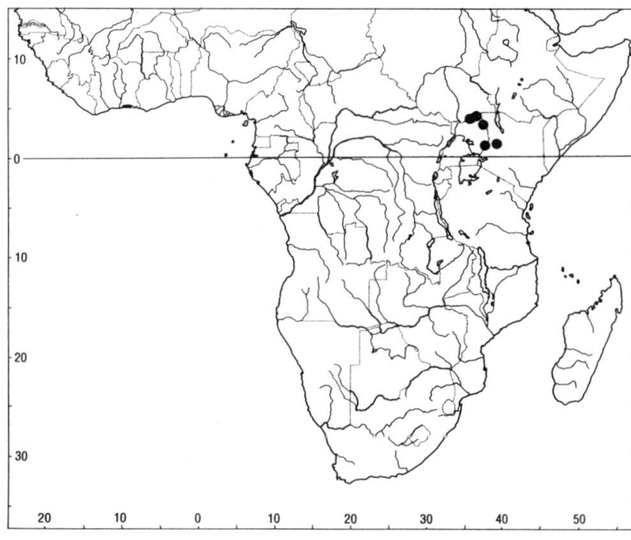

MAP 69. *Gladiolus dichrous.*

SELECTED SPECIMENS

Sudan. Equatoria: near Kipia, Imatong Mountains, stream edge, 2700 m, 26 July 1939, *Myers 11605* (K); Kipia, grassy clearing in *Podocarpus* and *Dombeya* forest, 4 July 1947, *MacLeay 127* (K); Torit, Mt. Kinyeti, near summit, 1 June 1950, *Jackson 1539* (K).

Uganda. Karamoja, Mt. Debasien, near summit, 2800 m, 16 June 1942, *Eggeling 5799* (K); Morongale, c. 2600 m, 16 June 1942, *Dale 289* (BR, K); Mt. Elgon, alpine zone, Aug. 1952, *Saunders & Hancock 34* (K); Mt. Elgon, Bulembuli, Aug. 1937, *Eggeling 3427* (K); Mt. Elgon near stream above bamboo ford, Aug. 1952, *Lind 120* (K).

Kenya. Rift Valley: Cherangani Hills, Kipto, plants flowered in cultivation in Britain, July 1971, *Mabberley & McCall 151* (K).

SECTION *DECORATUS* GOLDBLATT
New Section

Type: *Gladiolus decoratus* Baker.

Latin Diagnosis

Plantae grandes vel mediae, raro parvae, eramosae, foliis lanceolatis ad linearibus, spicis erectis, bracteis (15–)25–40 mm longis saepe tubis perianthii longioribus vel subaequalibus, floribus usitate grandibus (25–)35–90 mm longis, saepe rubris albis notatis, tubis perianthii (8–)15–40 mm longis, tepalis inaequalibus, antheris apiculatis.

Description

Plants medium, or sometimes small, unbranched. LEAVES lanceolate to linear, the upper smaller and sometimes entirely sheathing. STEMS normally straight and erect. SPIKES usually erect; BRACTS fairly long, (15–)25–40 mm long, generally slightly longer, or about as long as the perianth tubes. FLOWERS large to medium (small in *Gladiolus salmoneicolor*), (25–)35–90 mm long, the lower tepals with median white to cream nectar guides; PERIANTH TUBE (8–)15–40 mm long, usually exserted from, but slightly shorter than the bracts; TEPALS usually unequal, the dorsal tepals usually slightly longer than the tubes, occasionally slightly shorter; lower tepals much narrower but about as long as the dorsal. ANTHERS with short to long acute apiculate appendages.

Comprising eight species, section *Decoratus* occurs widely across tropical Africa. Species extend from Ivory Coast in western Africa eastward to Ethiopia and southward to central Mozambique. The plants are usually moderate in size and the flowers large, but both *Gladiolus serenjensis* and *G. salmoneicolor* are small and have moderate-sized flowers. All the species have red to pink or white flowers, usually with yellow or white markings. The leaf texture is unusually soft, and the margins and midribs are not at all, or barely, thickened. The veins are visible only because they are translucent. Anther appendages are present in all species of the section, as they are in section *Acidanthera*.

67. *Gladiolus decoratus* Baker

PLATE 36, FIGURE 55, MAP 70. Baker, J. Bot. 14: 337 (1876); Handbook Irideae 222 (1892); Fl. Trop. Africa 7: 370 (1898). Goldblatt, Fl. Zambesiaca 12(4): 100 (1993). Type: Mozambique, Zambésia, Morrumbala Mountain, from the foot to 600 m,

Dec. 1858, *Kirk s.n.* (K, lectotype designated by Goldblatt, 1993: 100); Morrumbala, 1863 and 1866, *Kirk s.n.* (K, MO, syntypes).

SYNONYMY

Gladiolus zanguebaricus Baker, Kew Bull. 1897: 282 (1897), in passing; Fl. Trop. Africa 7: 365 (1898), new name for *G. kirkii* Baker, Handbook Irideae 222 (1892), an illegitimate homonym for *G. kirkii* Baker (1890) (= *G. ochroleucus* Baker, a South African species). Type: Zanzibar, without date, *Kirk s.n.* (K, holotype; the filaments exserted c. 10 mm, anthers apiculate, and flowers rather tubular and unusually small, 45–50 mm long).

Gladiolus quilimanensis Baker, Fl. Trop. Africa 7: 577 (1898). Type: Mozambique, Zambésia, Quelimane, 10 Feb. 1889, *Stuhlmann s.n.* (B, holotype).

Gladiolus morumbalaensis de Wildeman, Pl. Nov. Herb. Hort. Thennensis 1: 17–20, pl. 5 (1904). Type: Mozambique, Zambésia, Morrumbala, Dec. 1900, *Luja 393* (BR, lectotype designated by Goldblatt, 1993: 100; BR, isolectotype).

EPONYMY

decoratus, "attractive," describing the large bright orange-red flowers with contrasting pale markings.

DESCRIPTION

Plants 45–80 cm high. CORMS 15–25 mm in diameter, tunics initially membranous, becoming more or less fibrous and matted with age, fibers brown, enclosing numerous small cormlets. CATAPHYLLS membranous, green or flushed purple above the ground. LEAVES four or five, fairly soft-textured, the midrib and margins not thickened, the lower three more or less basal and longest, lanceolate to linear, reaching to about the base of the spike, 8–18 (–24) mm wide, the upper two leaves cauline and sheathing in the lower half. STEM unbranched, c. 3 mm in diameter at the base of the spike.

SPIKE erect, 5- to 9-flowered; BRACTS green and soft-textured, the outer sometimes flushed red, 25–40(–50) mm long, the inner two-thirds as long as the outer. FLOWERS bright orange-red or dark red, the lower tepals each with a median white to yellow streak, the marks on the lower laterals spathulate and widest near the tepal apices; PERIANTH TUBE 25–35 mm long, the lower part slender and cylindric, 18–25 mm long, the upper part funnel-shaped, 8–10 mm long; TEPALS lanceolate, unequal, the dorsal largest, 45–55 mm long, 15–20 mm wide, arched over the stamens and style, the lower three united for 10–12 mm, c. 40 mm long, nearly horizontal or tilted toward the ground, in profile the lower equalling or exceeding the dorsal. FILAMENTS 35–40 mm long, exserted 20–25 mm; ANTHERS 10–12 mm long, reaching to about the upper third of the dorsal tepal, with a slender, usually inconspicuous apiculate appendage 0.5–1 mm long. OVARY c. 4 mm long; STYLE arched over the stamens, dividing opposite to or beyond the anther apices, the branches c. 4 mm long, widened and channeled above. CAPSULES elliptic, 18–27 mm long; SEEDS elliptic to oval, 10–12 × 5–6 mm. CHROMOSOME NUMBER $2n$ = 39.

FLOWERING TIME. December to February.

DISTRIBUTION & HABITAT

With a fairly wide distribution, *Gladiolus decoratus* extends from coastal Mozambique near Beira inland into the Shire Highlands of Malawi, and in Tanzania into the Ruaha Valley and the Uluguru Mountains. Along the coast it extends north to Lindi on the mainland and offshore on the large islands of Zanzibar and Pemba Island off the central and northern Tanzanian coast. Largely a species of woodland and forest, it favors rocky outcrops and screes and other more open sites. Its altitudinal range is from close to sea level to 2000 m in the Uluguru Mountains.

FIGURE 55. *Gladiolus decoratus.* Corm, leaves, and flowering spike, × 0.5; single flower, full size (*la Croix 292, 2246*).

Records from Pemba and Zanzibar, a good deal farther north than other records, seem strange, especially as *Gladiolus decoratus* evidently does not occur on Mafia, least disturbed of these islands and lying farthest to the south. The specimens from Pemba accord closely with *G. decoratus* from coastal southern Tanzania and Mozambique and with the species in general in having an arched dorsal tepal, well-exserted filaments, and nearly spathulate tepal markings. Specimens from Zanzibar consist only of the type of what Baker called *G. zanguebaricus.* Those plants have smaller flowers than normal for the species. The anthers are well exserted and the upper tepal is strongly arched. More puzzling are plants from the Ruaha Valley in the dry interior of Tanzania that also match fairly well with typical *G. decoratus* except for the dark red perianth and less pronounced spathulate markings. These differences do not seem to merit taxonomic separation but are not entirely surprising in plants from a habitat so different from that of the populations to the south that occur at either higher elevations or along the more humid coast.

DIAGNOSIS & RELATIONSHIPS

The large red flowers with white to yellow markings on the lower tepals, soft-textured leaves with distinctive venation, and matted fibrous outer corm tunics containing numbers of small cormlets place *Gladiolus decoratus* among a complex of tropical African species that extend from the central coastal belt of Mozambique to Ivory Coast and Ethiopia. Within the group, *G. decoratus* can be distinguished by a combination of bright red perianth, lower lateral tepals with spathulate markings, and strongly arched and hooded dorsal tepal. The anthers are also apiculate, but the rather small appendages are difficult to see and easily fall from the anthers. The similar *G. rupicola,* with which it is often confused, is widespread in moist montane habitats in Tanzania and occurs with *G. decoratus* in the Uluguru Mountains. *Gladiolus rupicola* has a dorsal tepal notably larger than the others, obtuse nonapiculate anthers, and appears to lack cormlets around the base. Also easily confused with *G. decoratus* is *G. oligophlebius,* which has pink flowers with narrow white markings and strongly apiculate anthers. That species can also be distinguished by its shorter filaments and dorsal tepal not or hardly arched so that the perianth does not have the inflated appearance of *G. decoratus.*

Gladiolus morumbalaensis, based on plants from the same part of Mozambique as the type of *G. decoratus,* was described by Emile de Wildeman in 1904. He described it as differing from *G. decoratus* in having broader leaves, 18 mm wide, and broader tepals, the dorsal measuring 20 mm in contrast to 12 mm in *G. decoratus.* Examination of the type specimens of *G. morumbalaensis* and *G. decoratus* reveals no significant differences between them and does not confirm the measurements noted by de Wildeman. The type material of *G. quilimanensis* (Baker, 1898) from coastal Mozambique also matches *G. decoratus* closely.

The western African *Gladiolus mirus,* a species sometimes confused with *G. decoratus,* is discussed in detail under *G. oligophlebius,* to which it appears

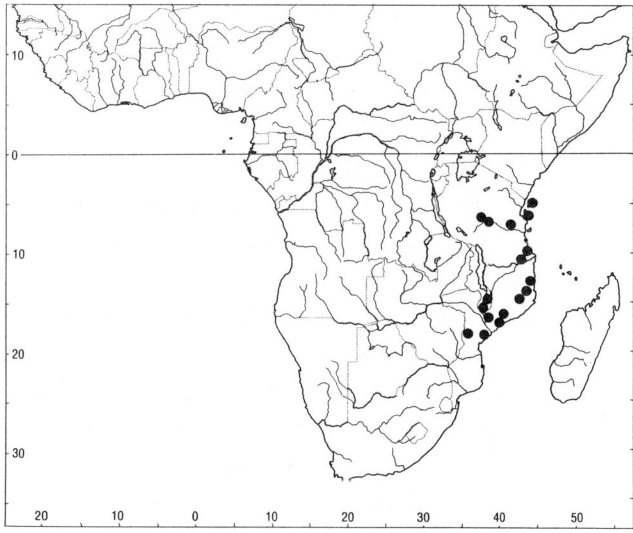

MAP 70. *Gladiolus decoratus.*

to be most closely related. *Gladiolus zanguebaricus,* reportedly from Zanzibar, has unusually small flowers for *G. decoratus,* but they otherwise correspond well with the species, especially in the arched dorsal tepal and filaments exserted about 10 mm.

HISTORY

The first records of *Gladiolus decoratus* were made by the explorer, John Kirk, during David Livingstone's Zambezi Expedition, in 1858. Plants collected near Morrumbala in interior Mozambique were named by J. G. Baker in 1876. Smaller-flowered plants from Zanzibar, also collected by Kirk (the date of collection is not recorded), were described as a separate species, *G. kirkii,* by Baker in 1892. That name had already been used for a southern African species (now *G. ochroleucus*) and Baker renamed the Zanzibar plants *G. zanguebaricus* in 1898. Plants collected by the Belgian collector, Edouard Luja, in 1900, also at Morrumbala, were described as *G. morumbalaensis* by Emile de Wildeman in 1904, and plants collected by Franz Stuhlmann near Quelimane, southeast of Morrumbala, were named *G. quilimanensis* by Baker. Neither of the two latter species differs in any significant way from the type specimens of *G. decoratus,* and it is difficult to see why they should ever have been regarded as separate species.

SELECTED SPECIMENS

Tanzania. Morogoro: Uluguru Mountains, north aspect, 450 m, 28 Jan. 1933, *Schlieben 3301* (B, BM, BR, G, LISC, P, S, Z); 24 km north of Morogoro, 1500 m, 1 Jan. 1931, *Haarer 2005* (K). Lindi: Lake Lutamba, 40 km west of Lindi, 250 m, 7 Jan. 1935, *Schlieben 5836* (B, BM, BR, G, HBG, K, LISC, MO, P, PRE, S, Z). Mtwara: Newala, 17 Dec. 1958, *Hay 9* (K). Pemba: Shengijuu–Pandani, 19 Feb. 1929, *Greenway 1490* (EA, K); Matangatwani, Nov. 1935, *Vaughan 2296* (EA). Iringa: Ruaha–Iringa road, woodland, 5 Feb. 1970, *Brown 1840* (K); Ruaha National Park, between Msembe and Mbage Camp, 840 m, 10 Jan. 1966, *Richards 20947* (K); Ruaha National Park, woodland north of Kilimatonge, 850 m, 5 Jan. 1987, *Lovett 1320* (DSM, K, MO).

Malawi. Southern: Zomba, 1897, *Whyte & McClounie s.n.* (K); Zomba. Old Naisi Road, 23 Jan. 1980, *Chapman 5501* (BR, K, MAL); Zomba, Likangala River, 2 Jan. 1957, *Banda 338* (BM, MAL, SRGH); Bvumbwe, Thyolo District, 5 Jan. 1982, *la Croix 292* (K), 2 Jan. 1983, *la Croix 2246* (BM); Mulange, Dec. 1893, *Scott Elliot 8673* (BM); Midima Mountain near Blantyre, 1 Jan. 1967, *Hilliard & Burtt 4103* (E, MAL).

Mozambique. Cabo Delgado: Porto Amélia to Ancuabe, 200 m, 21 Dec. 1963, *Torre & Paiva 9635* (LISC). Nampula: Eráti, c. 6 km from Namapa to Odinepa, c. 320 m, 12 Dec. 1963, *Torre & Paiva 9533* (LISC); near Nampula, 12 Jan. 1937, *Torre 1195* (LISC); Murrupula, Serra Marrutulo, 13 Jan. 1961, *Carvalho 429* (K, LMU). Manica e Sofala: Dondo near Beira, swamp, Dec. 1899, *Cecil 255* (K); woodland near confluence of the Haroni and Makurupini, 14 Jan. 1969 (fruit), *MacDonald 54* 6; 40 km north of Dondo, Inhaminga road, 120 m, 3 Dec. 1971, *Pope & Müller 503* (K, SRGH). Zambézia: Campo, 27 Dec. 1904, *le Testu 578* (BM, BR, P); km 7, road from Maganja to Maganja da Costa, c. 40 m, 25 Jan. 1966, *Torre & Correia 14,074* (LISC); Namagoa, Mocuba, Dec. 1943, *Faulkner 186* (PRE).

68. *Gladiolus oligophlebius* Baker

PLATE 37, FIGURE 56, MAP 71. Baker, Kew Bull. 73 (1895); Fl. Trop. Africa 7: 367 (1898). Hutchinson, Fl. West Trop. Africa 2: 379 (1936). Hepper, Fl. West Trop. Africa, Ed. 2, 3(1): 144 (1968). Goldblatt, Fl. Zambesiaca 12(4): 100 (1993). Type: Zambia, Mbala (Abercorn), 1893, *Carson 25/1893* (K, holotype).

SYNONYMY

Gladiolus caudatus Baker, Kew Bull. 1895: 74 (1895); Fl. Trop. Africa 7: 367 (1898). Type: Zambia, Fwambo (Lake Tanganyika), 1893, *Carson 19* (K, holotype; the specimens are depauperate and poorly pressed but clearly conspecific with *G. oligophlebius*).

EPONYMY

oligophlebius, "few-veined," referring to the soft-textured leaves that lack conspicuous venation.

DESCRIPTION

Plants 40–80 cm high. CORMS 15–25 mm in diameter, tunics initially membranous, becoming more or less fibrous and matted with age, fibers brown, enclosing numerous small cormlets. CATAPHYLLS membranous, often broken at flowering time, greenish or flushed purple above the ground. LEAVES four or five, fairly soft-textured, the midrib and margins not thickened, the lower three more or less basal and longest, lanceolate to linear, reaching to about the base of the spike, 8–18(–24) mm wide, the upper two leaves cauline and sheathing in the lower half. STEM erect, unbranched, 3–4 mm in diameter at the base of the spike.

SPIKE erect, (2–)5- to 9-flowered; BRACTS pale green and soft-textured, the outer sometimes flushed red, 25–40(–50) mm long, the inner two-thirds as long as the outer. FLOWERS pale to deep pink, the lower tepals each with a median white to yellow streak outlined in dark red or purple, the mark broadest in the center and largest on the lower laterals; PERIANTH TUBE narrowly funnel-shaped, the lower part, 30–40 mm long, slender and cylindric, the wider upper part c. 20 mm long; TEPALS lanceolate, more or less equal in length or the dorsal somewhat longer than the others, 38–45 mm long, the dorsal more or less horizontal, 15–20 mm wide, the lower three joined for c. 5 mm, tilted slightly below horizontal, 8–12 mm wide, in profile usually slightly exceeding the dorsal. FILAMENTS 15–20 mm long, exserted (3–)5–10(–15) mm from the tube; ANTHERS 10–15 mm long, reaching to about the middle of the dorsal tepal, with apiculate appendages c. 1.5 mm long. OVARY c. 4 mm long; STYLE arched over the stamens, dividing at or beyond the anther apices, the branches c. 4 mm long, wider and channeled above. CAPSULES more or less ellipsoid, c. 15 mm long; SEEDS not known.

FLOWERING TIME. Mostly December and January.

DISTRIBUTION & HABITAT

Gladiolus oligophlebius is centered in southern tropical Africa, where it extends from the Lake Malawi basin in southern Malawi, northward through western Tanzania to the southern end of the Lake Tanganyika valley in Zambia. Outlying populations occur in northern Tanzania near Shinyanga, and in the provinces of Shaba and Kasai Oriental, Zaire. These habitats are mostly at fairly low elevations, 500–600 m, in the Lake Malawi and Lake Tanganyika rift valleys, and relatively dry. Plants from the Ufipa Plateau, Tanzania, occur at elevations of 1200 to 1500 m, thus in cooler and wetter conditions. The species is often found in rocky sites in open woodland or sometimes in grassland. The closely related *G. decoratus* grows in similar habitats but in general grows in areas of higher rainfall, sometimes at considerably higher elevations, and also close to sea level in coastal south-eastern Africa.

DIAGNOSIS & RELATIONSHIPS

Gladiolus oligophlebius and *G. decoratus* can generally be distinguished by their soft-textured leaves, large flowers with a pink to red perianth, and anthers with prominent acuminate appendages. The flowers of *G. decoratus* are inflated because of the strongly arched dorsal tepal, and the perianth is a deep red with spathulate cream to yellow markings. Thus the pink perianth with elliptic markings and a weakly or not at all inflated perianth make it relatively easy to distinguish well-preserved specimens of *G. oligophlebius.* In addition, the stamens are generally shorter than in *G. decoratus,* and the filaments are typically only shortly exserted, for 5 to 10 mm, and they may sometimes be entirely included in the tube. In *G. decoratus* the filaments are exserted for 15 to 20 mm.

I am provisionally including an odd collection from Mporokoso, Zambia (*Bredo 5919*), in *Gladiolus oligophlebius.* The single plant has leaf blades with rough, shortly scabrid margins and vein edges, and densely scabrid leaf sheaths and cataphylls. In other respects the plant seems to accord with *G. oligophlebius,* as far as one can tell from the poorly pressed flowers, the anthers of which do have apiculate appendages typical of the species.

Closely allied to *Gladiolus oligophlebius* is the western African *G. mirus,* a species of rock outcrops in woodland or forest extending from Gabon to Ivory Coast. The two species are rather alike in general appearance, even to the large pink flowers with long-apiculate anthers and the rather

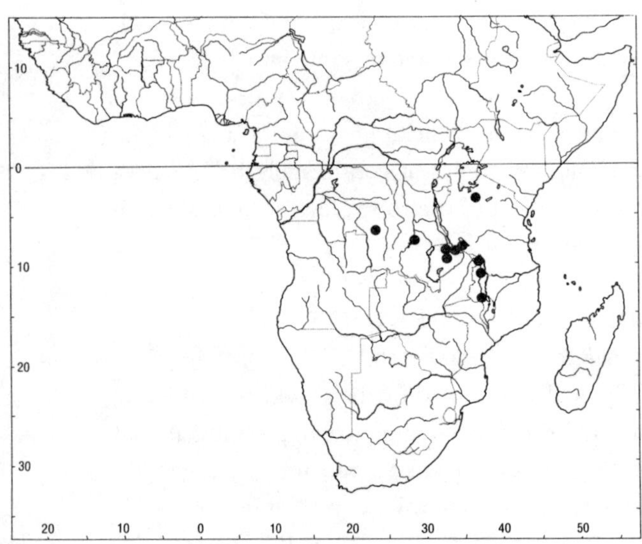

MAP 71. *Gladiolus oligophlebius.*

FIGURE 56. *Gladiolus oligophlebius.* Corm, leaves, and flowering spike, × 0.5; single flower, full size; anther with apiculate appendage, × 2 (*la Croix 2245*).

coarsely fibrous corm tunics that conceal several tiny cormlets. *Gladiolus oligophlebius* has slightly larger flowers and longer anthers than *G. mirus,* and the two can best be distinguished by their stamens, the filaments of *G. mirus* being only 12–15 mm long and extending 3–6 mm from the mouth of the tube, and the anthers, 8–9 mm long, reaching to just above the middle of the dorsal tepal. In *G. oligophlebius* the anthers are (10–)12–15 mm long, and the filaments, 15–20 long, may be exserted for up to 12 mm. Additional collections are needed to determine whether the two species truly merit separate status.

HISTORY

Gladiolus oligophlebius was first collected in 1893 by the Scottish naturalist, Alexander Carson, near Mbala (then Abercorn) at the southern end of Lake Tanganyika. These specimens became the type of the species, described by J. G. Baker in 1898. Plants from Fwambo, a short distance away, and collected by Carson in the same year were named *G. caudatus.* There are no differences of any taxonomic significance between the two collections, and the two species must be considered conspecific. The circumscription of *G. oligophlebius* was expanded by Hutchinson and Dalziel (1936) and Hepper (1968), who used the name for western African plants that had earlier been named *G. mirus* by Vaupel (1913). The two species are clearly very similar, but available material suggests that they are best considered separate species.

SELECTED SPECIMENS

Zaire. Kasai: Kazumba, Katende waterfall, forest edge, Nov. 1921, *Achten 609* (BR, C); Tchimboa, along the Lulua, 26 Dec. 1910, *Callewaert s.n.* (BR); c. 60 km southwest of Kananga (Luluabourg), above the falls at Katende, cultivated at Shangi, Rwanda, *Pierlot 2800* (BR). Shaba: Mitwaba–Manono, km 45, Kalumengongo, 1140 m, 3 Feb. 1986, *Bamps & Malaisse 8618* (BR, MO).

Tanzania. Rukwa: Sumbawanga, Ufipa Plateau, Malasa, rocky island, 780 m, 1 Jan. 1964, *Richards 18743* (BR, K, P, UPS). Shinyanga: Shinyanga, near Manyonga River, 1150 m, 17 Jan. 1937, *Burtt 6397* (BM, BR, K, P, PRE, S).

Zambia. Northern: Kalambo Falls, steep rocky slopes, 8 Feb. 1965, *Richards 19605* (K, MO); Mpulungu, steep bank above Lake Tanganyika, 16 Dec. 1954, *Richards 3650* (BR, K); Kambole Escarpment, 1500 m, 19 Feb. 1957, *Richards 8247* (BR, K); Mbala District, Chilongwelo Ravine, grassy scree, 1440 m, 27 Dec. 1954 *Richards 3753* (BR, K); Issi River Gorge, Mbala, 1200 m, 6 Jan. 1963, *Richards 17519* (EA, K, SRGH). Mporokoso, 6 Jan. 1944, *Bredo 5919* (BR).

Malawi. Northern: Karonga District, Vinthukulu Forest Reserve, 9 Jan. 1990, *Patel & Usi 4555* (MAL, MO); Nkhata Bay, Chikale Beach road, 6 Jan. 1976, *Pawek 10684* (K, MAL, MO, PRE); Nkhata Bay District, 8 km west of Chinteche, grassland, 500 m, 29 Dec. 1978, *Phillips 4487* (K, MO, WAG, Z); Rumphi District, Chiweta–Lura Escarpment, 3 Jan. 1973, *Pawek 6325* (K, MAL, MO, SRGH). Central: Salima, close to the lake shore, 1 Jan. 1963, *Chapman 1782* (K, PRE, SRGH); sand dune woodland, Johnson's Hotel, Salima, 7 Jan. 1953, *Jackson 1016* (BM, K, MAL).

69. *Gladiolus mirus* Vaupel

MAP 72. Vaupel, Bot. Jahrb. Syst. 48: 534 (1913). Type: Cameroon, Ebolowa District, Akumessin, 30 km west of Sangmelina, 2 June 1911, *Mildbraed 5542* (B, holotype; K, photo; HBG, isotypes).

SYNONYMY

Gladiolus staudtii Vaupel, Bot. Jahrb. Syst 48: 538 (1913). Type: Cameroon, Lolodorf, sunny rock faces, June 1895, *Staudt 328* (B, holotype; G, K, P, S, isotypes).

Gladiolus oligophlebius sensu Hutchinson, Fl.

West Trop. Africa 2: 379 (1936), and sensu Hepper, Fl. West Trop. Africa, Ed. 2, 3(1): 144 (1968).

EPONYMY

mirus, "wonderful, remarkable," alluding to the large, brightly colored reddish or pink flowers.

DESCRIPTION

Plants 40–80 cm high. CORMS (10–)18–25 mm in diameter, tunics initially membranous, becoming more or less fibrous externally, fibers straw-colored, mostly vertical. CATAPHYLLS membranous, often partly disintegrated, the uppermost reaching a short distance above the ground and purplish, or dry and brown. LEAVES four or five, the lower three more or less basal and longest, narrowly lanceolate, reaching to the base of the spike, (3–)6–9 mm wide, the upper two leaves cauline, becoming more or less bractlike, all fairly soft-textured, the midrib and margins lightly raised. STEM unbranched, c. 3 mm in diameter at the base of the spike.

SPIKE erect, 5- to 15-flowered; BRACTS green, the outer sometimes flushed red, 20–30(–40) mm long, the inner two-thirds as long as the outer.

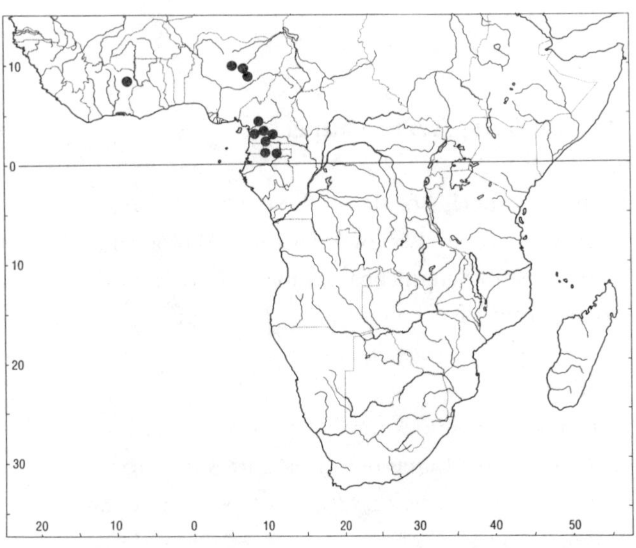

MAP 72. *Gladiolus mirus.*

FLOWER pink to red-purple, the lower tepals each with a median yellow streak outlined in purple; PERIANTH TUBE narrowly funnel-shaped, 30–36 mm long, the slender cylindric lower part 22–25 mm long, the wider upper part 8–10 mm long; TEPALS lanceolate, more or less equal in length, 30–35 mm long, the dorsal held apart, weakly ascending to almost horizontal, 10–12 mm wide, the lower three joined to the upper laterals for c. 3 mm, more or less horizontal, 6–8 mm wide, in profile slightly exceeding the upper. FILAMENTS 12–15 mm long, exserted 3–5 mm; ANTHERS 8–9 mm long, reaching to about the midline of the dorsal tepal, with apiculate appendages c. 1.5 mm long. OVARY 2–3 mm long; STYLE arched over the stamens, dividing near the apex of the anthers, branches c. 2.2 mm long, wider and channeled above. CAPSULE ellipsoid-obovoid, 20–25 mm long, SEEDS ovate, 8–9 × 5–6 mm, broadly winged.

FLOWERING TIME. Mostly June to August in the western part of the range, April to June in the east, and also November and December in Cameroon.

DISTRIBUTION & HABITAT

Widespread but infrequent across western Africa, *Gladiolus mirus* extends from Gabon in the east to Ivory Coast in the west. Plants favor higher elevations in open, usually rocky sites.

DIAGNOSIS & RELATIONSHIPS

Although closely related to the eastern and central African *Gladiolus oligophlebius* and so treated by Hepper (1968), *G. mirus* has somewhat smaller flowers, a less inflated perianth, and shorter filaments and anthers. *Gladiolus staudtii,* also described from southern Cameroon, is probably conspecific; it has smaller flowers, possibly due to shrinkage during drying. *Gladiolus mirus* and *G. oligophlebius* have similar finely fibrous, reticulate corm tunics that enclose numerous small cormlets, large red to pink flowers with a broad pale streak on each of the three lower tepals, and anthers with a slender mucro c. 1.5 mm long. The larger flowers of *G. oligophlebius* have tepals 40–45 cm long, filaments 15–20 mm long, and anthers (10–)15 mm long, all longer than in *G. mirus,* whereas the perianth tubes are the same length in both species. Some populations from northern Nigeria with very narrow leaves and relatively small flowers are probably depauperate forms of the species growing at the edge of its range and in degraded habitats.

HISTORY

Although first collected in the 1890s, both by French and German collectors, *Gladiolus mirus* was described in 1913 from specimens collected in 1911 by the German botanist, G. W. J. Mildbraed. At the same time that he described *G. mirus,* François Vaupel also named *G. staudtii,* this based on an 1895 collection, also from Cameroon. Both species were considered conspecific with the southern tropical African *G. oligophlebius* by Hutchinson and Dalziel (1936) and by Hepper (1968) in the two editions of the *Flora of West Tropical Africa.* As I have outlined above, the two species are indeed closely related, but at least on the basis of available evidence they seem best regarded as separate species.

SELECTED SPECIMENS

Ivory Coast: Upper Volta "Pays du Kong," 1898, *Bouet 2559* (P).

Nigeria. Kaduna: Jarawa Hills east of Federe, 1600 m, 25 Aug. 1962, *Lawler & Hall 353* (K). Bauchi: Panshanu Pass, Ziem Peak, 1300 m, 16 Aug. 1962, *Lawler & Hall 435* (K); Panshanu Pass, 17 July 1970, *Hall, Daramola, & Ekwuno s.n.* (FHI 67425, K, MO). Plateau: Jos Plateau, Amo, 1220 m, 23 Aug. 1956, *King sub Hepper 2870* (K); Mada Hills, Aug. 1926, *Hepburn 76* (K); Tof Peak, Jos Plateau, 1200 m, June 1962, *Williams s.n.* (K).

Cameroon: Akoakas Rock, N'Koemvone to Ambam, 650 m, 11 Nov. 1974, *de Wilde 7695* (YA); Central Province, Yaoundé, top of Akondoi hill, 960 m, 3 June 1987, *Manning 1906* (K, MO, P); Nkoltsia Hill near Gouap, 18 km northwest of Bipindi, 21 Apr. 1974, *Villiers 864* (P, YA).

Gabon: Bélinga, 1000 m, Dec. 1964 (fruit), *Hallé 3494* (K, LBV, MO, P, WAG); Bélinga, iron mines, June 1966, *Hallé 3696* (P); Bengo, meadows on the "rochers de Coss," 12 May 1933, *le Testu 9121* (BM, BR, MO, P); meadow on Ncolayop Rock, 14 May 1933, *le Testu s.n.* (BM, BR).

70. ***Gladiolus sudanicus* Goldblatt,** new species

PLATES 38, 39, FIGURE 57, MAP 73. Type: Ethiopia, Shoa, Blue Nile Gorge near km 205 on Addis Ababa–Debre Marcos road, c. 1050 m, 7 Aug. 1971, *Gilbert & Gilbert 2158* (ETH, holotype; C, EA, K, isotypes).

EPONYMY

sudanicus, "from the Sudan," more generally from sub-Saharan northeastern Africa.

LATIN DIAGNOSIS

Plantae 15–20 cm altae, cormis 10–12 mm diametro, foliis 4–5, inferioribus anguste lanceolatis vel linearibus 5–9 mm latis, caule simplici, spica 2–3 florum, floribus bubalinis tepalis flavoviridis rubrisque notatis, tubo perianthii c. 11 mm longis, tepalis lanceolatis 16—24 mm longis, filamentis 9–11 mm longis, antheris c. 6.5 mm longis violaceis.

DESCRIPTION

Plants 15–20 cm high. CORMS 10–12 mm in diameter, the tunics of medium-textured to fine netted fibers. CATAPHYLLS membranous, the uppermost green above the ground, or turning purple. LEAVES four or five, the lower narrowly lanceolate to linear and about as long as the stem, 5–9 mm wide, the uppermost smallest and partly to entirely sheathing, fairly soft-textured, the midvein and one pair of secondary veins evident but not noticeably thickened. STEM erect below, flexed outward above the sheath of the uppermost leaf, unbranched, c. 1.5 mm in diameter at the base of the spike.

SPIKE erect, 2- to 3-flowered; BRACTS green, 20–25 mm long, the inner slightly shorter than the outer. FLOWERS pale to deep pink, the lower three tepals each with a yellow-green median streak outlined in red; PERIANTH TUBE 16–20 mm long, arching outward and expanded above; TEPALS unequal, narrowly lanceolate, the three upper 20–24 mm long, 4–5 mm wide, the three lower joined to the upper laterals for c. 3 mm and to one another for c. 3 mm, 16–18 mm long, in profile exceeding the upper by 3–6 mm. FILAMENTS 10–12 mm long, exserted 2–3 mm from the tube; ANTHERS c. 6.5 mm long, violet, with short acute apiculi 0.5–1 mm long. OVARY 3–4 mm long; STYLE dividing opposite the middle of the anthers, the branches c. 2 mm long, not exceeding the anthers. CAPSULES and SEEDS unknown.

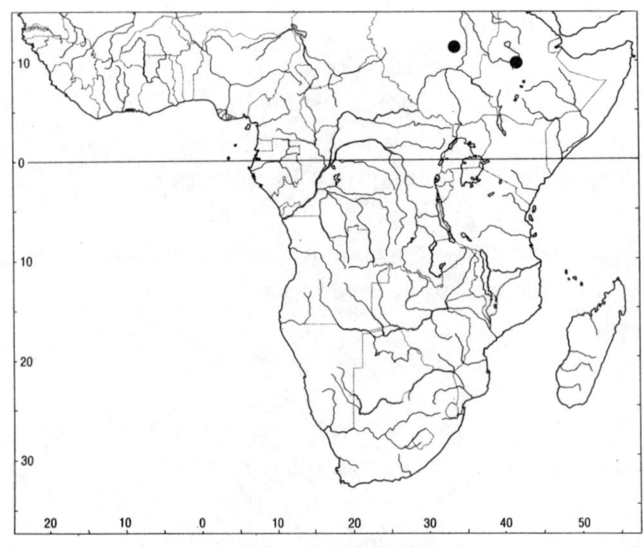

MAP 73. *Gladiolus sudanicus.*

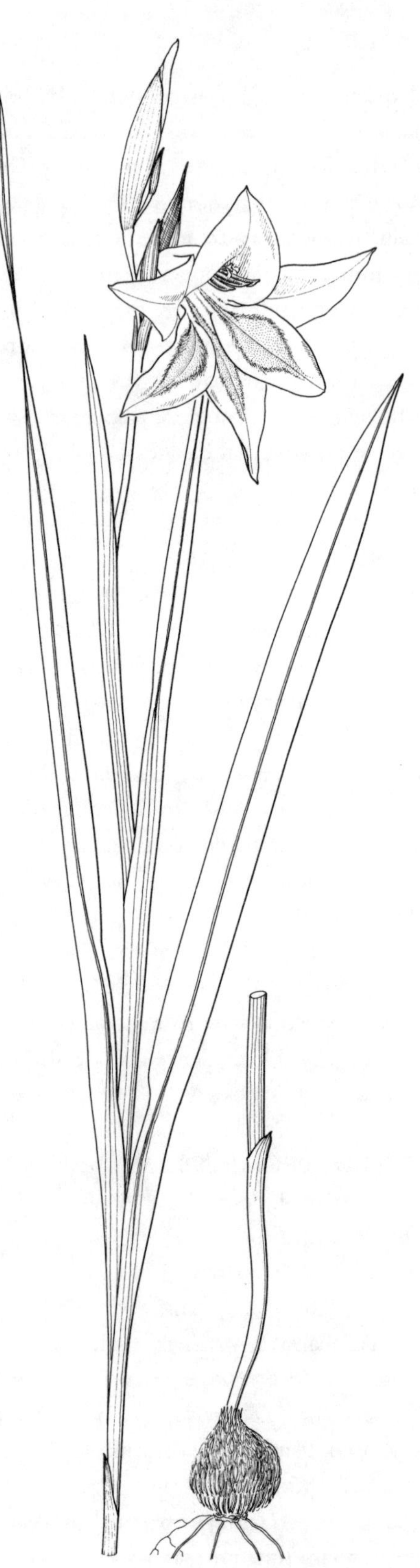

FIGURE 57. *Gladiolus sudanicus,* × 0.67 (*Gilbert & Gilbert 2155*).

Flowering Time. August, possibly also in September.

DISTRIBUTION & HABITAT

Gladiolus sudanicus is restricted to southern central Sudan and adjacent western Ethiopia. The species occurs at relatively low elevations of 1000 to 1200 m. Although it grows in relatively arid country, it occurs in grassland or among shrubs in places that are seasonally wet. The only record from Ethiopia is from the Blue Nile Gorge in the interior of the country, and it seems likely that *G. sudanicus* also occurs in the dry west. In Sudan, the only records of *G. sudanicus* are from the Nuba Mountains of southern Kordofan.

DIAGNOSIS & RELATIONSHIPS

Gladiolus sudanicus has a striking resemblance to the Zambian endemic, *G. serenjensis,* both in leaf and flower. The corms differ considerably, however, those of *G. serenjensis* being much larger and having coarsely fibrous tunics. The anthers of *G. sudanicus* have a short acute apiculus, suggesting a relationship with the largely southern tropical African *G. decoratus* complex, to which *G. serenjensis* also belongs. The soft-textured leaves, another characteristic of this complex, are surprising in species mostly adapted to fairly arid conditions. They suggest that the plants are adapted to grow and flower rapidly during the a wet season, after which they become dormant.

SELECTED SPECIMENS

Sudan. Kordofan: Grassland near Wadi Sheir, 8.3 km north of Kadugli, 15 Aug. 1983, *Musselman 6237* (E); grazed grassland near Wadi Sheir north of Kadugli, 20 Aug. 1983, *Musselman 6287* (E).

Ethiopia. Shoa: Blue Nile Gorge, near km 205 on Addis Ababa–Debra Marcos road, 7 Aug. 1971, *Gilbert & Gilbert 2158* (C, ETH, K).

71. *Gladiolus grantii* Baker

MAP 74. Baker, Handbook Irideae 206 (1892); Fl. Trop. Africa 7: 366 (1898). Type: Tanzania, "mountains east of Lake Tanganyika," 1200 m, probably 1860, *Speke & Grant s.n.* (K, holotype).

SYNONYMY

Gladiolus angustus sensu Baker, Trans. Linn. Soc. 29: 154 (1875), not *G. angustus* Linnaeus (1753), a South African species.

EPONYMY

grantii, named to commemorate Colonel James Grant of the Speke and Grant Expedition of 1864 to search for the source of the White Nile.

DESCRIPTION

Plants 30–45 cm high. Corms 15–20 mm in diameter, tunics initially membranous, becoming more or less fibrous externally, fibers straw-colored, mostly vertical, enclosing a few small cormlets. Cataphylls pale and membranous, the uppermost reaching up to 6 cm above the ground and then green or purple. Leaves four or five, the lower two or three more or less basal and longest, narrowly lanceolate, reaching to about the base of the spike, 4–7 mm wide, the upper two leaves cauline, becoming more or less entirely sheathing, all fairly soft-textured, the midrib and margins hardly thickened. Stem evidently erect, unbranched, 2–3 mm in diameter at the base of the spike.

Spike erect, 5- to 6-flowered; bracts green, soft-textured, the outer 33–38 mm long, the inner two-thirds as long as the outer. Flowers white to cream, fading to pale salmon or pink, the lower three tepals sometimes each with a median pale yellow mark in the midline; perianth tube narrowly funnel-shaped, 33–35 mm long, the lower cylindrical part c. 23 mm long, the wider upper part 10 mm long; tepals lanceolate, more or less equal in length, 28–30 mm long, the dorsal widest and horizontal, broadly lanceolate, c. 17 mm wide, the other tepals narrowly lanceolate, 5–8 mm wide, the lower three more or less horizontal, joined together for c. 2 mm, in profile appearing 3–5 mm longer than the dorsal. FILAMENTS c. 15 mm long, barely exserted from the tube for c. 2 mm; ANTHERS 12–13 mm long, reaching to about the middle of the dorsal tepal, with acute apiculate appendages c. 1–2 mm long. OVARY c. 5 mm long; STYLE arched over the stamens, dividing just below the apex of the anthers, branches c. 3.2 mm long, wider and channeled above. CAPSULES and SEEDS unknown.

FLOWERING TIME. Mid December to early February.

DISTRIBUTION & HABITAT

Gladiolus grantii is restricted to eastern central Tanzania. It is apparently fairly common locally in Shinyanga and Dodoma Regions, where it occurs in moist habitats near streams or in seasonally marshy sites at elevations of 1200 to 2000 m. It flowers fairly early in the wet season and may appear in vast numbers among young grasses, sometimes growing with *Lapeirousia schimperi,* which also has long-tubed white flowers.

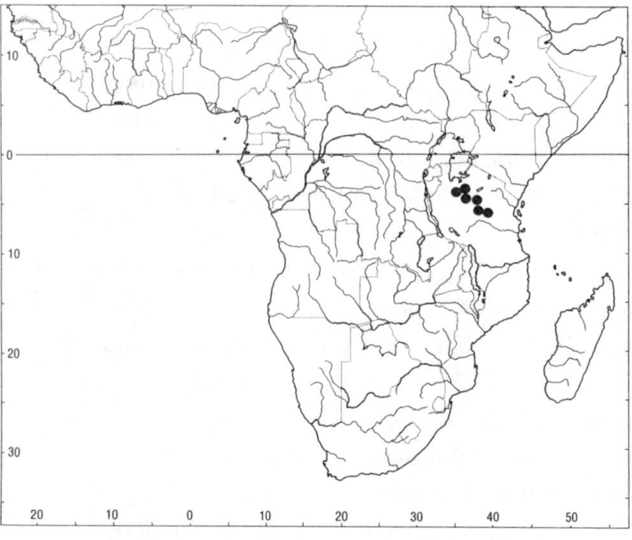

MAP 74. *Gladiolus grantii.*

DIAGNOSIS & RELATIONSHIPS

Gladiolus grantii resembles the southern tropical African *G. oligophlebius* in general aspect, size, and flower shape. The flower color of the two differs, for *G. grantii* has white to cream flowers, whereas *G. oligophlebius* has pink flowers with cream markings on the lower tepals. Careful examination of the flowers reveals other differences. The perianth tube of *G. grantii* is longer than the tepals, unlike other members of section *Decoratus*, and the filaments are unusually short, being exserted for only c. 2 mm from the tube. *Gladiolus grantii,* like other members of section *Decoratus,* has anthers with long-apiculate appendages and soft-textured leaves with only one pair of secondary veins normally evident although not thickened.

SELECTED SPECIMENS

Tanzania. Shinyanga: Huruhuru and Mantini road, swamp margin, 16 Jan. 1932, *Burtt 3511* (K); Huruhuru, margin of seasonal swamp, 1150 m, 16 Jan. 1932, *Burtt 3512* (BM, BR, K); Simba Hills, c. 1500 m, savanna, 8 Jan. 1928, *Burtt 1732* (K); near Shinyanga, open marshy ground, Jan.–Feb. 1933, *Bax 398* (K); Kahama District, Nyashimbi, 12 Jan. 1955, *Akiley s.n.* (K). Dodoma: Kondoa-Irange, Mangoloma from Siambo, 15 Dec. 1925, *Burtt 310* (K); 24 km south of Kondoa, Great North Road, dry bush near steam, 1300 m, 19 Jan. 1962, *Polhill & Paulo 1218* (BR, K, P, PRE); 70 km from Kibaya to Kongwa, 18 Jan. 1965, *Leippert 5469* (B, K).

72. *Gladiolus serenjensis* Goldblatt

MAP 75. Goldblatt, Fl. Zambesiaca 12(4): 101 (1993). Type: Zambia, Central Province, Serenje District, Kundalila Falls, crevices in sandstone, 1500 m, 4 Feb. 1973, *Strid 2834* (C, holotype; K, LD, MO, isotypes).

EPONYMY

serenjensis, named after the Serenje District of central Zambia, where the species occurs.

DESCRIPTION

Plants 15–25(–40) cm high. CORMS 14–25 cm in diameter, the tunics of coarse fibers. CATAPHYLLS pale and membranous, the upper purplish above the ground. LEAVES five or six, the lower three or four basal and longest, usually slightly exceeding the spike, linear to narrowly lanceolate, (2–)3–4 mm wide, the margins and midrib not thickened, the upper two leaves cauline and narrower and shorter than the basal. STEMS erect, unbranched, c. 1 mm in diameter at the base of the spike.

SPIKE erect, 3- to 6-flowered, internodes 10–15 mm long; BRACTS green, sometimes flushed purple above, attenuate, imbricate, 1.5–2 internodes long, 15–20(–25) mm long, the inner about half as long as the outer. FLOWERS pink, the lower tepals without contrasting markings; PERIANTH TUBE obliquely and narrowly funnel-shaped, curving outward between the bracts, c. 15 mm long, the lower part 6 mm long; TEPALS subequal, lanceolate, obtuse, 18–20 mm long, c. 8 mm wide in the midline, all similarly oriented, directed forward and curving outward distally. FILAMENTS unilateral and lying below the dorsal tepal, c. 11 mm long, usually shortly exserted from the mouth of the tube; ANTHERS c. 6.5 mm long, unilateral, with short acute apiculate appendages c. 0.4 mm long, apparently yellow. OVARY oblong, c. 4.5 mm long; STYLE arched over the stamens, dividing opposite the anther apices, the branches 4–5 mm long, extending beyond the anthers. CAPSULES and SEEDS unknown.

FLOWERING TIME. December.

DISTRIBUTION & HABITAT

Gladiolus serenjensis is known only from a small area of the Muchinga Mountains of central Zambia in the Serenje District. It occurs in rugged country, and records indicate that it grows in rock outcrops in thin soils or in crevices. The restricted distribution probably explains why it was recorded for the first time only in 1955.

DIAGNOSIS & RELATIONSHIPS

Short stature, rather soft-textured leaves, and attractive bright pink flowers with nearly subequal, unmarked tepals that extend forward for most of their length distinguish *Gladiolus serenjensis.* Its relationships are uncertain, but the soft leaf texture, pink perianth, and apiculate anthers suggest that it is allied to the *G. decoratus–G. oligophlebius* group of species. Members of the alliance have larger flowers, usually contrastingly marked with white or yellow on the lower tepals, and the anthers have conspicuous long-apiculate appendages. The possibility that the peculiar Zairian *G. salmoneicolor* is related to *G. serenjensis* is discussed under that species.

SELECTED SPECIMENS

Zambia. Central: Serenje District, Kundalila Falls, 1500 m, 4 Feb. 1973, *Kornas 3176* (K); Great North Road, 125 km south of Mpika, near Kapandashila,

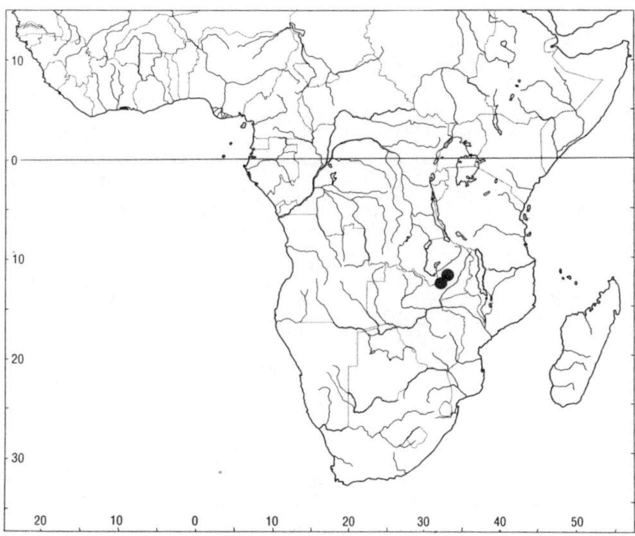

MAP 75. *Gladiolus serenjensis.*

rock shelves, 1600, 28 Feb. 1972, *Kornas 1304* (K); Serenje, granite rocks, 23 Jan. 1955, *Fanshawe 1854* (K); 148 km south of Mpika, quartzite rocks, 2 Mar. 1962, *Robinson 4974* (K, SRGH).

73. *Gladiolus stenolobus* Goldblatt, new species

MAP 76. Type: Tanzania, Mpanda, Buyenze escarpment 30 km south of Uvinza, c. 1250 m, Feb. 1956, *Proctor 447* (K, holotype).

EPONYMY

stenolobus, "narrow-lobed," referring to the unusually narrow, attenuate tepals.

LATIN DIAGNOSIS

Plantae c. 75 cm altae, cormis c. 10 mm diametro, foliis 8, inferioribus 4 basalibus caulem brevioribus, laminis 3–4 mm latis, marginibus costisque leviter incrassatis, caule eramoso erecto, spica 14 florum, bracteis viridibus exterioribus 14–25(–35) mm longis, floribus manifeste rubropurpureis, tubo perianthii c. 11 mm longo, tepalis subaequalibus anguste lanceolatis c. 33 × 3–3.5 mm, filamentis c. 5 mm longis in tubo inclusis, antheris c. 9 mm longis obtusis, stylo diviso 3–4 mm infra apices antherarum.

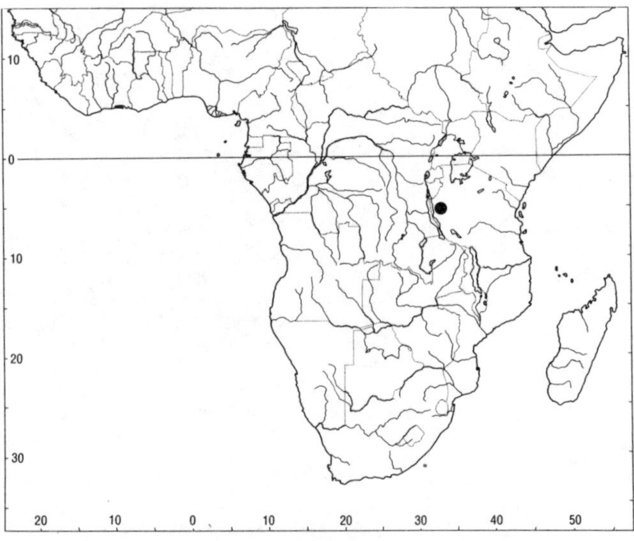

MAP 76. *Gladiolus stenolobus.*

DESCRIPTION

Plants c. 75 cm high. CORMS globose, evidently unusually small, c. 10 mm in diameter, the tunics membranous, light brown. CATAPHYLLS membranous, probably green above the ground. LEAVES eight, the lower four basal and longest, about two-thirds as long as the stem, and not reaching the base of the spike, the blades sublinear, 3–4 mm wide, the margins and midribs lightly thickened, the upper four leaves inserted on the lower part of the stem, sheathing for at least half their length. STEM erect, unbranched, c. 1.7 mm in diameter at the base of the spike.

SPIKE straight and erect, 14-flowered; BRACTS green, the outer 14–25(–35) mm long, the inner about half as long. FLOWERS evidently reddish purple, as far as known the tepals without contrasting markings; PERIANTH TUBE c. 11 mm long, the lower half straight, the upper half curving outward and becoming horizontal; TEPALS subequal, narrowly lanceolate and attenuate, c. 33 × 3–3.5 mm, somewhat shorter in dry material, evidently extended forward for most of their length and curving gently outward in the upper third. FILAMENTS c. 5 mm long, entirely included in the tube; ANTHERS c. 9 mm long, the lower 2–3 mm included in the tube, with acute apiculate appendages c. 0.6 mm long. OVARY c. 3 mm long; STYLE dividing 3–4 mm below the anther apices, the branches c. 2.3 mm long. CAPSULES and SEEDS unknown.

FLOWERING TIME. February, probably also in January.

DISTRIBUTION & HABITAT

Gladiolus stenolobus is known from just a single plant from the Mpanda Distict of western Tanzania, collected in 1956. It was recorded as growing in pockets of soil in weathered sandstone of the

Buyenze escarpment a short distance south of Uvinza. Additional collections are needed to understand the species more fully, and to determine its relationships and its patterns of variation.

DIAGNOSIS & RELATIONSHIPS

The reddish purple flowers with narrowly lanceolate, attenuate, subequal tepals, the short stamens with the filaments included in the tube, and the anthers with apiculate appendages readily distinguish *Gladiolus stenolobus.* The short tube, apiculate anthers, and reddish flower color place the species in section *Decoratus,* in which it seems most closely related to the Zambian *G. serenjensis.* That species also occurs on sandstone outcrops in shallow soil, but although it is a much smaller plant, it has longer stamens and broader obtuse tepals. The similarities between the two species are, however, striking. The structure of the flower of *G. stenolobus* recalls another Tanzanian species, *G. canaliculatus* (section *Gladiolus*), but the latter has anthers apically obtuse and only two leaves, these additionally channeled for their entire length. The floral similarity is probably due to convergence and presumably represents an adaptation to a particular pollination strategy, as yet unidentified.

SPECIMENS

Known only from the type, cited above.

74. *Gladiolus salmoneicolor* Duvigneaud & van Bockstal ex Còrdova

MAP 77. Còrdova, Bull. Jard. Bot. Nat. Belgique 60: 328 (1990). Type: Zaire, Shaba, Lupoto, open *Marquesia* and *Brachystegia* woodland, 6 Jan. 1960, *Duvigneaud 4938* (BRLU, holotype; BR, BRLU, isotypes).

EPONYMY

salmoneicolor, "salmon-colored," referring to the pale pinkish flower color.

DESCRIPTION

Plants (14–)30–60 cm high. CORMS 13–15 cm in diameter, the tunics of fairly fine fibers. CATAPHYLLS pale and membranous, the upper reaching well above the ground and then purplish. LEAVES five or six, the lower two or three basal and longest, reaching to about the base of the spike, narrowly lanceolate (to linear), 4–6 mm wide, soft-textured, the margins and midrib not thickened, upper two leaves cauline and narrower and shorter than the basal. STEMS unbranched, c. 1.3 mm in diameter at the base of the spike.

SPIKE erect, 5- to 7-flowered, internodes 10–20 mm long; BRACTS green, lanceolate, shortly imbricate, 1–1.5(–2) internodes and (15–)18–25 mm long, the inner half to two-thirds as long as the outer. FLOWERS evidently actinomorphic with a curved tube and white, but described as salmon-colored, evidently drooping slightly below horizontal, the tepals without contrasting markings; PERIANTH TUBE obliquely and narrowly funnel-shaped, curving outward between the bracts, 8–9 mm long, the lower part c. 3 mm long; TEPALS more or less equal, narrowly lanceolate, obtuse, 18–20 mm long, 3–5 mm wide in the midline, all similarly oriented, directed forward and curving outward distally. FILAMENTS c. 3.5 mm long, usually shortly exserted from the mouth of the tube, apparently symmetrically disposed; ANTHERS 5–6.5 mm long, the bases included in the tube, evidently symmetrically arranged around the style, with short acute apiculate appendages c. 0.15 mm long, apparently yellow. OVARY ellipsoid, c. 3.5 mm long; STYLE dividing opposite the upper third of the anthers, the branches 2–3 mm long, reaching to just beyond the anther apices. CAPSULES and SEEDS unknown (according to Còrdova, 1990, seeds winged and 3–4 mm in diameter but not present in material cited).

FLOWERING TIME. December and January.

DISTRIBUTION & HABITAT

Gladiolus salmoneicolor is known only from the type collection from Shaba Province, Zaire, near Lupoto, between Lubumbashi and Lukasi. It occurs on rocky hills in light *Brachystegia–Marquesia* woodland. It is presumably rare although its small, inconspicuous flowers may explain why it has not been collected more often.

DIAGNOSIS & RELATIONSHIPS

Everything about *Gladiolus salmoneicolor* is unusual, and at first glance the plant hardly seems to belong in the genus. The fine corm tunics and soft-textured leaves accord especially well with the eastern and southern African genus *Freesia,* but the simple style branches, slightly expanded toward the apices, in contrast to the divided style branches of *Freesia,* leave little room for doubt that it is correctly placed in *Gladiolus.* The flowers are remarkable in the genus in being evidently actinomorphic with symmetrically disposed stamens, slightly nodding orientation, and with a very short perianth tube 8–9 mm long. With all these peculiarities it is difficult to assess the relationships of *G. salmoneicolor.* It may be most closely related to the central Zambian *G. serenjensis,* which has somewhat larger pink flowers with similarly oriented subequal, unmarked tepals, soft-textured leaves, and short filaments. The stamens are unilateral in *G. serenjensis* and the filaments are usually exserted 1–2 mm from the tube, in contrast to the symmetrically disposed stamens and fully included filaments of *G. salmoneicolor.* I can offer no explanation for the specific epithet *salmoneicolor*'s presumably alluding to the perianth color, for there is little doubt that the tepals are white or possibly pale yellow while the tube is probably yellow.

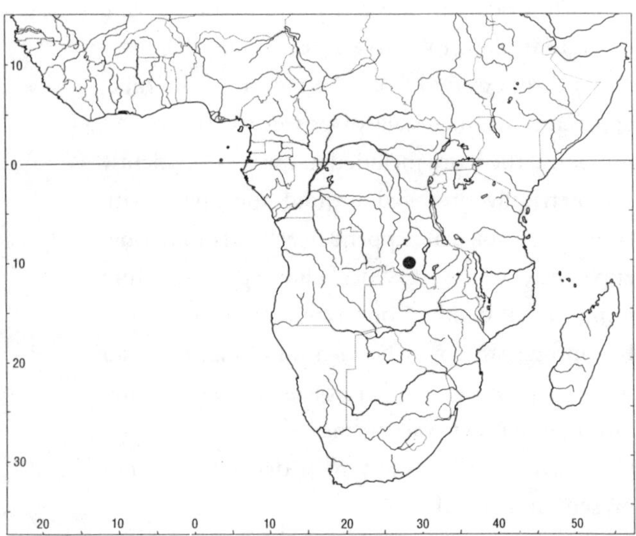

MAP 77. *Gladiolus salmoneicolor.*

SPECIMENS

Known only from the type, cited above.

75. *Gladiolus lithicola* Goldblatt, new species

FIGURE 58, MAP 78. Type: Ethiopia, Harerge, slopes above Mojjo River and south of Gara Mullata Mountains, c. 2100 m, 16 Aug. 1963, *Burger 3150* (K, holotype; ACD, isotype).

EPONYMY

lithicola, "growing among rocks," alluding to the habitat, rocky slopes and cliffs on the slopes of the Gara Mullata Mountains in Harerge Province, southeastern Ethiopia.

LATIN DIAGNOSIS

Plantae (8–)12–28 cm altae, cormo 8–10 mm in diametro, foliis 2–4, laminibus linearibus 1.5–2.5 caule excedente (2–)3–4 mm latis, spica (1–)2–3 florum, tubo perianthii c. 18 mm longo, tepalis lanceolatis inaequalibus, superioribus c. 18 mm longis, inferioribus c. 20 mm longis, filamentis c. 6 mm longis in tubo inclusis, antheris c. 5 mm longis breve apiculatis.

FIGURE 58. *Gladiolus lithicola*, × 0.67 (*de Wilde 5840*).

DESCRIPTION

Plants (8–)12–28 cm high. CORM 8–10 mm in diameter, the tunics of fine netted fibers. LEAVES two to four, the lowermost longest, 1.5–2.5 times as long as the stem, the blades linear, (2–)3–4 mm wide, the uppermost smallest and with oblong blades or largely to entirely sheathing. STEM erect below, flexed outward above the sheath of the uppermost leaf, or inclined to drooping, unbranched.

SPIKE (1–)2- to 3-flowered; BRACTS green, 15–30(–35) mm long, usually attenuate, the inner about two-thirds as long as the outer. FLOWERS lavender or mauve, the tepals evidently unmarked; PERIANTH TUBE narrowly funnel-shaped, c. 18 mm long; TEPALS lanceolate, unequal, the lower three longer than the upper, the dorsal and upper laterals c. 18 mm long, the three lower c. 20 mm long. FILAMENTS short, c. 6 mm long, included in the tube; ANTHERS c. 5 mm long, dark violet, the apices drawn into short acute appendages. OVARY oblong, c. 4 mm long; STYLE arching over the anthers, dividing at or 1–2 mm beyond the anther apices, the branches c. 2.5 mm long. CAPSULES and SEEDS unknown.

FLOWERING TIME. August to October.

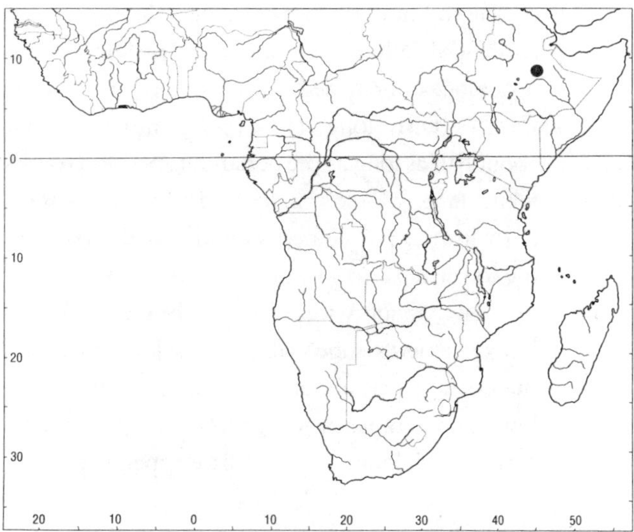

MAP 78. *Gladiolus lithicola*.

DISTRIBUTION & HABITAT

Gladiolus lithicola is apparently restricted to the slopes of the Gara Mullata Mountains of south-eastern Ethiopia. It grows on fairly steep rocky slopes and cliffs at c. 2750 m, originally fairly well covered with forest but today largely deforested. On the steep slopes that the species appears to favor, plants tend to grow out horizontally and the leaves generally droop, whereas the short flowering stem curves upward toward the apex, presenting the flowers in an upright position. The first collections of this plant appear to have been made by the American botanist, William Burger, in 1961.

DIAGNOSIS & RELATIONSHIPS

One of the most distinctive of the several Ethiopian species of *Gladiolus, G. lithicola* is readily recognized by the short stems, much shorter than the long drooping leaves, the relatively small lavender or mauve perianth, and short dark violet stamens borne at the mouth of the perianth tube.

SELECTED SPECIMENS

Ethiopia. Harerge: south face of Gara Mullata (Mullata Mountains), c. 2760 m, 18 Oct. 1969, *de Wilde 5840* (WAG); steep wet slope of Gara Mullata, in ravine below the pass, 2500–3000 m, 24 Sept. 1961, *Burger 1019* (ACD, ETH, K).

SECTION *ACIDANTHERA* (HOCHSTETTER) GOLDBLATT
New Combination and Rank

Basionym: *Acidanthera* Hochstetter, Flora 27: 25 (1844). Type: *A. bicolor* Hochstetter (= *Gladiolus murielae* Kelway).

Description

Plants medium to large, unbranched. LEAVES lanceolate to linear, the upper smaller and sometimes entirely sheathing. STEMS erect or inclined to drooping, unbranched. SPIKES erect or inclined; BRACTS fairly long, (30–)40–80 mm long, but about half as long as the perianth tubes. FLOWERS white, large, 75–180 mm long, the lower tepals often with red-purple nectar guides at the bases of the lower tepals; PERIANTH TUBE (60–)100–140 mm long, about twice as long as the bracts or even longer; TEPALS subequal, more or less lanceolate, the dorsal tepals about half as long as the tubes; lower tepals more or less equal to the dorsal. ANTHERS with long acute apiculate appendages.

Comprising seven species, section *Acidanthera* is widespread in tropical Africa. Species extend from Sierra Leone in western Africa eastward to Ethiopia and southward to Malawi and central Mozambique. The section does not occur in Angola or Zaire. One species, *G. candidus,* centered in eastern Africa, is shared with Oman in southern Arabia. The center of diversity for section *Acidanthera* is in the highlands of Guinea, Sierra Leone, and Ivory Coast, where there are three local endemic species.

76. *Gladiolus aequinoctialis* Herbert

FIGURE 59, MAP 79. Herbert, Bot. Reg. 28, Misc. 85, no. 97 (1842). Marais, Kew Bull. 28: 311 (1973). *Acidanthera aequinoctialis* (Herbert) Baker, J. Linn. Soc. 16: 160 (1877); Curtis's Bot. Mag. 121: pl. 7393 (1895); Fl. Trop. Africa 7: 358 (1898), in part, including *Gladiolus praecostatus* Marais. Hutchinson & Dalziel, Fl. West Trop. Africa 2: 376 (1936). Hepper, Fl. West Trop. Africa, Ed. 2, 3(1): 139 (1968). Type: Sierra Leone, the collector and date not recorded (CGE, holotype, not seen).

SYNONYMY

Acidanthera divina Vaupel, Notizbl. Bot. Gart. Berlin-Dahlem 7: 32 (also 375) (1920). *Gladiolus aequinoctialis* var. *divinus* (Vaupel) Marais, Kew Bull. 28: 312 (1973). Type: Fernando Po, north slope of Santa Isabella Peak, 2600 m, 16 Nov. 1911, *Mildbraed 7175* (B, holotype).

Gladiolus aequinoctialis var. *tomentosus* Marais, Kew Bull. 28: 312 (1973). Type: Sierra Leone, Konta-Bumban, bare granite hills, 300 m, 28 Aug. 1928, *Deighton 1244* (K, holotype).

EPONYMY

aequinoctialis, "from the tropics," its being one of the first tropical African species of *Gladiolus* to be discovered.

DESCRIPTION

Plants (40–)90–120 cm high. CORM 2–3 cm in diameter, the tunics coriaceous to papery, often becoming more or less fibrous with age, dark red-brown. CATAPHYLLS firm-membranous, green or purple above the ground. LEAVES 4–10, the lower two to five basal, reaching to about the base of the spike, sometimes slightly exceeding it, the blades lanceolate, (6–)12–17 mm wide, glabrous and sometimes glaucous, or occasionally densely tomentose, firm-textured, the midrib, margins, and often one or more pairs of secondary veins lightly thickened, the upper leaves progressively shorter, and the uppermost one or two sometimes entirely sheathing. STEM often inclined in the wild, unbranched, 3–4 mm in diameter at the base of the spike.

SPIKE (3–)5- to 8-flowered; BRACTS green entirely or flushed purple below, 50–80 mm long, the inner shorter than and concealed by the outer. FLOWERS white, the lower three tepals each with a narrow to broad dark purple streak in the lower midline; PERIANTH TUBE more or less cylindric and straight throughout, (85–)120–140 mm long; *tepals* lanceolate, more or less equal, 35–40 × 18–20 mm. FILAMENTS c. 25 mm long, exserted (2–)5–8 (–10) mm from the tube; ANTHERS 10–17 mm long, with acute apiculate appendages 0.5–1(–1.5) mm long, reaching to at least the middle of the tepals. OVARY oblong, 5–8 mm long; STYLE arching over the stamens, dividing at or beyond the anther apices, the branches 4–6 mm long, much expanded in the upper half. CAPSULES ellipsoid, 18–20 mm long, concealed in the bracts at least until mature; SEEDS 7–8 × c. 5 mm, broadly winged. CHROMOSOME NUMBER $2n = 30$.

FLOWERING TIME. Mainly October to December, the tomentose variant from interior Sierra Leone flowering in August.

DISTRIBUTION & HABITAT

Gladiolus aequinoctialis has a scattered distribution across western Africa, with centers in Sierra Leone in the west and the highlands of Cameroon in the east. Surprisingly, the species also occurs on the island of Fernando Po, southwest of Cameroon, and there it grows on Santa Isabella Peak. Plants are found mainly in montane habitats, usually above 450 m and up to 2600 m on Fernando Po,

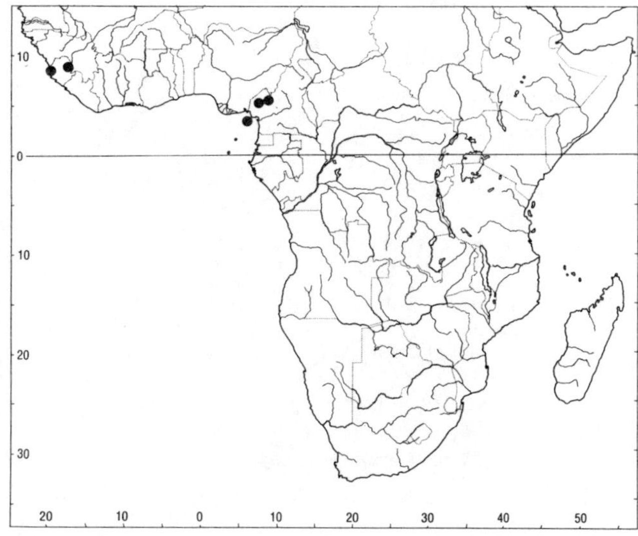

MAP 79. *Gladiolus aequinoctialis.*

FIGURE 59. *Gladiolus aequinoctialis*, × 0.5 (*Boussard sub Goldblatt 9583*).

where they grow in rocky situations, often on wet ledges on steep cliffs but also in stony grassland.

DIAGNOSIS & RELATIONSHIPS

The first of the several West African species of section *Acidanthera* to be discovered, *Gladiolus aequinoctialis* has the traits typical of the section, a white flower with a long perianth tube, typically 120–140 mm long, and anthers with long acute apiculate appendages. Possibly most closely related to the Ethiopian and eastern African montane *G. murielae,* and easily confused with it, *G. aequinoctialis* can be distinguished only with difficulty. According to Marais (1973), the main differences are the leaves, those of *G. aequinoctialis* being firmer in texture and having more conspicuous veins, particularly the secondary veins, which in *G. murielae* are very weakly developed. Another difference is the size of the anther apiculi, typically at least 2 mm long in *G. murielae* but only 0.5–1.5 mm in *G. aequinoctialis.* The filaments of *G. murielae* are also longer and exserted (5–)10–12 mm from the tube, whereas in *G. aequinoctialis* the filaments are exserted 2–8(–10) mm.

Particularly short filaments characterize the form of *Gladiolus aequinoctialis* from Cameroon and Fernando Po. In those plants the filaments are exserted for no more than 2 mm. These populations of *G. aequinoctialis* are also distinctive in being relatively short and in having very inconspicuous markings on the lower tepals. Plants from Sierra Leone to the west are often taller, have filaments exserted for 5 to 10 mm, and the markings on the tepals are larger. These differences prompted François Vaupel to name the Cameroon form, first collected by Mildbraed on Fernando Po, *G. divinus.* Marais (1973) considered that to be no more than a variety of *G. aequinoctialis,* and I do not recognize these plants taxonomically.

Another variant of *Gladiolus aequinoctialis* recognized by Marais, variety *tomentosus*, is also reduced to synonymy here. Tomentose plants occur locally in interior Sierra Leone, but they seem to differ in no other way from typical *G. aequinoctialis,* except possibly in their earlier flowering, in August. Flowering in other populations of *G. aequinoctialis* is mostly in October to December, but occasionally in late August and September. Additional collections of the poorly known tomentose variant are needed so that this treatment can be verified. The presence of pubescence alone is normally not considered sufficient to merit recognition of new taxa. Variants with pubescent or scabrid leaves are also known, for example, in the southern tropical African *G. oligophlebius* and *G. benguellensis,* and the development of pubescence on the leaves of *G. laxiflorus* is notably variable.

SELECTED SPECIMENS

Sierra Leone. Gumah, rock crevices, 450 m, 10 Nov. 1911, *Lane-Poole 81* (K); Sugar Loaf Mountain, 450 m, rock outcrops, 15 Oct. 1951, *Jones 250* (BR, K); top of Sugar Loaf Mountain, Mar. 1893, *Scott Elliot s.n.* (K); Picket Hill, 700 m, crevices on rock faces, 18 Nov. 1951, *Jones 205* (K, WAG); Mt. Loma, summit of Bintumani Peak, 26 Sept. 1945, *Jaeger 1990* (G, MO); Bumban, near Makeni, 27 Aug. 1962, *Harvey 81* (K); without precise locality, Dec. 1925, *Dawe 571* (K).

Cameroon. Gotel Mountains, 30 km north-northeast of Banyo, 13 June 1967, *Letouzey 8624* (BR, G, P, YA); Mt. Nshelle, 10 km east-southeast of Bamenda, 12 Aug. 1975, *Letouzey 14279* (BR, P, YA); Mt. Vokre, 30 Sept.1967, *Jacques-Felix 8493* (P, YA).

Fernando Po. Known from here only from the type of *Acidanthera divinus,* cited above.

77. *Gladiolus praecostatus* Marais

MAP 80. Marais, Kew Bull. 28: 314–315 (1973). Type: Guinea, Nzérékoré, top of Mt. Koiré, rock crevices, 8 Sept. 1939, *Baldwin 13285* (K, holotype; MO, NY, isotypes).

SYNONYMY

Acidanthera aequinoctialis (Herbert) Baker, in part, sensu Hepper, Fl. West Trop. Africa, Ed. 2, 3(1): 139 (1968). Adam, Fl. Descript. Monts Nimba, Part 5, 1642 (1981).

EPONYMY

praecostatus, "strongly ribbed," referring to the heavily thickened and raised veins of the foliage leaves.

DESCRIPTION

Plants 30–65 cm high. CORM 1.5–3 cm in diameter, tunics membranous, becoming irregularly broken, occasionally fibrous, light red-brown. CATAPHYLLS firm-membranous to herbaceous. LEAVES four to several, the lower three to five basal, narrowly lanceolate, half as long (perhaps just emerging) to slightly exceeding the spike, sometimes slightly exceeding it, 4–12 mm at the widest, stiff, with heavily thickened margins, midribs and at least two other veins, the thickenings raised and arching over the blade surface, remaining leaves cauline and smaller than the basal. STEM erect or inclined, unbranched, bearing one to two entirely sheathing bractlike leaves in the upper half.

SPIKE erect or inclined, (2–)4- to 6-flowered; BRACTS green, 3–7 cm long, the inner usually shorter than and concealed by the outer, occasionally nearly equal. FLOWERS white, the lower tepals each with a prominent dark purple streak in the lower midline; PERIANTH TUBE more or less cylindric, straight, slightly expanded at the throat, (80–)100–125 mm long; TEPALS more or less equal, obovate to broadly elliptic, 32–35 mm long, 15–18 mm wide, apiculate. FILAMENTS included or exserted for up to 5 mm from the tube; ANTHERS 12–16 mm long, with short apiculi less than 1 mm long. OVARY c. 5 mm long; STYLE arching over the stamens, dividing at or just beyond the apex of the anthers, the branches c. 5 mm long, broad and channeled in the upper half. CAPSULES narrowly ellipsoid, (16–)20–30 mm long; SEEDS 7–8 × 3.5–4 mm, broadly winged.

FLOWERING TIME. Mostly June to September, sometimes in October.

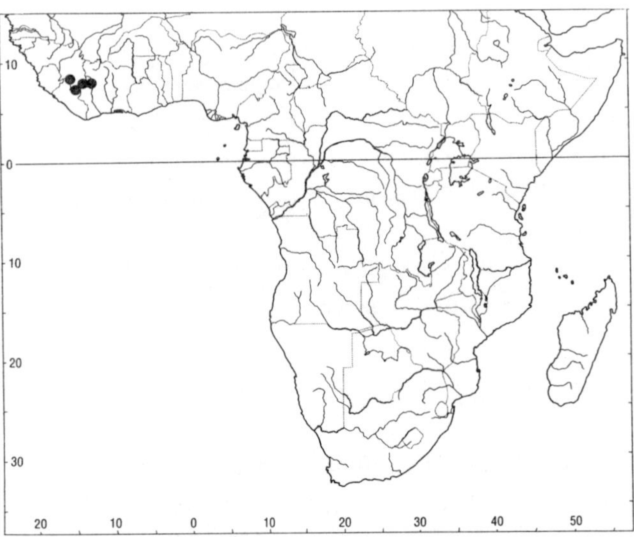

MAP 80. *Gladiolus praecostatus.*

DISTRIBUTION & HABITAT

Gladiolus praecostatus is endemic to the Nimba highlands of southern Guinea, northern Liberia, and neighboring Ivory Coast. Plants grow in open grassland, often in rocky sites, at elevations above 1500 m. They flower in the middle of the rainy season, from June to September, whereas the closely allied *G. aequinoctialis* flowers later, usually from October to December.

DIAGNOSIS & RELATIONSHIPS

Recognized as a distinct species only in 1973 but known since at least 1936, *Gladiolus praecostatus* is readily distinguished from the related *G. aequinoctialis* by the heavily ribbed leaves, at least the margins and three veins of which are thickened and raised, arching over the leaf surface. The flowers of the two species are similar, but those of *G. praecostatus* are slightly smaller, the tube seldom exceeding 10 cm and the tepals up to 35 mm long and 15–18 mm wide as compared to a tube 8–14 cm long and tepals c. 40 mm long and 18–20 mm wide in *G. aequinoctialis.* When it was first described, the difference between the leaves of *G. praecostatus* and *G. aequinoctialis* seemed absolute. Some collections from the Nimba Mountains of southeastern Guinea (e.g., *Adam 25,825, 29,018*), are, however, rather difficult to place. The leaves seem intermediate between those of typical *G. praecostatus* and those of *G. aequinoctialis,* having thickened but not strongly raised veins. I have included these plants in *G. praecostatus,* but it seems evident that the differences between the two species are less clear than originally thought by Marais in 1973. The situation needs to be investigated when more material becomes available.

The very similar *Gladiolus leonensis,* distinguished largely by its fewer and much reduced leaves with blades up to 8 cm long and only the midrib and margins thickened, is restricted to the mountains of Sierra Leone. Recognition of this as a distinct species is somewhat arbitrary since some specimens of *G. praecostatus* have rather short leaves. One plant of *Scaetta 3260,* for example, has leaves with blades less than 15 cm long. The species differ virtually in no other significant way. The purple markings on the lower tepals of *G. leonensis* are said by Marais (1973) to be poorly defined and the inner and outer bracts subequal. The latter feature can also be found in some spikes of *G. praecostatus* and is not consistent in *G. leonensis*; one spike of *Jaeger 9807* (P) has equal bracts, and in the other spike the inner bracts are appreciably shorter.

SELECTED SPECIMENS

Guinea: Bala Nord, Macenta, Mt. Ziama, 1383 m, 17 Aug. 1949, *Adam 5975* (MO, P). Nimba, Nzérékoré, 1800 m, Oct. 1937 *Jacques-Felix 1950* (P); grassland on the summits of the Nimba Mountains, 1700 m, July 1942, *Schnell 1561* (K, MO, P); Nimba Mountains, 15 Sept. 1942, *Schnell 1825* (P); crests of the Nimba Mountains, Aug. 1954, *Schnell 6269* (BR, K, MO, P).

Ivory Coast: Man, Mt. Tonkoui, Oct. 1936, *Jacques-Felix 1273* (P); Mt. Tonkoui, Sept. 1955, *Nozeran s.n.* (P); Nimba Mountains, without precise locality, 10 Aug. 1954, *Boughey X51* (BR); Mt. Momi, 3 Oct. 1966, *Aké Assi 9181* (K).

Liberia. Yekepia, 14 Oct. 1975, *Adam 29865* (MO).

78. *Gladiolus leonensis* Marais

MAP 81. Marais, Kew Bull. 28: 315 (1973). Type: Sierra Leone, Loma Mountains, 15 June 1966, *Morton s.n.* (SL 3556 in K, holotype).

EPONYMY

leonensis, "from Sierra Leone," celebrating the country in which the type collection was made.

DESCRIPTION

Plants 25–30 cm high, not including the flowers. CORM evidently c.15 mm in diameter, the tunics papery, becoming irregularly broken and sometimes finely fibrous with age, light brown. CATAPHYLLS firm-textured, green above the ground. LEAVES three or four, the lower one or two basal, the blades always very short, 2–6(–12) cm long, linear, 2–3 mm wide, oval in transverse section, the margins and midrib heavily thickened, thus with two narrow grooves on each surface, the upper one or two leaves without blades and entirely sheathing. STEM erect, unbranched, bearing one or two sheathing bractlike leaves in the upper half, c. 3 mm in diameter at the base of the spike.

SPIKE evidently inclined at the top of the stem, 2- to 3-flowered; BRACTS green, 30–35(–55) mm long, the inner only slightly shorter than the outer, occasionally nearly equal. FLOWERS apparently uniformly white; PERIANTH TUBE more or less cylindric, straight, slightly expanded at the throat, 130–140 mm long; TEPALS more or less equal, broadly lanceolate, 30–35 × c. 12 mm. FILAMENTS c. 15 mm long, exserted for 3–5 mm from the tube; ANTHERS 12–14 mm long, with short apiculate appendages c. 1.5 mm long. OVARY c. 5 mm long; STYLE arching over the stamens, dividing at or just beyond the apex of the anthers, the branches c. 5 mm long, broad and channeled in the upper half. CAPSULES and SEEDS unknown.

FLOWERING TIME. April to June.

DISTRIBUTION & HABITAT

Gladiolus leonensis is known from just a handful of collections, all from the interior mountains of Sierra Leone. Plants flower early in the season and usually complete fruit development by the end of May, although the type collection, of plants in full flower, was made in June. Plants evidently grow in thin rocky soil and flower before the surrounding vegetation has grown up. Whether plants continue to grow after flowering, perhaps producing longer leaves from separate shoots, is unknown. The very short leaves seen on flowering material seem too small to provide enough nutrition for the plants.

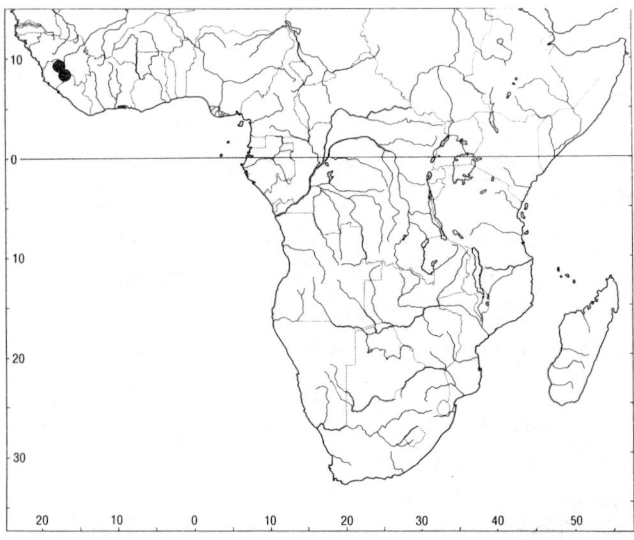

MAP 81. *Gladiolus leonensis.*

DIAGNOSIS & RELATIONSHIPS

The flowers of *Gladiolus leonensis* are white and have the long perianth tube typical of section *Acidanthera.* Both the bracts and flowers correspond in size almost exactly to those of *G. praecostatus,* and it is to that species that *G. leonensis* is probably most closely related. The two differ mainly in their foliage. Leaves of *G. praecostatus* have well developed and relatively broad blades, distinctive in having strongly primary and secondary veins, as well as thick margins. The leaves are also usually about as long as the stems, reaching at least to the base of the spikes. This is quite different from the short leaves of *G. leonensis,* which are less than half as long as the stems and have just the main vein and the margins thickened. Although said by Marais in the protologue to have flowers without markings, some specimens show traces of a dark median streak on each of the lower tepals, thought by Marais to have perhaps developed on drying. The flowers of *G. praecostatus* have conspicuous purple streaks on the lower tepals. Subequal floral bracts are also said to be characteristic of *G. leonensis,* but the bracts of *G. praecostatus* are also sometimes nearly equal, whereas the inner bract in some specimens of *G. leonensis* is as much as 10 mm shorter than the outer. *Gladiolus leonensis* and *G. praecostatus* are separated both by geography and flowering time. *Gladiolus praecostatus* flowers from June to October and occurs in the Nimba Mountains of southern Guinea and adjacent Ivory Coast and Liberia, some distance to the east of the range of *G. leonensis,* which flower in April to June.

ADDITIONAL SPECIMENS

Sierra Leone. Loma Mountains, 1600 m, 13 Apr. 1966, *Jaeger 9807* (K, P), 2 Apr. 1966, *Jaeger 9708* (K); Tingi Mountains, summit of Sankanbiriwa, 12 April 1965, *Morton & Gledhill 1904* (K); Bintumane Peak, c. 1800 m, 2 May, 1949, *Deighton 5087* (K).

79. *Gladiolus chevalieranus* Marais

MAP 82. Marais, Kew Bull. 28: 313 (1973), as new name for *Acidanthera amoena* (A. Chevalier) Jacques-Felix, Adansonia, Sér. 2, 9: 131 (1969), not *Gladiolus amoenus* Salisbury. *Dortania amoena* A. Chevalier, Bull. Mus. Nat. Hist. Nat. Paris, Sér. 2, 9: 402 (1937). Type: Guinea, Benna, 1000 m, June 1937, *Jacques-Felix 1755* (P, lectotype designated here; P, isolectotype); between Limbou and Lontonta, July 1937 (fruit), *Jacques-Felix 1812* (P, syntype).

EPONYMY

chevalieranus, named in honor of Auguste Chevalier, 19th and early 20th century botanist and explorer in northern and western Africa.

DESCRIPTION

Plants 40–60 cm high. CORM ovoid, c. 15 mm in diameter, tunics firm-membranous, red-brown, becoming more or less fibrous by decay. CATAPHYLLS pale and membranous. LEAVES four to six, all with rather short blades and thus grading imperceptibly into entirely sheathing bractlike leaves, the lower one or two basal, the others inserted above the ground, soft-textured, blades narrowly lanceolate, 3–16 cm long, 4–8 mm at the widest. STEM erect, unbranched, c. 3 mm in diameter at the base of the spike.

SPIKES erect, 1- to 2-flowered; BRACTS green, very unequal, the outer 40–50 mm long, the inner 18–24 mm. FLOWER pale yellowish green, evidently without markings on the tepals; PERIANTH TUBE cylindric, 60–70 mm long, widening toward the upper 20 mm; TEPALS subequal, lanceolate, c. 20 × 5 mm, their orientation unknown. FILAMENTS c. 20 mm long, included in the tube; ANTHERS c. 8 mm long, with acute apiculate appendages c. 1 mm long. OVARY oblong, c. 4 mm long; STYLE evidently shortly exserted from the tube, dividing just above the base of the anthers, the branches diverging, c. 4.5 mm long. CAPSULES narrowly obovoid, 20–25 mm long; SEEDS broadly winged, 10–12 × 5–6 mm.

FLOWERING TIME. June (collected only once in bloom).

DISTRIBUTION & HABITAT

Gladiolus chevalieranus is known from just two collections, both made by the French botanist, H. Jacques-Felix, in the Fouta Djallon highlands of western Guinea in western Africa. The species is evidently a local endemic, restricted to the Benna Plateau, where it grows in grassy meadows on rock terraces at elevations of c. 1000 m (Chevalier, 1937).

DIAGNOSIS & RELATIONSHIPS

Gladiolus chevalieranus appears to be most closely allied to *G. aequinoctialis*, and it has the long-tubed flower and soft-textured leaves of that species. As noted by Jacques-Felix (1969), however, it is easily distinguished by its shorter leaf blades, seldom exceeding 15 cm, single-flowered spikes, pale yellow-green flower with a tube 60–70 mm long (described as 12–15 cm in the protologue), and tepals

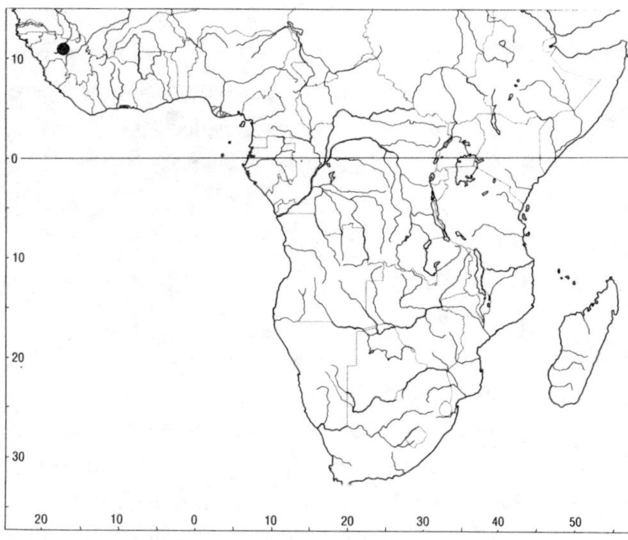

MAP 82. *Gladiolus chevalieranus.*

c. 2 cm long. This is quite different from *G. aequinoctialis,* which has long leaf blades, often longer than the stems, spikes of four to eight flowers, a perianth tube 80–140 mm long, and tepals c. 40 mm long.

The type species of the genus *Dortania, Gladiolus chevalieranus* was renamed by Marais (1973) when he transferred the tropical African species of *Acidanthera* to *Gladiolus.* The epithet *amoena* given to the species by Chevalier had already been used in *Gladiolus,* hence the need for a new name. Although still poorly known, there seems no doubt that it is a good species, and one of the few endemic Iridaceae of western Africa.

SPECIMENS

Known only from the type, cited above.

80. *Gladiolus iroensis* (A. Chevalier) Marais

MAP 83. Marais, Kew Bull. 28: 313 (1973).

SYNONYMY

Acidanthera iroensis A. Chevalier, Bull. Mus. Nat. Hist. Nat. Paris, Sér. 2, 9: 401 (1938). Type: Chad, Koulfé, marshy banks of Lake Iro, 29–30 June 1903, *Chevalier 8799* (P, lectotype designated here; B, BM, BR, K, MO, P, WAG, isolectotypes); 29–30 June 1930, *Chevalier 8899* (P, syntype).

EPONYMY

iroensis, "from Lake Iro," in southern Chad, the type locality.

DESCRIPTION

Plants 40–60 cm high. CORMS 14–18 mm in diameter, tunics membranous, becoming finely fibrous, red-brown to straw-colored. CATAPHYLLS firm-membranous, green and leaflike above the ground. LEAVES four to six, the lower two to four basal and longest, one-third to about half as long as the stem, the upper cauline and shorter, all with comparatively short blades, narrowly lanceolate, 6–9 mm at the widest, the margins and two or three other veins lightly thickened and hyaline. STEM unbranched, 2–3 mm in diameter at the base of the spike.

SPIKE erect, 1- to 2-flowered; BRACTS green, the outer 5–7 cm long, the inner 1–2 cm shorter than and hidden by the outer. FLOWERS greenish white, sometimes flushed with mauve, the lower three tepals each with a reddish band in the lower midline; PERIANTH TUBE cylindric and straight, widening somewhat near the throat, 120–140 mm long; TEPALS lanceolate, the upper c. 35 mm long, the lower c. 30 mm long, width uncertain. FILAMENTS exserted for c. 10–12 mm; ANTHERS 8–11 mm long, with well-developed apiculi 4–5 mm long. OVARY 5–6 mm long; STYLE and branches not seen. CAPSULES and SEEDS not known.

FLOWERING TIME. June and July.

DISTRIBUTION & HABITAT

Known from just three collections, *Gladiolus iroensis* is evidently restricted to a small portion of

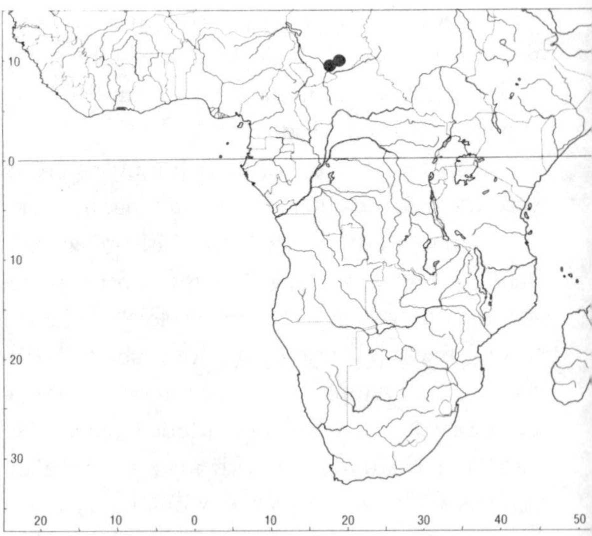

MAP 83. *Gladiolus iroensis.*

southern Chad. It is very unusual in section *Acidanthera* in favoring marshy habitats. The type locality, the marshes around Lake Iro, is seasonally inundated but dries out during the dry season.

DIAGNOSIS & RELATIONSHIPS

Gladiolus iroensis is clearly related to *G. aequinoctialis* and *G. murielae* and somewhat intermediate between them, but it is distinguished by its rather short leaf blades and, from the former, also by the very long anther appendages 4–5 mm long. The species bears a close resemblance to the western Guinean endemic, *G. chevalieranus,* which also has short leaf blades. *Gladiolus chevalieranus,* however, has single-flowered spikes, smaller flowers with perianth tubes 70–80 mm long, and filaments not exserted from the tubes. More material in good flowering condition is needed to assess the possible close relationship between the two species. The habitat, at least, of the two appears quite different. *Gladiolus iroensis* grows in seasonally flooded or marshy areas, whereas *G. chevalieranus* occurs in meadows in thin soil on rock terraces.

ADDITIONAL SPECIMENS

Chad: Djouman, seasonally inundated waterlogged soil, 8 July 1969, *Fotius 1527* (ALF, G).

81. *Gladiolus murielae* Kelway

PLATE 40, FIGURE 60, MAP 84. Kelway, Gard. Chron., Ser. 3., 92: 107 (1932). Type: Ethiopia, exact locality unknown, 1931, *Erskine s.n.* (location of specimens unknown).

SYNONYMY

Acidanthera bicolor Hochstetter, Flora 27: 25 (1844). Bouché & Wittmack, Monatschr. Ver. Beförd. Gartenb. Berlin 19: pl. 1 (1876). Baker, Handbook Irideae 188 (1892); Fl. Trop. Africa 7: 358 (1898). *Gladiolus callianthus* Marais, Kew Bull. 28: 311–317 (1973), as new name for *A. bicolor* Hochstetter, which cannot be transferred to *Gladiolus* because of the existence of *G. bicolor* Baker (1878). Goldblatt, Fl. Zambesiaca 12(4): 101 (1993). Type: Ethiopia, Shoa, hills and valleys on wet rocks, "in collibus vallium districtus Schoata ad rupes humentes," 24 July 1838, *Schimper s.n.* (no type material has been located and it may have been destroyed during World War II, but protologue leaves no doubt about the identity of the species).

Ixia quartiniana A. Richard, Tent. Fl. Abyssinia. 2: 310 (1851), not *Gladiolus quartinianus* A. Richard (1851). *Sphaerospora gigantea* Klatt, Linnaea 34: 699 (1866), an illegitimate superfluous name for *Ixia quartiniana.* Type: Ethiopia, fields near Adoa, "pratis circa Adoua," Aug. 1939, *Quartin-Dillon s.n.* (P—Herb. Richard, lectotype designated here; P (3), isolectotypes); Beless plains, "in planitie Beless, Prov. Shiré," Feb. 1839, *Quartin-Dillon s.n.* (P, syntypes).

EPONYMY

murielae, named in honor of Muriel Erskine, wife of the collector of the plants on which the description was based.

DESCRIPTION

Plants 30–65 cm high. CORM 15–22 mm in diameter, tunics firm to softly membranous, fragmenting irregularly, sometimes becoming subfibrous, dark red-brown. CATAPHYLLS firm-membranous, green or purple above the ground. LEAVES four to eight, the lower three to five basal, narrowly lanceolate, reaching at least to the base of the spike, sometimes slightly exceeding it, 5–12 mm at the widest, relatively soft-textured and without thickened margins or midrib. STEM erect or inclined, unbranched, 3–4 mm in diameter at the base of the spike.

SPIKE often inclined, 3- to 5-flowered; BRACTS

FIGURE 60. *Gladiolus murielae.* Corm, leaves, and flowering spike, × 0.5; single flower with bract and anther showing apiculate appendage, full size; cross-section of leaf, × 2 (*la Croix 2728*).

green, the outer 5–8(–10) cm long, the inner shorter than and concealed by the outer. FLOWERS white, with a prominent dark purple streak in the midline of the lower three or all of the tepals, sweetly scented, more strongly in the evenings; PERIANTH TUBE cylindric and straight, slightly wider near the throat, (90–)120–150 mm long; TEPALS more or less equal, lanceolate, 35–45 mm long, 17–22 mm wide. FILAMENTS exserted for 10–15 mm; ANTHERS c. 15 mm long, with rigid filiform apiculi 2–4 mm long. OVARY 6–8 mm long, style arching over the stamens, dividing beyond the anthers, the branches c. 5 mm long, much expanded in the upper half. CAPSULES oblong-ellipsoid, 20–25 mm long; SEEDS c. 8 × 5 mm, broadly winged. CHROMOSOME NUMBER $2n = 30$.

FLOWERING TIME. Mostly January to March in Tanzania and Malawi, July to September in Ethiopia.

DISTRIBUTION & HABITAT

Gladiolus murielae has a scattered distribution from northeastern Africa southward to Malawi and central Mozambique. It is most common in northern Ethiopia and has been collected frequently there. The plant is also well known in the southern half of Malawi, but the record is poor between the northern and southern ends of the range. There are no records from either southern Ethiopia or Kenya, where there are suitable habitats for the plant, and the few collections from Tanzania are from the Mufindi District and farther south near Songea. The Mozambique record is from near Domué, not far from central Malawi, where it seems to be common in the mountains around Dedza, only a short distance from Mozambique. The distribution pattern is unusual. As far as it is possible to tell, there are no differences between the Malawian and Ethiopian forms of the species. *Gladiolus murielae* favors rocky, partly shaded sites, growing on cliffs and in rocky outcrops where the soil is thin and there are few other plants.

DIAGNOSIS & RELATIONSHIPS

The tall plants with long-tubed white flowers splashed with dark purple on the lower tepals and well-developed soft-textured leaves readily distinguish *Gladiolus murielae* from most species of tropical African *Gladiolus,* including all those from the eastern half of the continent. Western tropical African *G. aequinoctialis* appears on morphological grounds to be most closely related to *G. murielae,* and without knowledge of the source of plants the two are often difficult to separate. Both have large white flowers with a tube usually 12–15 cm long, elliptic tepals 35–45 mm long, the lower marked with dark purple, and long anthers with prominent apiculate appendages. Marais (1973) distinguished the two on the basis of leaf texture, softer in *G. murielae,* and length of the anther apiculi, 4–5 mm long in *G. murielae* compared with 2–3 mm in *G. aequinoctialis.* In the main, these criteria work and it seems useful to maintain the two as

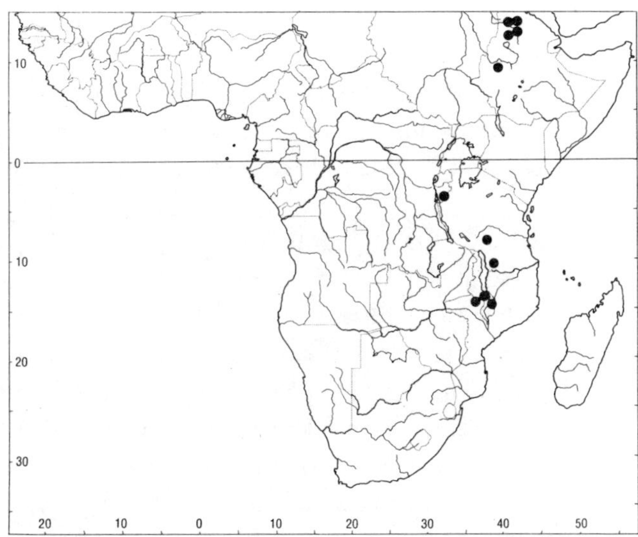

MAP 84. *Gladiolus murielae.*

separate species. Their relationship is, however, very close. In *G. murielae* the soft leaves generally have only one pair of secondary veins in addition to the lightly raised midrib; the remaining veins are all very fine. *Gladiolus aequinoctialis* has firmer-textured leaves, and the secondary pairs of veins are well developed. In cultivation, the two flower at different times of the year, *G. murielae* in the summer and *G. aequinoctialis* in the late autumn.

HISTORY

One of the earliest species of tropical African *Gladiolus* to be collected, *G. murielae* was first recorded by Georg Heinrich Schimper in northern Ethiopia in 1838, and a year later, also in northern Ethiopia, by the botanist of a French expedition to that country, Richard Quartin-Dillon. The latter collection became the type of *Ixia quartiniana,* described in 1851, whereas Schimper's specimens formed the basis for *Acidanthera bicolor,* the type and only species of a new genus *Acidanthera,* recognized because of the elongate perianth tube. The existence of *G. bicolor* (a synonym of *G. villosus,* now *Sparaxis villosa*) and *G. quartinianus* (a synonym of *G. dalenii*) prevents the transfer of either name to *Gladiolus.* When the genus *Acidanthera* was merged with *Gladiolus* in 1973, Marais gave it the new name, *G. callianthus.* Marais was evidently unaware that the species had already been validly described as *G. murielae* in 1932.

SELECTED SPECIMENS

Ethiopia. Wollega: Didessa, shade near stream, 15 July 1975, *Ash 3096* (BR, K, WAG). Tigre: 2 km from Axum along road to Adowa, 2090 m, 25 Aug. 1973, *Aweke & Gilbert 802* (K); 10 km north of Adowa, on cliffs of volcanic rock, 2000 m, 12 Sept. 1962, *Mooney 9564* (ETH); 18 km north of Adowa on the road to Asmara, 1950 m, 20 Aug. 1973, *Gilbert & Getachew 2758* (ETH, K); 26 km southwest of Inda Sellasie on the Gondar road, 29 Aug. 1973, *Gilbert & Getachew 2916* (K); swampy ground on Tekkeze side of Inda Selassie, Sept. 1972, *Tewolde 882* (ETH); Tigre, 7 Sept. 1970, *de Wilde 7029* (ACD, BR, HBG, MO). Shiré, without date, *Quartin-Dillon & Petit 112* (K, P). Without Precise Locality: Arwassa, 1800 m, 4 Aug. 1852, *Schimper 434* (P); Bellgass near Taserotsch, 1200–1800 m, 8 Aug. 1852, or without locality and dated 1854, *Schimper 434* (G, P); Simen, 1852, *Schimper 434* (BR, P); plains in the Walcha Mountains, "in planitie montana Walcha, provinciae Sana," 6 Aug. 1841, *Schimper 1634* or *s.n.* (B, BR, G, K).

Burundi. Kagera waterfalls, 28 Dec. 1968, *Delarge 2* (BR).

Tanzania. Iringa: Mufindi District, Kigogo Forest Reserve near old forest station, 1800 m, 2 Mar. 1987, *Keeley s.n.* 6. Rovuma: Matagoro Hill just south of Songea, crevices in steep rock face, 3 Feb. 1956, *Milne-Redhead & Taylor 8473* (B,BR, EA, K).

Malawi. Central: Dedza, Chincherere Hill, Chongoni Forest, 11 Mar. 1967, *Salubeni 581* (K, LISC, MAL, PRE); Dedza, Kalichero Hill, foot of wet rocks, 1700 m, 21 Jan. 1959, *Robson & Jackson 1289* (K, LISC, MAL, PRE, SRGH). Southern: Zomba Mountain, near Chingwe's Hole, 7 Feb. 1985, *Salubeni & Nachamba 4006* (K, MAL, MO); Mt. Malosa, opposite Domasi Valley, 3 Jan. 1980, *Salubeni, Banda, & Masiye 2690* (MAL); Ndirandi, 5 Dec. 1967, *Abrams et al. s.n.* (MAL); banks of the Mtungusi, Shire Highlands, Oct. 1880, *Buchanan 5* (E, K).

Mozambique. Tete: Angonia, north of Mt. Domué, date unknown, *Macuácua & Mateus 1092* (B).

82. *Gladiolus candidus* (Rendle) Goldblatt

PLATE 41, FIGURE 61, MAP 85. Goldblatt in Stork & Lebrun, Enumeration Plantes Fleurs Afrique Tropicale 3: 89 (1995). *Acidanthera candida* Rendle, J. Linn. Soc. Bot. 30: 404 (1895). Baker, Fl. Trop. Africa 7; 360 (1898). J. D. Hooker, Curtis's Bot.

Mag. 129: pl. 7879 (1903), not *Gladiolus candidus* Herbert, a name without description (e.g., see Baker, J. Linn. Soc. Bot. 16: 177 (1878)). Type: Kenya, Athi Plains, Lanjaro, *Gregory s.n.* (BM, holotype).

SYNONYMY

Acidanthera laxiflora Baker in Johnston, Kilimanjaro Exped. 346 (1886), name without description; Trans. Linn. Soc., Ser. 2, 2: 350 (1887); Handbook Irideae 188 (1892). Type: Kenya, Taita, Maungu, 1884, probably May, *Johnston s.n.* (K, holotype; B, BM, isotypes), not *Gladiolus laxiflorus* Baker (1878b).

Acidanthera gracilis Pax, Bot. Jahrb. Syst 15: 154 (April 1892). *Acidanthera zanzibarica* Baker, Handbook Irideae 188 (October 1892), illegitimate superfluous name, based on the same type, for *A. gracilis*. Type: Kenya, mainland near Mombasa, July 1876, *Hildebrandt 2015* (B, holotype; B, BM, isotypes), not *Gladiolus gracilis* Jacquin, from South Africa.

Acidanthera ukambanensis Baker, Fl. Trop. Africa 7: 359 (1898). *Gladiolus ukambanensis* (Baker) Marais, Kew Bull. 28: 313–314 (1973). Miller & Morris, Pl. Dhofar (Southern Oman) 150 (1988). Type: Kenya, Ukamba, on rocks, 1500–1800 m, *Scott Elliot 6434* (K, holotype).

Gladiolus ukambanensis var. *alatus* Marais, Kew Bull. 28: 314 (1973). Type: Kenya, Naivasha District, Kedong Valley, Mt. Margaret Estate, *Bally 1060* (K, holotype, not seen; G, isotype).

EPONYMY

candidus, alluding to the pure white flowers without contrasting markings.

DESCRIPTION

Plants 20–40 cm high. CORM globose, 12–25 mm in diameter, tunics firm-papery, breaking into vertical fibers above and below. CATAPHYLLS membranous, the uppermost extending above the ground and green. LEAVES two or three, all more or less basal, narrowly lanceolate, about half as long as the stem, 5–10 mm wide, the margins and midribs evident but not noticeably thickened. STEM erect, unbranched, c. 2.5 mm in diameter below the first flower.

SPIKE erect, 2- to 4-flowered; BRACTS (25–)40–50(–80) mm long, the inner shorter and narrower than the outer. FLOWERS white, rarely pink, occasionally with purple median streaks in the lower midline of the lower tepals, sweetly scented; PERIANTH TUBE (70–)80–100 mm long, more or less straight and cylindric; TEPALS subequal, broadly lanceolate to elliptic, (20–)25–30 mm long, c. 15 mm wide. FILAMENTS c. 20 mm long, included in the tube or, rarely, exserted 1–2 mm; ANTHERS 8–10 mm long, with acute apiculate appendages 1.3–1.8 mm long. OVARY oblong, c. 5 mm long; the style dividing opposite the anther apices, the branches 5–7 mm long, often unusually broad and fringed above. CAPSULES narrowly elliptic to obovate, 18–22 mm long; SEEDS fully to partly winged, to angled or rounded and wingless.

FLOWERING TIME. Mainly May and June in southern Ethiopia, March to May in Kenya and Tanzania, October to December in Somalia, July and August in Oman.

DISTRIBUTION & HABITAT

Gladiolus candidus is a widespread eastern African species, evidently most common in interior highland Kenya but extending south to the Arusha District of Tanzania, north into Bale and Sidamo Provinces of Ethiopia, and east through Somalia to Djibouti. Outlying populations occur in the highlands of Dhofar, southwestern Oman, in the southern Arabian Peninsula. The distribution appears rather scattered and includes a range of habitats from well-watered highlands, where plants seem to favor rocky outcrops, to woodland and fairly dry bushland. Nothing has been reported about the biology of the species. Unusual for *Gladiolus,* the corms are eaten when available in Dhofar (Miller & Morris, 1988). Corms are reported to be eaten as a remedy for snake bite in northern Kenya (*Gillett 12833*)!

DIAGNOSIS & RELATIONSHIPS

Gladiolus candidus is a fairly representative member of section *Acidanthera,* and it has the long perianth tube and white flower characteristic of the section. It is distinctive in generally having a uniformly white flower (occasionally the lower tepals show light purple median streaks) and fairly slender habit. The stems are erect and bear two to three or, occasionally, four flowers with tubes 70–100 mm long. The other eastern African member of the section, *G. murielae,* is typically a robust plant and has large flowers with tubes 120–150 mm long and the lower tepals always with conspicuous dark purple markings. Another notable feature of *G. candidus* is the short filaments that reach only to the mouth of the perianth tube or are exserted for only 1 to 2 mm. The anthers are thus lodged at the base of the tepals.

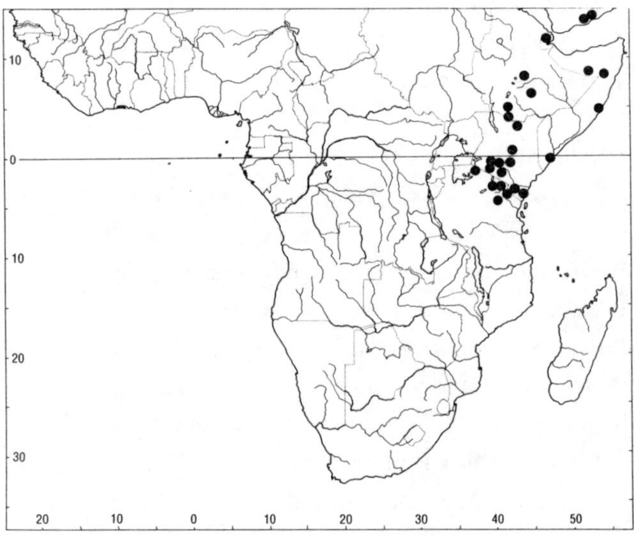

MAP 85. *Gladiolus candidus.*

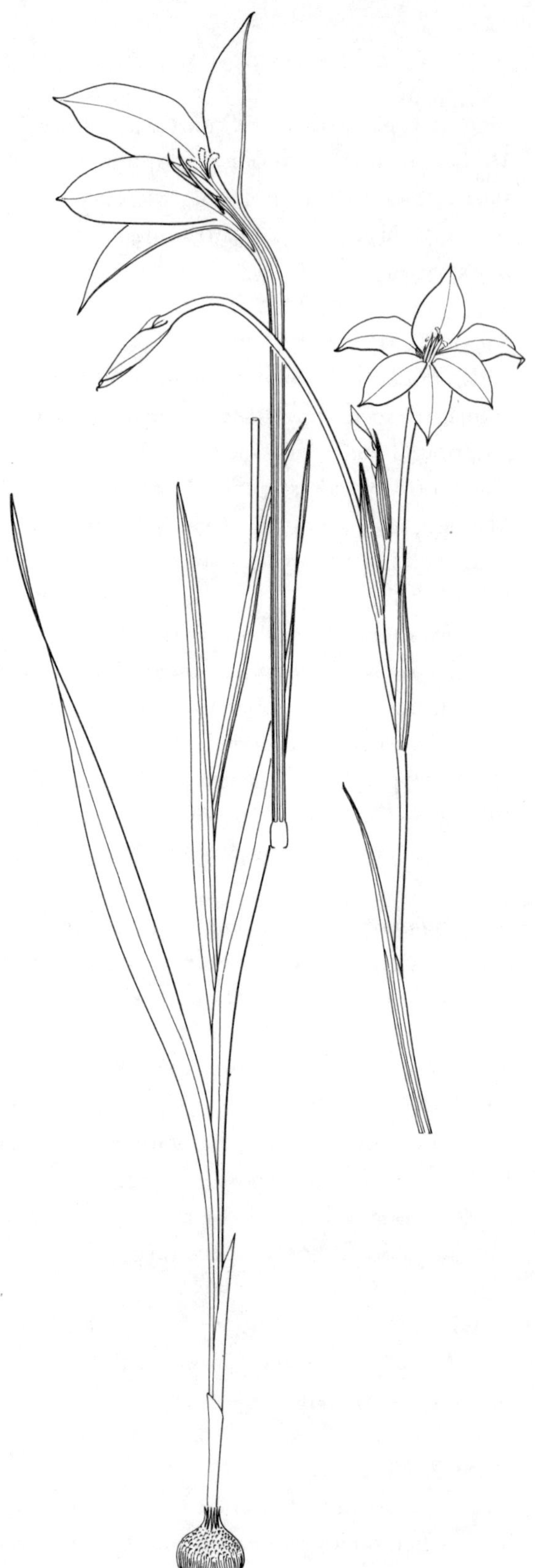

FIGURE 61. *Gladiolus candidus.* Corm, leaves, and flowering spike, × 0.5; single flower, full size (*Faden et al. 74/556*).

One unusual aspect of the *Gladiolus candidus* is the occasional absence of a wing on the seeds. Noting this variability, Marais (1973) provided the name *G. ukambanensis* var. *alatus* for plants with broadly winged seeds (the type specimens of *G. ukambanensis* have wingless seeds). This characteristic does not seem to be correlated with any other. Most specimens have been collected without fruits so it is difficult if not impossible to determine to what degree the winged and wingless states intergrade. It seems best not to attach too much weight to the feature, for at least in some populations, seeds may range from nearly wingless to nearly entirely winged. A decision about the recognition of formal varieties must be postponed until more is known of the variation in the character.

HISTORY

There is a surprisingly confused history for *Gladiolus candidus,* first collected by the German, J. M. Hildebrandt, in 1876. It was first named *Acidanthera laxiflora* by Baker in 1886, based on plants from Maungu in southeastern Kenya, collected on H. H. Johnston's Kilimanjaro Expedition of 1884. Then in early 1892 Pax described *A. gracilis* from the plants collected by Hildebrandt near Mombasa. Later that same year and based in part on the same collection, Baker named *A. zanzibarica,* neither author mentioning any differences between his species and *A. laxiflora.* In 1895, Rendle described *A. candida* from plants from interior Kenya, and Baker named *A. ukambanensis* in 1898 based on a collection from the same area.

There is no reason to doubt that all the above are conspecific and they were so treated by Marais (1973) when he transferred the tropical African species of *Acidanthera* to *Gladiolus.* Marais chose to transfer *A. ukambanensis* to *Gladiolus,* after considering the epithets of all earlier synonyms preoccupied in *Gladiolus.* Certainly, *G. laxiflorus* Baker (dating from 1876) and *G. gracilis* Jacquin (dating from 1792) prevent transfer of *A. laxiflora* and *A. gracilis,* and Baker's *A. zanguebaricus* is illegitimate, being based on the same type collection as *A. gracilis.* That leaves *A. candidus* the next available synonym for use in *Gladiolus.* Marais thought that the existence of "*G. candidus*" (actually a hybrid cultivar listed by Baker, 1878b, without description) prevented future use of that epithet in *Gladiolus.* Nomina nuda, names published without description, however, are not considered valid under the *International Code of Botanical Nomenclature*; the epithet *candidus* may therefore be used in *Gladiolus.*

SELECTED SPECIMENS

Ethiopia. Bale: Web Gorge, 15 km west of Goro on the road to Ginir and Sof Omar to Robe, 2250 m, 1 June 1983, *Gilbert, Ensermu, & Vollesen 8024* (ETH, K). Harerge: Ogaden, without precise locality, Nov. 1958, *Ashall 5/58* (EA, ETH, K). Sidamo: c. 15 km south of Yavello, old airfield, 1700 m, 17 May 1976, *Gilbert & Jefford 4587, 4587A* (K); 13 km north of Negeli, road to Debre Mengist, 1300 m, 9 May 1982, *Friis, Tadesse, & Vollesen 2674* (C, K, UPS); 22 km from Mega turnoff on road from Negeli, 11 May 1980, *Thulin, Hunde, & Tadesse 3529* (BR, K, MO, UPS); 9 km from Negeli toward Filtu, 11650 m, 9 May 1980, *Thulin, Hunde, & Tadesse 3407* (ETH, K, UPS); under Mega Mountain, 1200 m, 16 May 1959, *Thesiger 2060* (BM). Arusi: Arusi Plateau, Oct. 1908, *Drake-Brockman s.n.* (K).

Djibouti. Adonta, open forest, 16 Apr. 1983, *Blot 251* (P).

Kenya. Central: Karen, Ngong, 2000 m, Apr. 1950, *Babault s.n.* (P); Nairobi National Park, shallow soil over rock, 30 Apr. 1967, *Agnew 9196* (C, MO); Thika Roadhouse, wet grassland, 5 Apr. 1951, *Verdcourt 458* (MO, PRE); behind Nairobi Golf Range, 1700 m, 14 May 1974, *Faden & Faden 74/556* (MO); Ngong Escarpment near Nairobi, 7

June 1947, *Bally 5033* (G); Athi Plains, Mar.–Apr. 1929, *Benson s.n.* (BM). Eastern: north of Mt. Kenya, c. 20 km south of Isiolo, 1685 m, 28 June 1979, *Linsen & Giesen 9* (B, G, HBG, MO, NY); 24 km south of Isiolo, road from Meru, 25 Dec. 1969, *Gillett 18935* (EA, PRE). Moyale, cracks in granite rocks, 17 Apr. 1952, 1080 m, *Gillett 12833* (B, BM, BR, EA, LISC, MO, P, PRE, S). Coastal: Taita, Nov. 1895, *Sacleux 2279* (P).

Somalia. Nugaal: Taleh, 600 m, 25 Oct. 1961, *Hemming 2309* (K, PRE); between Bihen and Las Anod, 27 Oct. 1944, *Glover & Gilliland 250* (BM); Northern Rangelands, 8°56′ N, 48°23′ E, edge of seasonal watercourse, 700 m, 2 Dec. 1980, *Beckett 643* (EA). Mudug: 11–14 km north of Xarardheere (Harardere) to Hobyo (Obbia), coastal plain with open grassland, 150 m, 29 May 1989, *Thulin & Dahir 6697* (K, UPS). Jubbada Hoose: 21 km west of Kismayo, evergreen bushland on red sand, 5 July 1983, *Gillett, Hemming, Watson, & Julin 25326* (K); Kismayo, 1926, *Gorini 58* (FI).

Oman. Dhofar, Jebel Qara near Kaftawt, escarpment woodland, 790 m, 1 Aug. 1985, *Miller 7194* (K); Hagif to Ayun road, c. 8 km west of turn-off from Salalah road, c. 870 m, 17 Sept. 1985, *Miller 7647* (K); Khadrafi, Sarfait area, c. 700 m (2200 ft), 16°42′ N, 53°10′ E, 26–27 Sept. 1976, *Mandaville 6915* (BM).

Tanzania. Mara: Shirati to Utegi, 23 May 1957, *Gane 120* (EA). Arusha: west of Arusha, grassland, c. 1200 m, Apr. 1965, *Beesley 112* (K); Mt. Meru, Nasala track, Arusha National Park, 26 Mar. 1968, *Greenway & Kanuri 13249* (BR, EA, K, PRE); Ngurdoto National Park, Lake Tulusia, grassland, 1650 m, 11 Apr. 1965, *Richards 20168* (MO); Monduli District, Essimingor, 1100 m, 13 Apr. 1967, *Carmichael 1359* (PRE); below Marangu, 1000 m, Apr. 1894, *Volkens 2131* (BM, G).

ADDITIONAL OR INCOMPLETELY KNOWN SPECIES

Gladiolus somalensis Goldblatt & Thulin

Goldblatt & Thulin, Novon 5 (1995). Type: Somalia, Sanaag, escarpment south of Laasqoray near Mirci, evergreen bush on limestone, 16 Jan. 1995, *Thulin, Dahir, & Hassan 9079* (UPS, holotype).

EPONYMY

somalensis, from Somalia.

DESCRIPTION

Plants (7–)12–30 cm high. CORM obconic, c. 12 mm in diameter, the tunics of softly textured layers, these decaying with age into fine netted fibers. LEAVES four or five, the lower three basal and longest, reaching at least to the base of the spike and one or more often slightly exceeding the spike, the blade more or less linear, (1–)2–4 mm wide, the upper one or two leaves inserted on the lower half of the stem, smaller than the basal leaves. STEM erect, simple or with one or two branches, c. 1.2 mm in diameter below the base of the spike. SPIKE lightly flexuose, 2- to 10-flowered; BRACTS green and soft-textured, the outer (7–)12–17(–23) mm long, the inner about two-thirds as long as the outer. FLOWERS zygomorphic, orange, the lower lateral or lower median tepals bright yellow in the lower half; PERIANTH TUBE funnel-shaped, 6–8 mm long; TEPALS unequal, lanceolate, the upper three larger than the lower, the dorsal inclined over the stamens, 16–18 × 8 mm, the upper laterals about as long, the lower tepals more or less parallel to the ground, the lower lateral tepals c. 15 × 5.5 mm, the lower median c. 12 × 5 mm, the margins raised below and the surface channeled in the lower half. FILA-

MENTS 8–10 mm long, exserted 4–6 mm from the tube; ANTHERS 3–5 mm long, yellow. OVARY ovoid, 2–3 mm long; STYLE arching over the stamens, the style dividing c. 1.5 mm beyond the anther apices, the branches c. 2.5 mm long, filiform, evidently not expanded apically. CAPSULES and SEEDS unknown.

FLOWERING TIME. January and February.

DISTRIBUTION & HABITAT

Although first collected in the 1950s, *Gladiolus somalensis* remained too poorly known to be described or even to be assigned with confidence to genus. A new collection made in January 1995 by the Swedish botanist Mats Thulin and his Somalian colleagues, however, is well preserved and is accompanied by photographs and spirit material. This has made it possible to draw up a description and formally name the species. The new collection reached me too late for the species description to be placed in the body of this volume or in the key to species, hence its inclusion at the end of the treatment.

Gladiolus somalensis is restricted to northeastern Somalia. According to Thulin, the species grows in montane habitats in rocky situations, and it may be restricted to limestone-derived soils. The population found by Thulin occurred in evergreen bushland dominated by a native boxwood (*Buxus*) and *Cadia* (Fabaceae), at an elevation of 1350 m.

DIAGNOSIS & RELATIONSHIPS

Gladiolus somalensis is readily distinguished from all other tropical African species of the genus by its small, bright orange flowers and the conspicuous yellow nectar guide located on the lower lateral or lower median tepals. The flowers are fairly short-tubed and have a superficial resemblance to some species of the genus *Tritonia*. Species of that genus usually have bright orange flowers but the nectar guides typically have a large, central toothlike callus. *Gladiolus* and *Tritonia* differ radically in their seeds and in the nature of the floral bracts. The bracts of *Tritonia* are typically fairly short, are more or less membranous or are dry, and have bifurcate or trifurcate apices. The capsules and seeds of *G. somalensis* are unknown and cannot be used to assist in generic placement. The bracts, however, are relatively large and are soft-textured and green, hence quite typical of *Gladiolus*. The style branches are linear and do not appear to be apically expanded as is the case with most species of *Gladiolus*, but I do not attach much significance to style branch structure in this case. The relationships of *G. somalensis* with other species of the genus are unclear. It bears a fair resemblance to the Ethiopian species, *G. calcicola*, and to the Eritrean *G. mensensis*. The possibility exists that its affinities may lie with southern African species of subgenus *Gladiolus*, which have short-tubed flowers of comparable size. Until more is known about the species, its affinities remain speculative at best.

ADDITIONAL SPECIMENS

Somalia. Wadi Hantara, Candala, 6 Feb. 1956, *Azzaroli 6* (FI); Azienda Uar Mahan, Jan.–Feb. 1959, *Sacco s.n.* (FI).

Gladiolus aff. *pretoriensis* O. Kuntze

A single specimen (*Fanshawe 11940*) from the Mafinga Hills in northern Malawi at the herbarium (SRGH) in Harare, Zimbabwe, appears to match no known species of *Gladiolus* from tropical Africa. It resembles most closely *G. pretoriensis* from the Transvaal, South Africa. The specimen, lacking a corm, was said to have reddish flowers, unlike *G. pretoriensis*, which has pale pink or lilac flowers. The leaves of the two species correspond in being terete to square in transverse section, with four narrow longitudinal grooves and strongly thickened and winged margins and midribs. The Fanshawe specimen has six leaves, all with long leaf blades. Additional specimens, including capsules and seeds, are needed to confirm to identity of the plant from the Mafinga Hills. If it has wingless seeds and ovoid capsules, there will be no remaining doubt that it is *G. pretoriensis.* Such a wide disjunction in the range of a species of *Gladiolus* is unexpected but not without precedent. For example, *G. murielae* occurs in northern Ethiopia, Burundi, and southern Tanzania and Malawi, and *G. aequinoctialis* has populations in Sierra Leone and Fernando Po and the Cameroon highlands.

Glossary

actinomorphic—a regular flower, i.e., radially symmetric
anatropous—of an ovule curved back on itself
androecium—the male part of the flower (filaments and anthers)
angiosperm—informal term for a flowering plant
anther—part of the male organ, consisting of two lobes (thecae), each with two chambers that contain the pollen
anthesis—time of opening of the flower
aperture—of a pollen grain, the opening or pore in the wall through which the pollen tube emerges
apiculus (apiculate)—small, usually sharp-pointed structure borne at the apex of an organ, e.g., leaf, tepal, anther
apomorphic—specialized or derived characteristic; synapomorphy, a shared derived characteristic
attenuate—with an elongated and tapering apex
axil (axillary)—the angle between the leaf and stem axis, also lateral as opposed to terminal

basifixed—attached basally, subbasifixed, attached shortly above the base
bifid—divided into two for a short distance
bract—modified leaf associated with a flower or inflorescence, sometimes much reduced in size, having a leafy or nonleafy texture

capsule—a dry fruit, as found in *Gladiolus* and most other Iridaceae, composed of fused carpels (three in all Iridaceae), and containing two or more seeds per chamber
cataphyll—modified leaf produced at the base of a plant, often underground, and without a blade

cauline—relating to the stem
centric—of a leaf, radially symmetric in transverse section
chalaza (chalazal)—anatomical part of the ovule and seed, the point at which the vasculature enters the ovule, often prominent in the mature seed
chlorenchyma—plant tissue, usually in the leaf, containing chloroplasts, the main site of photosynthesis
chromosome—cell organelle located in the nucleus that contains the genetic information
clade—evolutionary lineage defined by derived characters
columellate—in pollen exine, with small pillars, the pillars supporting the tectum (roof layer)
connective—sterile tissue between the pollen-bearing chambers of the anther, sometimes extended apically into a distinctive structure
coriaceous—leathery texture
corm—underground storage organ consisting of tissue derived from the base of the flowering stem
corm tunic—dry fibrous, papery, or leathery covering of the corm, derived from the bases of cataphylls and leaves
cuticle—thin impermeable layer external to the epidermal cells

dambo—term used in southern tropical Africa for poorly drained land, usually moist throughout the year and marshy or inundated in the wet season
dehiscence—method of splitting, e.g., of capsules, anthers
dorsal—upper
dysploid—chromosome number differing from the norm or basic number for a group

edaphic—relating to the soil
emarginate—with the apex notched
endemic—restricted to a particular geographic area, sometimes used as a noun
endothecium—tissue in the anther with distinctive wall thickenings
epidermis—distinct plant tissue layer, usually one cell layer thick
exine—outer wall of the pollen grain, usually with distinctive sculpturing

falcate—shaped like a sickle
filament—part of the male organ of the flower, the stalk bearing the anther, usually threadlike

filiform—threadlike, i.e., very slender
flavone—class of flavonol compounds of simple structure
flavonol—group of chemical compounds in plants, often studied by botanists

genome—entire genetic constitution of an organism
girder—strengthening structure of the leaf in which thickened cells extend from the vascular trace to the epidermis
glabrous—smooth, without hairs
glaucous—with a waxy or whitish bloom on the surface
glycoside—primitive type of flavonol
gynoecium—the female part of the flower, composed of one or more units called carpels (ovary, style, and stigma)

herbarium—museum collection of preserved plant specimens
herkogamous—referring to spatial separation of receptive part of the stigma from pollen-bearing structures
homonym—specific name matching exactly one already given to a plant in the same genus
hypocrateriform—shaped like a salver, of a flower with a long narrow tube and tepals spread at right angles to the tube

inaperturate—of pollen grains without pores or openings of any kind
inflorescence—cluster of flowers on a stem
isobilateral—both surfaces of a leaf identical

karyotype—appearance of the chromosome complement

lanceolate—shaped like a spear
linear—narrow and with sides parallel
locule—chamber of the ovary

meristem—area of active cell division, giving rise to cells that form tissues and organs
mesophyll—green, chloroplast-filled tissue
monophyletic—taxonomic group sharing a common ancestor
morphology—external appearance, usually at macroscopic level
muri—walls, often of the pollen wall sculpturing

ontogeny—developmental process of an organ
operculum—literally, lid, a discrete covering of the pollen grain aperture
ovary—part of the gynoecium that contains the placentas and ovules
ovule—organ within the ovary that bears the embryo sac and egg cell

papilla (papillate)—small round protuberance extending shortly above the cell surface
papyraceous—papery texture
parenchyma—tissue consisting of large, rounded thin-walled cells
pedicel (pedicellate)—stalk of an individual flower
peduncle—stalk of an inflorescence
periclinal—parallel to the external surface
pericycle—tissue between the epidermis and vasculature, usually in stems
petaloid—with a texture like a petal
phylogeny (phylogenetic)—relating to evolutionary relationships
plesiomorphic—unspecialized or primitive characteristic; symplesiomorphy, a shared primitive characteristic
polyploid—having more than two sets of the basic chromosome complement
proboscis (proboscid)—elongated mouth part of some insects, enclosing the tongue
protologue—the formal description of a species
pseudomidrib—the central and main vein of the unifacial leaf of many Iridaceae, so named to indicate that it is not homologous with the midrib of bifacial leaves of other plants
puberulous—with minute hairs
pubescent—with hairs

sagittate—shaped like an arrow
scabrae (scabrid)—small raised surface sculpturing, usually in a regular pattern
scarious—dry and papery
sclerenchyma—tissue consisting of cells with heavily thickened walls
sensu—in the sense of, or as interpreted by
septal nectary—a nectar-producing gland located within the septa (walls separating the locules) of the ovary
sessile—without a stalk, stemless
stamen—male organ within the flower
stigma (stigmatic)—apical part of the style and receptive surface for pollen grains

stolon—slender process produced from the base of the plant, arising from the stem and terminating in a bud, bulb, or cormlet
style—portion of the gynoecium linking the stigma(s) to the ovary, hollow, allowing passage of pollen tubes
sulcus—narrow groove, more specifically, the germination pore of a monocot pollen grain
symplesiomorphic—*see* plesiomorphic
synapomorphic—*see* apomorphic

tannin—plant chemical, located in specific dark-colored cells
taxonomy (taxonomic)—study of nomenclature, relating to classification
tectate—particular pollen exine structure with a roof layer (tectum) supported by pillars
tepal—term for petaloid part of the flower when the calyx (green outer segments of some flowers) is lacking, as in the lilioid monocots
terete—round in cross-section
trichome—surface outgrowth, including hairs of various types
truncate—of a rounded object abruptly flattened at one end

unifacial—of leaves in which the two surfaces have the same anatomical origin

vicariant—immediately related species
vlei—term used in southern tropical Africa and South Africa for a seasonally or permanently marshy area

windowed—for a *Gladiolus* flower in which there is a wide gap between the bases of the dorsal and upper lateral tepals so that in profile one can see through the flower

xeric—arid or semiarid, of habitats
xeromorphic—feature related to adaptation for dry habitats
xylem—plant tissue responsible for water conduction
xylem vessel—cell in the xylem with perforations in the walls allowing unrestricted water passage

zygomorphic—an irregular flower, i.e., bilaterally symmetric

Bibliography

Anderton, E. W., & R. Park. 1989. Growing Gladioli. Timber Press, Portland, Oregon.

Baker, J. G. 1876. New Gladioleae. J. Bot. 14: 333–339.

Baker, J. G. 1878a. Report on the Liliaceae, Iridaceae, Hypoxidaceae and Haemodoraceae of Welwitsch's Angolan herbarium. Trans. Linn. Soc. London, Bot., Ser. 2, 1: 245–273.

Baker, J. G. 1878b. Systema Iridacearum. J. Linn. Soc. Bot. 16: 61–180.

Baker, J. G. 1892. Handbook of the Irideae. George Bell & Sons, London.

Baker, J. G. 1896. Iridaceae. Pp. 7–171 *in* W. T. Thiselton-Dyer (editor), Flora Capensis, Volume 6. Reeve & Co., Ashford.

Baker, J. G. 1898. Irideae. Pp. 337–376 and 573–578 *in* W. T. Thiselton-Dyer, Flora of Tropical Africa, Volume 7. Reeve & Co., London.

Baker, J. G. 1901. Beiträge zur Kenntniss der Afrikanischen Flora—Iridaceae. Bull. Herb. Boissier, Sér 2, 1: 862–868.

Bamps, P. 1982. Flore d'Afrique Centrale. Répertoire des Lieux de Récolte. Jardin Botanique National de Belgique, Meise.

Bentham, G., & J. D. Hooker. 1883. Genera Plantarum, Volume 3(2). Reeve & Co., London.

Bolus, H. M. L. 1928. Plants new and noteworthy. S. African Gard. 18: 213.

Bolus, H. M. L. 1929. Novitates Africanae. J. Bot. (London) 67: 132–139.

Bolus, H. M. L. 1933. Plants new and noteworthy. S. African Gard. 23: 47.

Brown, N. E. 1932. Contributions to a knowledge of the Transvaal Iridaceae. Trans. Roy. Soc. S. Africa. 20: 261–280.

Bullock, A. A. 1930. *Oenostachys,* a new genus of Iridaceae from East Africa. Kew Bull. Misc. Inform. 1930: 465–466.

Chevalier, A. J. B. 1920. Exploration Botanique de l'Afrique Occidentale Française. Lechevalier, Paris.

Chevalier, A. J. B. 1937. Deux nouvelles Iridées de l'Afrique tropicale. Bull. Mus. Nat. Hist. Nat., Sér. 2, 9: 401–404.

Chiovenda, A. 1911. Plantae novae vel minus notae e regione Aethiopica. Ann. Bot. Roma 9: 125–152.

Còrdova, S. Ponce. 1990. *Gladiolus* nouveaux du Zaire (Iridaceae). Bull. Jard. Bot. Nat. Belgique 60: 325–329.

Cufodontis, G. 1974. Iridaceae. Pp. 1584–1592 *in* Conspectus Florae Africae, Volume 2. Jardin Botanique National de Belgique, Meise.

Dalziel, J. M. 1937. The Useful Plants of West Tropical Africa. Crown Agents for the Colonies, London.

De la Roche, D. 1768. Plantae Aliquot Novarum. Leiden.

De Vos, M. P. 1972. The genus *Romulea* in South Africa. J. S. African Bot., Suppl. 9.

De Vos, M. P. 1976. Die Suid-Afrikaanse species van *Homoglossum*. J. S. African Bot. 42: 301–359.

De Wildeman, E. 1913. Decades novarum specierum florae Katangensis. XII–XIV. Feddes Rep. Spec. Nov. Regni Veg. 12. 289–298.

Duvigneaud, P., & S. Denaeyer-de Smet. 1963. Cuivre et végétation au Katanga. Bull. Soc. Roy. Bot. Belgique 95–96: 93–231.

Ecklon, C. F. 1827. Topographisches Verzeichniss der Pflanzensammlung der C. F. Ecklon. Reise Verein, Esslingen.

Farris, J. S. 1988. Hennig86, Version 1.5. Program and software documentation. Published by the author, Port Jefferson Station, New York.

Garrity, J. B. 1975. Gladioli for Everyone. David & Charles, London.

Geerinck, D. 1972. Révision du genre *Gladiolus* L. (Iridaceae) au Zaire, au Rwanda et au Burundi. Bull. Jard. Bot. Nat. Belgique 42: 269–287.

Goldblatt. P. 1971. Cytological and anatomical studies on the southern African Iridaceae. J. S. African Bot. 37: 317–460.

Goldblatt, P. 1981. Notes on the cytology and distribution of *Anapalina*, *Tritoniopsis* and *Sparaxis*, Cape Iridaceae. Ann. Missouri Bot. Gard. 68: 562–564.

Goldblatt, P. 1984. New taxa and notes on southern African *Gladiolus* (Iridaceae). J. S. African Bot. 50: 449–459.

Goldblatt, P. 1985. Systematics of the southern African genus *Geissorhiza* (Iridaceae—Ixioideae). Ann. Missouri Bot. Gard. 72: 277–447.

Goldblatt, P. 1989. Systematics of *Gladiolus* (Iridaceae) in Madagascar. Bull. Mus. Nat. Hist. Nat., Sér. 4, Sect. B, Adansonia 11: 235–255.

Goldblatt, P. 1990a. Phylogeny and classification of Iridaceae. Ann. Missouri Bot. Gard. 77: 607–627.

Goldblatt, P. 1990b. Status of the southern African *Anapalina* and *Antholyza* (Iridaceae), genera based solely on characters for bird pollination, and a new species of *Tritoniopsis.* S. African J. Bot. 56: 577–582.

Goldblatt, P. 1991. An overview of the systematics, phylogeny and biology of the southern African Iridaceae. Contrib. Bolus Herb. 13: 1–74.

Goldblatt, P. 1993. Iridaceae. Pp. 1–106 *in* G. V. Pope (editor), Flora Zambesiaca, Volume 12(4). Flora Zambesiaca Managing Committee, London.

Goldblatt, P., & M. P. de Vos. 1989. The reduction of *Oenostachys, Homoglossum* and *Anomalesia*, putative sunbird pollinated genera, in *Gladiolus* L. (Iridaceae—Ixioideae). Bull. Mus. Hist. Nat., Sér. 4, Sect. B, Adansonia 11: 417–428.

Goldblatt, P., & J. C. Manning. 1990. *Devia xeromorpha*, a new genus and species of Iridaceae—Ixioideae from the Cape Province, South Africa. Ann. Missouri Bot. Gard. 77: 359–364.

Goldblatt, P., & J. C. Manning. 1995. Phylogeny of the African genera *Anomatheca* and *Freesia* (Iridaceae—Ixioideae), and a new genus *Xenoscapa*. Syst. Bot. 20: 161–178.

Goldblatt, P., & J. H. J. Vlok. 1989. New species of *Gladiolus* (Iridaceae) from the southern Cape and the status of *G. lewisiae*. S. African J. Bot. 55: 259–264.

Goldblatt, P., A. Bari, & J. C. Manning. 1991. Sulcus variability in the pollen grains of Iridaceae subfamily Ixioideae. Ann. Missouri Bot. Gard. 78: 950–961.

Goldblatt, P., M. Takei, & Z. A. Razzaq. 1993. Chromosome cytology in tropical African *Gladiolus* (Iridaceae—Ixioideae). Ann. Missouri Bot. Gard. 80: 461–470.

Grant, J. A. 1875. The botany of the Speke and Grant Expedition, an enumeration of the plants collected during the journey of the late Captain J. H. Speke and Captain (now Lieut.-Col.) J. A. Grant from Zanzibar to Egypt. Trans. Linn. Soc. 29: 1–61.

Gunn, M. & L. E. Codd. 1981. Botanical Exploration of Southern Africa. Balkema, Cape Town.

Harms, H. 1903. *In* H. Baum, Kunene-Sambesi-Expedition. Kolonial-Wirtschaftlichen Komitees, Berlin.

Harper, F. J. 1949. Notes on certain species of the botanical genera *Gladiolus* and *Eulophia*, Nigerian Field 14: 68–69.

Hedberg, O. 1957. Afroalpine vascular plants. Symb. Bot. Upsaliensis 15: 1–411.

Hepper, F. N. 1968. Flora of West Tropical Africa, Edition 2, Volume 3(1). Crown Agents for Oversea Governments, London.

Herbert, W. 1842. *Gladiolus aequinoctialis.* Bot. Reg. 28, Misc. 85, no. 97.

Herbert, W. 1847. On the hybridization amongst vegetables. J. Hort. Soc. London 2: 81–107.

Hiern, W. P. 1896. Catalogue of the African Plants Collected by Dr. Friedrich Welwitsch in 1853–1861. British Museum, London.

Hilliard, O. M., & B. L. Burtt. 1979. Notes on some plants of southern Africa, chiefly from Natal (VIII). Notes Roy. Bot. Gard. Edinburgh 37: 284–325.

Hilliard, O. M., & B. L. Burtt. 1983a. Notes on some plants of southern Africa, chiefly from Natal (X). Notes Roy. Bot. Gard. Edinburgh 41: 299–319.

Hilliard, O. M., & B. L. Burtt. 1983b. Notes on some plants of southern Africa, chiefly from Natal (XIII). Notes Roy. Bot. Gard. Edinburgh 43: 189–228.

Hitchcock, A. S., & M. L. Green. 1929. Standard species of Linnaean genera of Phanerogamae. Pp. 111–199 *in* Nomenclature proposals by British botanists. International Botanical Congress, Cambridge (England), 1930. HMSO, London.

Hooker, J. D. 1830. *Gladiolus psittacinus.* Curtis's Bot. Mag. 57: pl. 3032.

Hooker, J. D. 1831. *Gladiolus natalensis.* Curtis's Bot. Mag. 58: under pl. 3084.

Hulton, P. H., F. N. Hepper, & J. Friis. 1991. Luigi Balugani's Drawings of African Plants. Yale Center for British Art, New Haven, Connecticut.

Hutchinson, J., & J. M. Dalziel. 1936. Flora of West Tropical Africa, Volume 2. Crown Agents for the Colonies, London.

Irvine, F. R. 1930. Plants of the Gold Coast. Oxford University Press, London.

Jacot-Guillarmod, A. F. M. 1971. Flora of Lesotho. J. Cramer, Lehre, Germany.

Jacques-Felix, H. 1969. Sur un *Acidanthera* (Iridaceae) de Guinée. Adansonia, Sér. 2, 9: 131–133.

Johnson, S. D. 1992. Plant animal relationships. Pp. 175–205 *in* R. Cowling (editor), The Ecology of Fynbos: Nutrients, Fire and Diversity. Oxford University Press, Cape Town.

Johnson, S. D., & W. A. Bond. 1994. Red flowers and butterfly pollination in the fynbos of South Africa. Pp. 137–148 *in* M. Arianoutsou & R. Grooves (editors), Plant Animal Interactions in Mediterranean-Type Ecosystems. Kluwer Academic Press, Dordrecht.

Kassner, T. 1911. From Rhodesia to Egypt. Hutchinson & Co., London.

Klatt, F. W. 1864. Iridaceae. Pp. 515–517 *in* W. C. H. Peter (editor), Reise nach Mossambique—Botanik, Volume 6(2). Berlin.

Klatt, F. W. 1882. Ergänzungen und Berichtigungen zu Baker's Systema Iridacearum. Abh. Naturforsch. Ges. Halle 15: 44–404.

Klatt, F. W. 1895. Iridaceae. Pp. 143–230 *in* T. Durand & H. Schinz (editors), Conspectus Florae Africae, Volume 5. Jardin Botanique de l'État, Brussels.

Kuntze, O. 1898. Revisio Generum Plantarum, Volume 3. Arthur Felix, Leipzig.

Lewis, G. J. 1954a. Some aspects of the morphology, phylogeny and taxonomy of the South African Iridaceae. Ann. S. African Mus. 40: 15–113.

Lewis, G. J. 1954b. Iridaceae—new species and miscellaneous notes. Ann. S. African Mus. 40: 115–135.

Lewis, G. J., A. A. Obermeyer, & T. T. Barnard. 1972. A revision of the South African species of *Gladiolus.* J. S. African Bot., Suppl. 10: 1–316.

Linnaeus, C. 1753. Species Plantarum. L. Salvius, Stockholm.

Malaisse, F. & G. Parent. 1985. Edible wild vegetable products in the Zambezian woodland area: a nutritional and ecological approach. Ecol. Food & Nutrition 18: 43–82.

Marais, W. 1973. Notes on African Iridaceae. Kew Bull. 28: 311–317.

Marloth, R. 1917. Flora of South Africa, Volume 5. Juta, Cape Town.

Miller, A. G. & M. Morris. 1988. Plants of Dhofar. Office of the Advisor for Conservation of the Environment, Oman.

Nixon, K. 1992. CLADOS 1.2 IBM PC-Compatible Character Analysis Program. Published by the author, Ithaca, New York.

Oates, F. 1889. Matabeleland and the Victoria Falls. London.

Obermeyer, A. A. 1989. A new species of *Gladiolus*. Bothalia 14: 78.

Persoon, H. 1805. Synopsis Plantarum, Volume 1. Cramer, Paris.

Pourret, P. A. 1789. Description de deux nouveaux genres de la famille des Liliacées, désignés sous le nom *Lomenia* e de *Lapeirousia*. Observ. Phys. 35: 425–432.

Raamsdonk, L. W. D. van, & T. de Vries. 1989. Biosystematic studies in European species of *Gladiolus* (Iridaceae). Pl. Syst. Evol. 165: 189–198.

Radelescu, D. 1970. Recherches morphopalynologiques sur les éspeces d'Iridaceae. Lucr. Grăd. Bot. Bucureşti 1968: 311–350.

Rendle, A. B. 1911. A contribution to our knowledge of the flora of Gazaland—Monocotyledons. J. Linn. Soc., Bot. 40: 207–245.

Richard, A. 1851. Tentamen Florae Abyssinicae. Bertrand, Paris.

Rudall, P., & P. Goldblatt. 1991. Leaf anatomy and phylogeny of Ixioideae (Iridaceae). Bot. J. Linn. Soc. 106: 329–345.

Salisbury, R. A. 1866. The Genera of Plants. Van Voorst, London.

Schulze, W. 1971. Beiträge zur Pollenmorphologie der Iridaceae—Ixioideae. Wiss. Z. Friedrich-Schiller-Univ. Jena, Math.-Naturwiss. Reihe 19: 437–445.

Shneyer, V. S. 1990. A serotaxonomical study of the tribe Ixieae (Iridaceae). Bot. Zhurn. 75: 1657–1668. (In Russian.)

Thunberg, C. P. 1823. Flora Capensis. Schultes, Stuttgart.

Vaupel, F. 1913. Iridaceae Africanae novae. Bot. Jahrb. Syst. 48: 533–549.

Vaupel, F. 1920. Neue afrikanische Iridaceen. Notizbl. Bot. Gart. Mus. Berlin-Dahlem. 7: 376–380.

Vogel, S. 1954. Blütenbiologische Types als Elemente der Sippengliederung. Bot. Stud. 1: 1–338.

Watt, J. M., & M. G. Breyer-Brandwijk. 1962. Medicinal and Poisonous Plants of Southern and Eastern Africa, Edition 2. Livingstone Ltd., Edinburgh.

Index of Scientific Names